PLINY THE ELDER'S WORLD

Pliny the Elder's World offers readers a translation of the *Natural History*'s opening books unprecedented for its completeness, accuracy and accessibility. Here, in quirky, often breathless style, Pliny lays the foundation of a hugely influential encyclopedia with coverage of the universe, stars, planets and moon, followed by earth's climate and then its physical and human geography. From Rome as ruling centerpoint, Pliny surveys the known world and its countless peoples in a vast arc from the Atlantic to Sri Lanka, embracing the Danube, Euphrates and Nile lands, Atlas and Caucasus mountains, Germany, Africa, Arabia, India. Passages from later books further illustrating his geographical grasp are appended, on topics as varied as wine, water, trees, birds and fish. Throughout, Pliny's frank expression of strong opinions about religion, distorted human values, abuse of the environment (and more) reveals uncannily modern preoccupations. His work remained an inspirational resource through the Renaissance, and still fascinates today.

BRIAN TURNER is Associate Professor of History at Portland State University, Oregon. He is former director (2009–2011) of the Ancient World Mapping Center (https://awmc.unc.edu/wordpress), and associate editor of the Pleiades project (https://pleiades.stoa.org). Alongside ancient geography and worldview, his research focuses on the culture of warfare in the ancient world. He is co-editor of *Brill's Companion to Military Defeat in Ancient Mediterranean Society* (2018).

RICHARD J. A. TALBERT is Research Professor of History at the University of North Carolina, Chapel Hill. He is the editor of the *Barrington Atlas of the Greek and Roman World* (2000). His other books include *Rome's World: The Peutinger Map Reconsidered* (Cambridge 2010), *Roman Portable Sundials: The Empire in Your Hand* (2017), *Challenges of Mapping the Classical World* (2019), *World and Hour in Roman Minds* (forthcoming), and a translation *Plutarch: On Sparta* (revised 2005).

PLINY THE ELDER'S WORLD

Natural History, Books 2–6

BRIAN TURNER
Portland State University

RICHARD J. A. TALBERT
University of North Carolina, Chapel Hill

CAMBRIDGE
UNIVERSITY PRESS

CAMBRIDGE
UNIVERSITY PRESS

University Printing House, Cambridge CB2 8BS, United Kingdom

One Liberty Plaza, 20th Floor, New York, NY 10006, USA

477 Williamstown Road, Port Melbourne, VIC 3207, Australia

314–321, 3rd Floor, Plot 3, Splendor Forum, Jasola District Centre, New Delhi – 110025, India

103 Penang Road, #05–06/07, Visioncrest Commercial, Singapore 238467

Cambridge University Press is part of the University of Cambridge.

It furthers the University's mission by disseminating knowledge in the pursuit of education, learning, and research at the highest international levels of excellence.

www.cambridge.org
Information on this title: www.cambridge.org/9781108481755
DOI: 10.1017/9781108592758

© Cambridge University Press 2022

This publication is in copyright. Subject to statutory exception and to the provisions of relevant collective licensing agreements, no reproduction of any part may take place without the written permission of Cambridge University Press.

First published 2022

Printed in the United Kingdom by TJ Books Limited, Padstow Cornwall

A catalogue record for this publication is available from the British Library.

ISBN 978-1-108-48175-5 Hardback

Cambridge University Press has no responsibility for the persistence or accuracy of URLs for external or third-party internet websites referred to in this publication and does not guarantee that any content on such websites is, or will remain, accurate or appropriate.

For Kathryn, Erin and Zandra

Contents

Preface	*page* ix
Maps	xi
Introduction	1
Book 1 Pliny's Detailed Table of Contents	11
Book 2 The Universe, Astronomy, Climate	25
Book 3 Europe: Spain to Italy	99
Book 4 Europe: East, North	137
Book 5 Africa, Levant, Asia Minor	169
Book 6 The East, India	209
Books 7–37 Notable Geographical Passages	261
Appendix 1 Titles and Technical Terms	301
Appendix 2 Units of Measurement	302
Appendix 3 Latin Editions Translated (Books 2 to 6)	304
Modern Works Cited	306
Index	308

Preface

Pliny's encyclopedic *Natural History* has always both fascinated and challenged its readers. Few works from antiquity have so substantial and varied a manuscript tradition, few have been so praised and so criticized over the centuries. The ambition to translate the geographical books of the *Natural History* originated at the University of North Carolina's Ancient World Mapping Center in 2011, while Brian Turner was its Director. There, and with enthusiasm, Richard Talbert supported Turner's vision that such a work would make a valuable addition to the growing number of translated geographical texts. Having proved a tirelessly responsive reader of initial drafts, in early 2017 Talbert agreed to become co-translator at Turner's invitation.

By that date, Pliny had come to dominate our conversations, whether on a train from Portland, Oregon en route to one conference, or on buses rolling across Bulgaria and Serbia during others, and later in countless phone and (eventually) zoom calls. In 2018 we learned of a project similar to ours being considered by Duane W. Roller, who then welcomed our proposal to advance in parallel: Turner and Talbert to produce a translation and maps (with the help of the Ancient World Mapping Center), Roller a scholarly commentary.

The translation is a collaborative one throughout; we did not split books or passages between us. Our aim has been to achieve freshness and accuracy, and so far as possible to capture Pliny's own voice. Although the result may seem surprising in some respects, it faithfully reflects his distinct perspective on the world. No doubt we have introduced errors and perhaps even annoyances, but – to paraphrase Philemon Holland, the first translator of the *Natural History* into English at the beginning of the seventeenth century – we shall still consider ourselves successful if these unintended flaws prompt closer attention to Pliny's encyclopedia, and consequently foster renewed interest in it and in ancient geography.

Our thanks go to Michael Sharp, Sophie Taylor, Liz Davey and Malcolm Todd at Cambridge University Press. We are most grateful to the many editors of volumes in the Budé and Sammlung Tusculum series, whose Latin texts provide the foundation for our translation. Thanks also to Kai Brodersen, Alexander Jones, Duane W. Roller, Graham Shipley, and the anonymous readers whose comments greatly improved our own understanding of Pliny's Latin, in Book 2 especially. Like Alan Bowen (Bowen and Rochberg 2020, 102), we have often found Pliny here "either very confused or maddeningly confusing." Further thanks to the librarians and staff at Portland State's Millar Library and Chapel Hill's Davis Library. Ekaterina Ilyushechkina introduced Talbert to a Russian project to translate the *Natural History* and invited him to Moscow to speak about our translation in 2018. Portland State University and the *Friends of History* generously funded preparation of the index, which was greatly assisted by Jamie Ditzel. Lindsay Holman, Gabriel Moss, Ryan Horne and the staff of the Ancient World Mapping Center have, as always, provided outstanding maps. Tom Elliott at New York University's Institute for the Study of the Ancient World facilitated hosting of the volume's website and digital map. Finally, we express the greatest thanks for the patience and support of our families, especially Brian's wife Kathryn and their daughter Erin, and Richard's wife Zandra.

Maps

Pliny's World: overview	*page* 10
Book 3 Europe: Spain to Italy	98
Book 4 Europe: East, North	136
Book 5 Africa, Levant, Asia Minor	168
Book 6 The East, India	208

Digital map: https://isaw.nyu.edu/research/pliny-the-elder

Introduction

Gaius Plinius Secundus (23/24–79 CE), known today as Pliny the Elder, was born in Novum Comum (modern Como, Italy). An unfailingly proud member of Rome's upper-class equestrian order, Pliny spent his life serving the empire and his own insatiable curiosity; he evidently never married. As a military officer in the German provinces during the reign of Claudius (41–54), Pliny wrote works of military history and biography. The perilous reign of Nero (54–68) saw him mostly avoid imperial posts as he concentrated on further research and writing. After Nero's suicide in 68, and the subsequent civil war of 69–70, Pliny resumed his career, serving as an imperial administrator and advisor to the victorious claimant Vespasian and his son Titus. In 79, as commander of the fleet at Misenum, just across the bay from Pompeii and Herculaneum, Pliny witnessed the eruption of Vesuvius. Curious to experience the power of Nature and dedicated to saving the lives of the doomed, he sailed into the falling ash. He died on the seashore near Stabiae.[1]

According to his nephew (Pliny the Younger), Pliny was such a voracious reader and devoted scholar that he often fell asleep as he was working.[2] More than one hundred volumes by him went into circulation, including: one on throwing a javelin from horseback, two on the life of Pomponius Secundus (a friend and patron), twenty on Rome's German wars, three on the education of orators, eight on bad grammar, thirty-one on history and, finally, the thirty-seven Books of his *Natural History*, a massive collection of facts, "no less varied than Nature herself."[3] Along with these he bequeathed to his nephew 160 more volumes, filled with

---

[1] For Pliny's biography, see *Prosopographia Imperii Romani*² P 373; Cornell (2014); for his life and work, Murphy (2004); Doody (2015).
[2] Pliny the Younger, *Letters* 3.5.8.
[3] Ibid. 3.5.3–6.

notes scribbled on both sides and in minuscule script.[4] This publication record reflected a work ethic that proved challenging to match. Because walking did not afford the opportunity to read, Pliny once reprimanded his nephew for walking rather than riding in a litter.[5] He even criticized a member of his reading group who had demanded the rereading of a mispronounced word. There was no justification to waste time with repetition, Pliny argued; the point was clear enough.[6] For him, the pursuit of knowledge did not necessarily rest upon the quality of the presentation, but rather on the quantity of the material. No truer statement could be made to describe his surviving magnum opus.

Pliny's World

The *Natural History* collates data on a vast range of topics including cosmology, astronomy, geography, ethnography, anthropology, zoology, botany, dendrology, pharmacology and geology. In total, Pliny claims to have recorded 20,000 facts in thirty-six Books (not including Book 1, the table of contents) from his reading of some 2,000 volumes by 100 authors.[7] His own extensive bibliography proves his maxim that no book was so bad that something useful could not be found in it.[8] Even though the Books of the *Natural History* – especially those focused on the universe and the geography of the world – are primarily a product of Pliny's reading rather than of his original research and experimentation, they do consolidate the collective knowledge about the natural world available in the first century CE. As a result, they reflect wonderfully the Roman worldview that Pliny shared with many of his contemporaries, as well as their enthusiasm for assembling and organizing knowledge of all kinds.[9] These Books are also invaluable for the summaries they provide of much that is otherwise lost.

After the table of contents the first five of Pliny's Books cover the size, form and character of: the known universe (Book 2), western and southern Europe, including Spain and Italy (Book 3), eastern and northern Europe as well as Britain (Book 4), north Africa, the Levant

---

[4] Ibid. 3.5.17.
[5] Ibid. 3.5.15–16.
[6] Ibid. 3.5.12–13.
[7] *NH* Preface 17; Suetonius, *Life of Pliny*.
[8] Pliny the Younger, *Letters* 3.5.10.
[9] On contemporary worldviews, see Beagon (1992); on the tendency to assemble data, see König and Whitmarsh (2007) 3–39; Riggsby (2019).

and Asia Minor (Book 5), and the regions of the East beyond the Black Sea, including India and Sri Lanka, as well as what in antiquity was understood to be southern Africa (Book 6). Alongside Book 2, the 'geographical' Books have regularly been treated as a distinct group by scholars and editors alike.[10] Yet despite the good reason to do this, at the same time we should not fail to recognize that geography and ethnography continue to play a vital part in the subsequent Books of the *Natural History* – hence our inclusion of some passages from these.

Pliny's presentation of geography fits within a long tradition of Greek and Roman geographical and ethnographic writing.[11] During the past twenty years or so, English translations have appeared of much of this work. They have contributed to a surge in studies on ancient geography, travel and worldview. A fresh translation of Pliny's geographical Books boosts this surge, and also benefits medieval and Renaissance studies, given the persistent enthusiasm shown for the *Natural History* during those periods.

For readers' convenience Pliny summarized his sources for each Book and the total number of facts recorded in it, and modern editors usually place all this information in Book 1.[12] In our translation, however, we have placed these records for Books 2–6 at the end of each. In them Pliny consistently differentiates between Roman and foreign authors; none of the records is to be regarded as all-inclusive. For whatever reason, authors drawn upon for a Book may be omitted; equally, others listed are not specifically cited in the Book itself.

Beyond his reading, Pliny relied on other types of evidence, including what he had seen and experienced himself. Reports from traders and envoys are cited in Book 6 especially. In Pliny's time trade could occur under conditions which might encourage exploration. He praises quests for the Nile's sources (5.51, 6.181), and he admires the "unarmed men" who earlier were able to calculate the earth's dimensions (6.208). On the other hand, he is also aware that peace might not be invariably beneficial to the acquisition and spread of knowledge: "Nowadays, however, in such blissful peace and with a *princeps* so delighted by productivity in sciences and the arts, nothing more is being learned through new research, and really there is not even a full grasp of the ancients' discoveries" (2.117). Indeed, for Pliny and the Romans, war served as a means to acquire

---

[10] Note Doody's (2015) section dedicated to bibliography on Pliny's geography.
[11] For surveys, see Dueck (2012); Roller (2015); Irby (2016).
[12] *NH* Preface 33.

knowledge. Rome's armies repeatedly penetrated unknown parts and peoples of the world. During war with the Oeensians in Africa a shortcut to the territory of the Garamantes emerged (5.38); expeditions in northern Europe revealed some twenty-three islands off the Cimbrian cape, Glaesaria being the most prominent and a major source for amber (4.97). Preparations for war could also be productive. Soldiers on an intelligence-gathering mission reported upon Aethiopia's wilderness (6.181). Isidorus wrote a description of the world on instructions from Augustus "to make a complete record when his elder son [Gaius Caesar] was going to Armenia to deal with the Parthians and Arabians" (6.141). Not every report was reliable of course, as Pliny sharply points out when describing labeling errors found on "drawings of the landscape" around the Caucasian Gates in Hiberia sent back by Domitius Corbulo's men (6.40).

Furthermore, Pliny recognizes monuments as sources of geographical knowledge. He regularly refers to Marcus Vipsanius Agrippa's work, and specifically mentions Agrippa's presentation of "the world for the world to see" in the Porticus Vipsania at Rome (3.17). This does not survive, and so its appearance remains controversial. Elsewhere, Pliny knows the trophy erected by Augustus – which survives, albeit damaged, above modern Monaco – commemorating the emperor's pacification of Italy's northern neighbors. He reproduces its list of "all the peoples of the Alps extending from Upper [Hadriatic] sea to Lower [Tyrrenan]" (3.136).

The Text

The choice of which editor's Latin text to translate presents difficulties. The manuscript tradition and editorial disagreements have prevented any one edition from becoming the universally recognized standard. The *Natural History* was repeatedly copied, abridged and paraphrased during the Late Roman Empire, throughout the Middle Ages and on into the Renaissance.[13] Altogether, the mass of what survives is remarkable, including well over one hundred manuscripts of the complete work, an astonishing total given its length and density. If further testimony to continuing interest in the *Natural History* were needed, there are in addition two hundred or so excerpts, some copied earlier than

---

[13] For the post-classical reception of Pliny note, for example, Lozovsky (2000); Doody (2010); McHam (2013); Hiatt (2020).

Introduction 5

Charlemagne's reign around 800.[14] From the mid-fifteenth century, printed editions appeared in rapid succession, and these in turn raise problems.[15] The editions to be considered today owe much to indefatigable efforts by talented nineteenth-century German scholars. Ludwig von Jan and Karl Julius Sillig led the way by producing the first complete scholarly edition between 1831 and 1836 (published by Teubner). The two later parted company: Sillig published his second edition between 1851 and 1858 for Perthes, and Jan his between 1854 and 1860 for Teubner. Next, between 1866 and 1873, Detlef Detlefsen offered a rival edition published by Weidmann. Karl Mayhoff then emerged as a further rival, with his (Teubner) edition starting to appear in 1875 and continuing (with various revisions meantime) to completion in 1906. Mayhoff's edition (readily accessible online) could still serve as the basis for our translation, but that would be to ignore all subsequent efforts over the past century and more, in particular those by French scholars who have contributed to an edition for Budé (with detailed commentary) which now lacks only parts of Books 5 and 6.[16] Otherwise in this edition the geographical books appeared between 1950 and 2015, and it is their texts that we translate.[17] For the parts missing, we translate the text of the edition produced (in the 1990s) by Roderich König and others for Sammlung Tusculum.[18]

The Translation: Background

"English is less well served than other languages when it comes to translations," regrets Aude Doody in her authoritative 2015 overview of scholarship on Pliny. Relatively recent, reliable and scholarly translations of the *Natural History* are available in French, German, Italian and Spanish.[19] By contrast, English translations – with the exception of Mary Beagon's of Book 7 (2005) – are old, outdated, or drastically abridged. An early one by Philemon Holland published in 1601 – perhaps used by

---

[14] For the manuscript tradition, see Reeve (2007); Healy (1999) 380–392.
[15] Healy (1999) 388–389.
[16] Budé's format is for Latin text on the left page of a doublespread to be matched by French translation of it on the right page.
[17] See Appendix 3. Inevitably, the editors for Budé vary in the attention they give to the manuscript tradition: see Reeve (2007) 115–116.
[18] The format matches Budé's, with concise commentary, a list of variants in the Latin text, and detailed indices.
[19] A Russian translation – the first ever – has now started to appear: see Ilyushechkina et al. (2021).

Shakespeare[20] – was only supplanted by that of John Bostock and Henry Thomas Riley in 1855, with the translators' apologies to Holland in their Preface because his work was deemed "unsuited to the requirements of the nineteenth century." The same might have been said about their translation in the twentieth century by the Cambridge classicist Harris Rackham, who began an English translation for the Loeb Classical Library in the 1930s.[21] Strikingly, however, in a *Prefatory Note* Rackham warns that "[t]his translation is designed to afford assistance to the student of the Latin text; it is not primarily intended to supply the English reader with a substitute for the Latin."[22] The sole translator into English since Rackham – John Healy for Penguin in 1991 – was certainly intent upon being more helpful to those without Latin, but he offers merely a drastic abridgement. For example, he translates less than 40 of Book 5's 151 Paragraphs, with several of these abridged and even parts of some sentences omitted.

While we too are tackling only a portion of Pliny's *Natural History*, there can be no doubt about the potential value of a replacement for Rackham's translation in order to provide a coherent, self-standing addition to the expanding range of translated ancient texts about geography. Moreover, we append to Books 2 to 6 translations of a substantial range of varied passages from Books 7 through 37. Although the choices made are inevitably subjective, all in one way or another relate closely to geography and illustrate its importance to Pliny's understanding of the natural world.[23] We mean our audience to include classical scholars, but our translation is intended mainly for readers who will not consult a Latin text (which we do not include) and are not specialists in any area of ancient studies. In view of the commentary to be provided by Duane W. Roller, only the bare minimum of explanation accompanies our translation.

The Translation: Character

'Coherent' may not be the appropriate description of Pliny's notoriously challenging Latin.[24] Compressed constructions often stitch together breathless, rambling sentences. Tedious, though important, lists – "the

---

[20] Gillespie (2001) 421–425.
[21] The Loeb translation, like the Budé, faces a Latin text, but there is next-to-no commentary.
[22] Rackham (1942) vii, an approach curiously out of step with Loeb's avowed mission. This note no longer appears in reprints of the volume after 1969.
[23] Again we follow the Budé Latin text in each instance.
[24] Healy (1987); Pinkster (2005); Travillian (2015).

bare names of places," in Pliny's own phrase[25] – are interspersed with rhetorical descriptions and frank personal outbursts. Thus to strike a balance between readable English and a faithful representation of Pliny's sentence structure is no easy task. Accordingly, a brief explanation of the principles that we have developed and normally followed for our translation is called for in order to prepare readers.[26]

It has been our aim throughout to remain as consistent and accurate as possible in rendering Pliny's Latin and in capturing his tone, which switches repeatedly. Put another way, we always seek to reflect Pliny's style, and we emphatically reject the temptation to 'improve' his prose. This is often liable to resemble a terse, opaque tweet, with few or no verbs, drafted in haste more to himself rather than for any audience. Nonetheless, for our readers' sake we do split some interminable sentences into two or more, and we make liberal use of the colon and semi-colon. Also, our use of the comma is very deliberate, again to aid the reader. As one means of conveying the clipped, staccato character of Pliny's 'memo to self' style without impairing its comprehensibility, we mostly omit the definite article (which does not exist in Latin) before a name or term in contexts where really it is superfluous, even though expected in standard English. We believe that, after an initial jolt, readers are unlikely to be concerned by its absence.

An especially unsatisfactory feature of Rackham's translation is his inconsistent and unexplained handling of placenames. Our approach is to keep these names for the most part in the form used by Pliny (as does Rackham sometimes), with only limited conversion to English equivalents. Rackham by contrast is prone to substitute the current (1930s) placename for Pliny's, one liable to reflect the now long outdated (and often offensive) preference of European colonialists. So in 5.9 Pliny's *Rutubis* Rackham translates Mazagan, and Pliny's *promunturium Solis* he translates Capo Blanco. Alternatively and unpredictably, Rackham may opt to substitute a literal English translation of an ancient name's meaning: hence Hiera Sycaminos becomes Holy Mulberry (6.184), and Zeugma becomes Bridgetown (5.67). His muddled handling of *Autoteles* ("Free State") sows serious confusion. In 5.5 he retains Pliny's name. Later, in 5.17, he translates it "Free State"; but then, further into the same sentence when he needs to refer to this people by name (to clarify the pronoun used here by Pliny), he switches to "Autoteles", without

---

[25] *NH* 3.2.
[26] See further Talbert (2020).

explaining that these are in fact the "Free State" just mentioned rather than a different people, as readers of the translation alone would reasonably infer. We on the contrary reject the temptation to substitute modern names for Pliny's, or to use literal translations. A modern equivalent for an ancient name (where there is one) should be easy to establish by consulting pleiades.stoa.org. In addition, all places and features mentioned in our translation that can be located may be viewed on the accompanying digital map.[27]

Again, however, we do not go to extremes. We do use a few very common equivalent English placenames, such as Athens, Egypt, or Rome, but we still retain, for example, Danuvius, Germania, Graecia, Italia, and (because it is *not* the equivalent of modern Ethiopia) Aethiopia. We anglicize the names of well-known ancient individuals and peoples: so Pompey not *Pompeius,* Greeks not *Graeci,* Scythians not *Scythae.* On the other hand, for the less well-known we retain Pliny's form of their name. We do anglicize many adjectival forms, such as Atlantic sea for *Atlanticum mare* and Persian gulf for *Persicus sinus.* These last examples as presented here also illustrate our rendering of feature names (sea, gulf) uncapitalized, an attempt to soften the visual impact of a translation that teems with upper-case letters.

As commonly in English usage, we opt for J rather than I when a choice is called for (so Juba, not Iuba), and likewise U when needing to decide between U and V (so Ubii, not Vbii; but Andecavi, not Andecaui). More awkward is the issue of how to present names that Pliny (or his manuscripts at least) spells more than one way, such as Epiros (4.1) and Epirus (6.215). Normally the variants reflect Pliny's use of both Greek and Latin sources, and their retention may confuse readers. In such cases therefore, we adhere consistently to one spelling only. Our choice is the form that Pliny uses more often or, failing that, the form used in the *Barrington Atlas.*

There is a case for maintaining that the translation should take care to reflect at least some of Pliny's vocabulary choices. Consider, for example, his unexpected reference to the leading city of Caesarea in Mauretania as an *oppidum* (5.20). We determined therefore to translate certain Latin terms as consistently as possible. So the Latin *oppidum* is normally rendered as town, *urbs* city, *gens* people, *populus* community, *civitas* state, *litus* shore, *ora* coast, *portus* harbor, *sinus* bay; *Persicus sinus,* however, we

---

[27] See also in print or as App for i-pad Talbert's *Barrington Atlas of the Greek and Roman World* (2000, 2013).

translate by the very familiar Persian gulf. In a few cases we do not differentiate Latin terms because suitable distinct English terms are lacking: consequently we render *flumen, fluvius* and *amnis* river, both *regio* and *tractus* region.

For various directional terms, too (such as *adversus, ante, contra, intra, sub, super*), we translate Pliny's prepositions literally. In a context where *sub*, for example, could be taken to signify 'to the south', we just retain 'below'. We do, however, translate *septentrio* (and cognates) as north, *meridies* south, *oriens* east, *occasus* west. Some Roman technical terms, like *conventus* or *princeps*, we leave in Latin but present in italics; such terms are explained in Appendix 1.

We treat numerals with similar restraint. Mostly, Pliny (or his manuscripts) state them in figures, although some are written out in full, and there are even occasional formulations like 'two short of 80' for 78 (3.62). The highest number our translation normally writes out in full is ten; above that, we use figures. In rendering units of length or distance (cubit, mile, *schoenus*, stade, etc.), we retain whatever Pliny states; Appendix 2 offers some possible conversions. Where Pliny gives a number in the thousands – with no unit stated, though presumably paces – we divide by 1,000 (the number of paces in a Roman mile). So, $\overline{XV}$, literally 15,000, we translate '15', miles being understood. However, where Pliny writes $\overline{XV}$ p., we translate '15 miles'.

[Square brackets] denote an addition by the translators, a cross-reference, translation (where Pliny uses a Greek word, for example), or short editorial explanation. Cross-references marked with an asterisk signify a passage in Pliny's text *not* included in our translation [e.g. 18.156*].

<Angled brackets> indicate a passage where some words have evidently gone missing from the Latin and cannot be recovered, or a passage which makes sense only if it is corrected.

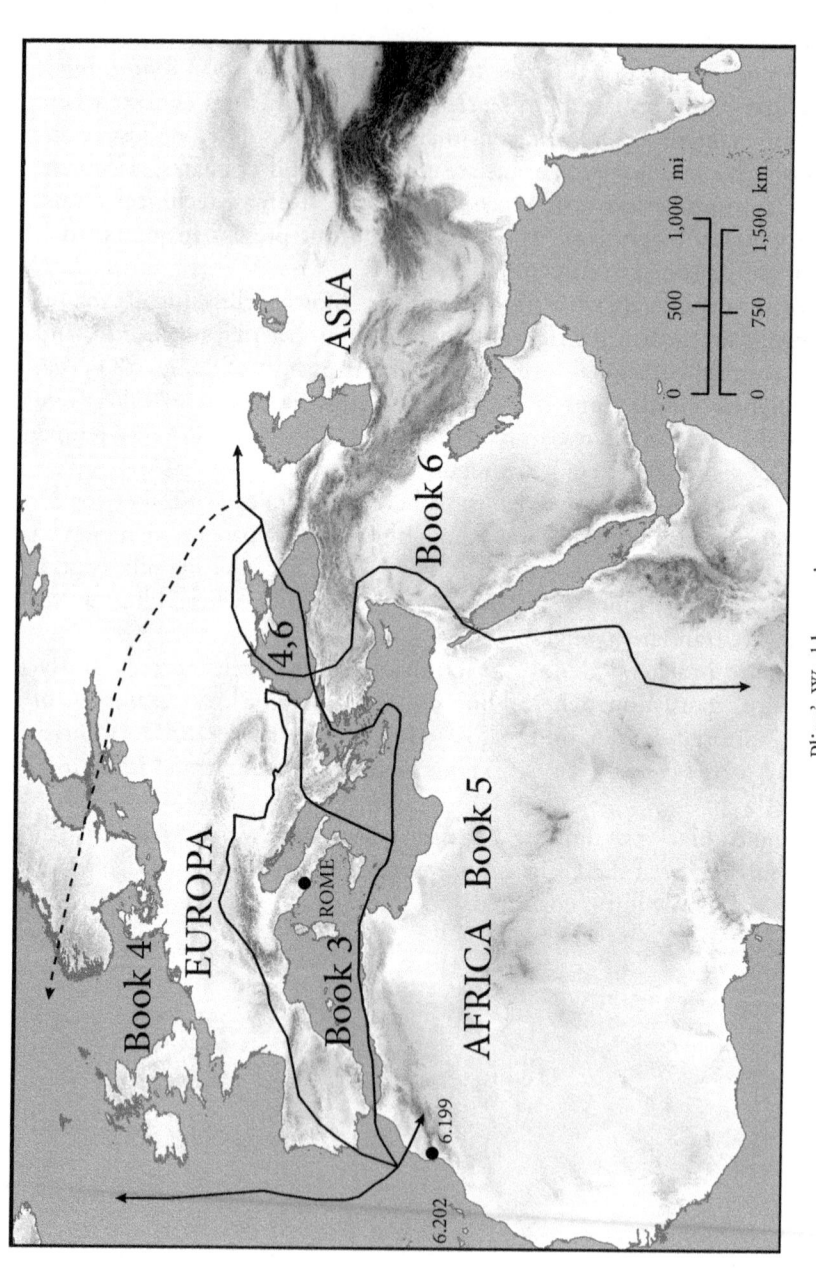

Pliny's World: overview

BOOK 1

Pliny's Detailed Table of Contents

We divide Pliny's text into Books, Chapters and Paragraphs. The thirty-seven Books, and within each the Chapters (usually with a header), are in fact Pliny's own divisions, with his first Book offering a prefatory letter to the emperor Titus followed by a detailed table of contents. For Books 3 to 6, in each instance the table begins with a general header:

> **Locations, peoples, seas, towns, harbors, mountains, rivers, distances, communities present or past.**

Modern editors have divided the text further into short, numbered Paragraphs – usually between five and ten lines of Latin. Their Paragraph numbers are mostly consistent across modern editions and are the best means (with the Book number) by which to cite the *Natural History*.[1]

Book 2
The Universe, Astronomy, Climate

Chapter(s)	Paragraph(s)	
I	1–4	The world: Is it finite? Is it one?
II	5	About its shape
III	6–9	About its motion. Why it should be called the universe
IV	10–13	About the elements
V	14–27	About God
VI	28–46	About the nature of the wandering stars

---

[1] Pliny's text has also been divided in other ways, which are not taken into account here.

Chapter(s)	Paragraph(s)	
VII	47–48	About lunar and solar eclipses. About darkness
VIII	49–52	About the size of stars
IX–X	53–57	What has been discovered by observing the sky
XI	58	About the moon's movement
XII	59–61	Movements of planets and patterns of their radiance
XIII–XIV	62–76	Why they should appear more distant and at other times closer
XV	77–78	General principles of planets
XVI	79–80	Reasons why their colors should change
XVII	81	The sun's movements. Theory of unequal days
XVIII	82	Why thunderbolts are assigned to Jupiter
XIX	83	Distances between planets
XX	84	About the planets' music
XXI	85–88	About the universe's geometry
XXII–XXIII	89–94	About shooting stars. About the nature, position and types of comets
XXIV	95	About Hipparchus' identification of stars. About celestial prodigies, with historical examples
XXV	96	Torches, fireballs [i.e. meteorites]
XXVI	96	Celestial beams, opening of the sky
XXVII	97	About the sky's colors. About celestial blazes
XXVIII–XXIX	98	About celestial crowns. About suddenly formed circles
XXX	98	\<Protracted solar eclipses\>
XXXI	99	Numerous suns
XXXII	99	Numerous moons
XXXIII	100	Daylight at night
XXXIV	100	Burning shields
XXXV	100	A celestial marvel noted once
XXXVI	100	About stars running this way and that

Chapter(s)	Paragraph(s)	
XXXVII	101	About the stars called Castores
XXXVIII	102–104	About air
XXXIX	105–106	About fixed weather conditions
XL	107	About the rising of Canicula
XLI	108–110	Force of annual seasons
XLII	111	About unpredictable weather conditions. About rains and why it should rain stones
XLIII	112–113	About thunder and thunderbolts
XLIV–XLVIII	114–130	Why echoes should occur. Winds: Their types, natures and observations about them
XLIX	131–132	Cloudburst, typhoon
L	133–134	Tornados, *presteres,* vortexes and other monstrous types of storms
LI–LVI	135–146	About thunderbolts:
		Lands in which they do not fall and why
LII	137	Thunderbolt types and marvels
LIII	138–139	Etruscan and Roman observations about them
LIV	140–141	About calling down thunderbolts
LV	142–145	General principles of thunderbolts
LVI	146	What may never be struck
LVII	147	Showers of milk, blood, flesh, iron, wool and fired bricks
LVIII	148	<Clashing arms and trumpet sounds heard from the sky>
LIX	149–150	About stones falling from the sky. Anaxagoras' view of them
LX	150–151	Rainbows in the sky
LXI	152	Nature of hail, snow, frost, mist, dew. Shapes of clouds
LXII	153	Distinct characteristics of the sky according to location
LXIII	154–159	Earth's character

Chapter(s)	Paragraph(s)	
LXIV	160	About her form
LXV	161–165	Whether there may be antipodes
LXVI	166	How water is bound up with earth. The reason for rivers
LXVII	167–170	Whether ocean circles earth
LXVIII	171–175	What part of earth may be habitable
LXIX	176	That earth is in the center of the universe
LXX	177	About the zones' slant
LXXI	177–179	About variations in earth's tilt
LXXII	180	Where eclipses may not be seen and why
LXXIII	181	The reason for daylight on earth
LXXIV	182	Relevant gnomonics
LXXV–LXXVI	183–185	Where and when there are no shadows; where twice a year. Where shadows may be cast in opposite directions
LXXVII	186–187	Where days are longest, where shortest
LXXVIII	187	About the first sundial
LXXIX	188	How days may be defined
LXXX	189–190	Differences between peoples according to the world's principles
LXXXI–LXXXIII	191–196	About earthquakes. About chasms in the earth. Signs of an impending quake
LXXXIV	197–198	Precautions against future quakes
LXXXV	199	Unique predictions of earthquakes
LXXXVI	200	Miraculous earthquakes
LXXXVII	201	Locations where the sea has receded
LXXXVIII	202	The creation of islands explained
LXXXIX	202–203	Which ones have been created and when
XC	204	Lands broken up by sea
XCI	204	Islands joined to mainland
XCII	205	Lands changed completely by sea

Chapter(s)	Paragraph(s)	
XCIII	205	Lands that have shrunk spontaneously
XCIV	206	Cities swallowed up by the sea
XCV	206–208	About airholes
XCVI	209	About land that never stops shaking. About islands always in motion
XCVII	210	Locations where there is no rain
XCVIII	210–211	Collection of earth's marvels
XCIX	212–218	On what principles the sea's tides should rise and fall
C	219	Where tides may equally occur according to no principle
CI–CV	220–224	Sea marvels:
CII	221	The moon's power over land and sea
CIII	222	The sun's [power]
CIV	222–223	Why the sea is salty
CV	224	Where the sea is deepest
CVI	224–234	Marvels from springs and rivers
CVII–CX	234–238	Related marvels of fire and water:
CVIII	235	About fossil pitch
CIX	235	About naphtha
CX	236–238	Places which may always be burning
CXI	239–241	Marvels of fire itself
CXII	242–248	Size of the whole world
CXIII	248	Harmony theory of the universe

Book 3
Europe: Spain to Italy

	1–4	[Division of the world into parts]
I	5	[Europa]
II	6	[Hispania Further]
III	7–17	Baetica

Chapter(s)	Paragraph(s)	
IV	18–30	Hispania Nearer
V	31–37	Narbonensis province
VI–X	38–75	Italia to Locri:
IX	53–70	Tiberis, Rome
XI–XIV	76–94	Sixty-four islands, including:
XI	76–79	Baliares
XII	80–82	Corsica
XIII	83–85	Sardinia
XIV	86–94	Sicilia
XV–XX	95–114	Italia from Locri to Ravenna
XX	115–122	About the Padus
XXI–XXII	123–128	Italia across the Padus
XXIII	129–132	Histria
XXIV	133–138	Alpes and Alpine peoples
XXV	139–140	Illyricum, Liburnia
XXVI	141–145	Delmatia
XXVII	146	Norici
XXVIII	147–148	Pannonia
XXIX	149–150	Moesia
XXX	151–152	Islands of the Ionian and Hadriatic [seas]

Book 4
Europe: East, North

I–X	1–22	Epirus, Achaia
XI–XIII	23–28	Graecia
XIV–XV	28–31	Thessalia
XVI	32	Magnesia
XVII	33–39	Macedonia
XVIII	40–51	Thracia

Chapter(s)	Paragraph(s)	
XIX	52–57	Islands off these lands, including:
XX	58–62	Creta
XXI	63–64	Euboea
XXII	65–67	Cyclades
XXIII	68–74	Sporades
XXIV	75–79	Hellespont, Pontus, Maeotis
XXV–XXVI	80–91	Dacia, Sarmatia, Scythia
XXVII	91–97	Islands of Pontus
XXVIII–XXIX	98–101	Germania
XXX	102–104	Ninety-six islands in the Gallic ocean, including Britannia
XXXI	105–106	Belgica Gallia
XXXII	107	Lugdunensis Gallia
XXXIII	108–109	Aquitanica Gallia
XXXIV	110–112	Nearer Hispania along the ocean
XXXV	113–118	Lusitania
XXXVI	119–120	Islands in the Atlantic sea
XXXVII	121–122	Dimensions of Europa in its entirety

Book 5
Africa, Levant, Asia Minor

I	1–21	Mauretaniae
II	22	Numidia
III	23–25	Africa
IV	26–30	Syrtes
V–VI	31–40	Cyrenaica
VII	41–42	Islands off Africa
VIII	43–46	Africa's hinterland
IX–XI	47–64	Egypt: Chora [district of Lower Egypt], Thebaid, Nile

Detailed Table of Contents

Chapter(s)	Paragraph(s)	
XII	65	Arabia near the Egyptian sea
XIII–XIV	66–69	Idumaea, Syria, Palaestine, Samaria
XV	70–73	Judaea
XVI	74	[Decapolitan region]
XVII	75–78	Phoenicians
XVIII–XIX	79–82	Syria Coele, Syria Antiochia
XX–XXI	83–90	Euphrates
XXII	91–93	Cilicia and neighboring peoples
XXIII	94	Isaurica, Omanades
XXIV	94	Pisidia
XXV	95	Lycaonia
XXVI	96	Pamphylia
XXVII	97–99	Mount Taurus
XXVIII	100–102	Lycia
XXIX	103–109	Caria
XXX	110–111	\<Lydia\>
XXXI	112–120	Ionia
XXXII	121–123	Aeolis
XXXIII	124–127	Troas and neighboring peoples
XXXIV–XXXIX	128	212 islands off Asia, including:
XXXV	129–131	Cyprus
XXXVI	132–134	Rhodos, Cous
XXXVII	135	Samus
XXXVIII	136–138	Chius
XXXIX	139–140	Lesbos
XL	141–144	Hellespont, Mysia
XLI	145	Phrygia
XLII	146–147	Galatia and neighboring peoples

Chapter(s)	Paragraph(s)	
XLIII	148–150	Bithynia
XLIV	151	[Islands in Propontis]

Book 6
The East, India

I	1–4	Pontus, Mariandyni
II	5–7	Paphlagones
III	8–9	Cappadoces
IV	10–14	Themiscyrene region and its peoples. Heniochi
V	15–17	Colica region and peoples. Achaean peoples. Other peoples in the same region
VI	18	Cimmerian Bosporus
VII	19–22	Maeotis. Peoples around Maeotis
VIII	23–24	Cappadoces
IX	25	Lesser Armenia, Greater Armenia
X	26–28	Cyrus river, Araxes river
XI	29	Albania, Hiberia and nearby peoples
XII	30–31	Caucasian Gates
XIII	32	Islands in the Pontus
XIV	33–35	Peoples around the Scythian ocean
XV	36–40	Caspian and Hyrcanian sea
XVI	41–42	Adiabene
XVII	43–45	Media, Caspian Gates
XVIII	46–49	Peoples around the Hyrcanian sea
XIX	50–52	Scythian peoples
XX	53–55	Places after the Eoan ocean. Seres
XXI	56–64	Indians
XXII	65–70	Ganges
XXIII	71–80	Indus
XXIV	81–91	Taprobane
XXV	92–95	Ariani and neighboring peoples

Chapter(s)	Paragraph(s)	
XXVI	96–106	Voyages to India
XXVII	107	Carmania
XXVIII	107–111	Persian gulf
XXIX	112–116	Parthians' kingdoms
XXX	117–126	Mesopotamia
XXXI	126–141	Tigris
XXXII	142–162	Arabia
XXXIII	163–168	Red Sea bay
XXXIV	169–177	Trogodytice
XXXV	178–197	Aethiopia
XXXVI	198–201	Islands in the Aethiopic sea
XXXVII	202–205	About the Fortunatae islands
XXXVIII	205–210	Earth's comparative dimensions
XXXIX	211–220	Division of the lands into parallels and equal shadows

Books 7–37 Notable Geographical Passages

Book 7

XXVI	95–99	Clemency, magnanimity
LVI	191–206	About creativity, about the mind

Book 8

LXXXIII	225–228	Creatures possibly in one location but not another
LXXXIV	229	Which [creatures] may only harm outsiders and where; which may [harm] only locals and where

Book 9

XVII	44–53	Which fish are the largest [*not including 47–48*]

Book 10

XXXVII–XXXVIII	74	Memnonides, meleagrides

Chapter(s)	Paragraph(s)	
XXXIX–XL	75	Seleucides, the ibis
XLI	76–79	Birds possibly in one location but not another
LXVII	132	Foreign birds: phalerides, pheasants, numidicae
LXVIII	133–134	Foreign birds: flamingos, partridges, cormorants, choughs, ptarmigans
LXIX	135	'New' birds, *vibiones*

Book 12

Chapter(s)	Paragraph(s)	
XXX	51–54	About the frankincense region
XXXI	55–65	About trees that produce frankincense [*not including 58–62*]
XLI	82–84	About Arabia's good fortune
XLIX	107	*Hammoniacum*
L	108	*Sphagnos*
LI	109	*Cypros*

Book 14

Chapter(s)	Paragraph(s)	
VIII	59–72	Fifty fine wines
IX	73–76	Thirty-eight fine overseas wines

Book 16

Chapter(s)	Paragraph(s)	
I	2–4	Peoples who are without trees
II	5–6	Tree marvels in the northern region
LXV	159–162	Reeds, twenty-eight types of bamboo
LXXXVIII	238	Trees planted by Agamemnon. Trees from the first year of the Trojan War. Trees at Troy named after Ilium older than the Trojan War
LXXXIX	239–240	The same at Argos. Ones planted by Hercules. Ones planted by Apollo. A tree older than Athens

Book 18

Chapter(s)	Paragraph(s)	
LVII	210–216	The constellations' pattern day by day and signal on earth of agricultural work to be done

Chapter(s)	Paragraph(s)	
		Book 19
I	2–6	Nature and remarkable features of flax
II	7–15	Seventeen outstanding varieties
		Book 27
I	1–3	[Medicinal plants]
		Book 31
II	4–5	Different kinds of water
III	6	Types of waters beneficial to the eyes
IV	8	Types producing fertility. Waters curing insanity
V	9	Types for gallstones
VI	10	Types for wounds
VII	10	Types to protect the fetus
VIII	11–12	Types to remove acne
IX	13	Ones for coloring fleeces
X	14	Ones for humans
XI	15	Ones for memory, ones for forgetfulness
XII	15	Ones for heightened senses, ones for sloth, ones for melodious voices
XIII	16	Ones to induce distaste for wine, ones for drunkenness
XIV	17	Ones to substitute for oil
XV	18	Salty and bitter ones
XVI	19	Ones that toss out rocks, induce laughter or tears, or allegedly cure love
XVII	20	Ones that, once drawn, stay hot for three days
XVIII	21–24	Water marvels: Ones in which everything is liable to sink, in which nothing [sinks]

Chapter(s)	Paragraph(s)	
XIX	25–28	Deadly waters, poisonous fish
XX	29–30	Ones that may turn into stone or make stone

Book 32

VI	15	Fishes' mental abilities. Fishes' remarkable qualities
VII	16	Where fish may eat out of the hand
VIII	17	Where there may be responses from fish. Where they may recognize a voice
IX	18–19	Where fish may be pungent, where salty, where freshwater, where not mute. Also their feeling, or lack of feeling, for locations
XI	21	Coral

Book 37

XI	30–46	About amber
LXXVII	201–205	Nature's works compared by region. Its products compared by price

BOOK 2

The Universe, Astronomy, Climate

I The world: Is it finite? Is it one?

1. The world [i.e. universe] and this – whatever other name be preferred to call the sky under whose arc everything occurs – is rightly believed to be a deity, eternal, boundless, neither created nor ever to perish. Humans are neither concerned to investigate what is beyond, nor are mortal minds capable of imagining that.

2. It is sacred, eternal, boundless, all in all, yet really itself all, infinite and yet seeming finite, sure of everything and yet seeming unsure, drawing to itself everything without and within, a work of the natural order while also being itself the natural order.

3. To have pondered its extent – as some have done – and dared to make it known is craziness, as it is for others to have then taken the opportunity (or been given it by them) to claim that there are countless worlds, obliging us to believe that there are just as many natural orders or, if one does encompass all, that there are as many suns and as many moons as there are already in a single one, as well as huge stars beyond counting; as if, despite the wish to conclude them, these same issues would not always occur at the end of our deliberations, or, if this boundlessness could be attributed to Nature the creator of all, that same fact would not be easier to understand in a single work, especially one so large.

4. It is craziness, pure craziness, to rove out from there in order to examine what lies beyond, as if everything within was already clearly known; it is as if someone ignorant of their own dimensions could be preoccupied with those of anything else, or as if the human mind should see what the world itself does not comprehend.

II About its shape

5. Its name in the first place, together with mankind's general agreement to call it a 'globe', inform us that its form has the rounded appearance of a perfect sphere. So also do the facts of the situation, not only because with all its parts such a shape turns upon itself and – with the need to be self-supporting – cordons itself and holds together without the need for joints; not only, too, the sense that none of its parts has a start or finish, as well as its [shape] being ideal for constant rotation (as will shortly emerge). But there is also confirmation from our own eyes because from wherever it be viewed it is an arc centered [on the viewer], an impossibility with any other shape.

III About its motion. Why it should be called the universe

6. The sun's rising and setting leave no doubt, then, that this is its shape and that in the space of 24 hours it is rotated in an eternal, unceasing orbit at indescribable speed. I, for one, cannot easily say whether such a great mass rotating with its incessant spin makes a loud noise that is beyond our hearing ability – any more, by Hercules, than I can say whether at the same time the circling stars jingle as they revolve in their own orbits – or whether there is some soft, incredibly charming harmony. For those of us living in it, the world glides silently both day and night.

7. Marked on it are any number of likenesses of animals and everything else, but it does not have – what we observe in birds' eggs – a slippery body, smooth all over, as very famous authors have said. Evidence on the ground shows this, since likenesses are produced from the seeds of all that drops down there, beyond count especially in the sea, and typically freakish when crossbred; our own eyes confirm this too – here the figure of a bear, there of a bull, elsewhere of a letter [of the alphabet], while the center of the circle through the pole is brighter.

8. I, for one, am also influenced by mankind's consensus. Thus Greeks gave it the name 'ornament' [Greek *kosmos*], and ours is *mundus* because of its perfect, complete elegance. There is certainly no doubt that we have called it *caelum* from the fact that it is engraved,[1] as Marcus Varro explains.

---

[1] Latin *caelare*, to engrave in relief, or carve.

9. The orderly structure of things helps, since its circle called the zodiac is divided into the likenesses of 12 creatures, and the sun has plotted its course accordingly through them for age after age.

IV About the elements

10. I see no reason to doubt that there are four elements: uppermost is fire, the source of all those stars that shine like eyes; next is breath which Greeks and we have the same word for, 'air', the means of life that passes through everything and is bound up with all. Suspended by its force along with the fourth element water, the earth is poised in mid-space.

11. Hence a bond is created by the mutual incorporation of different [substances]: light ones are prevented from flying up by the weight of heavy ones, while on the other hand the heavy, suspended by light ones with upward tendencies, do not fall down. So, because of equal pressure from opposite directions each one remains in its place, bound together by the world's unceasing orbit, and since this is always running back on itself, the earth is at the bottom and in the center overall. It remains suspended at the pivot of the universe and balances what makes its suspension possible, so it alone is fixed while the universe rotates around it; it is connected to everything, and everything rests on it.

12. Between it and the sky are suspended – by the same breath, at fixed intervals – the seven stars we call 'wanderers'[2] because of their courses (although nothing else wanders less than them). Prominent in their midst is the sun, greatest in size and power as well as ruler not only of seasons and lands, but also of the stars themselves and sky.

13. To judge by its efforts, we may fairly believe [the sun] to be the whole world's spirit, or more clearly its mind, as well as the principal controller and deity of Nature. It provides light for activity and removes darkness; it dims or brightens the rest of the stars; it regulates the changes of season and the invariable rebirth of the year in line with Nature's practice; it dispels dismal skies, and even lightens gloom in human minds; it lends its light to other stars too, in addition to being brilliant, outstanding, all-seeing, even all-hearing – its qualities exclusively, as I see Homer, the foremost of authors, favored.

---

[2] *Errantia*, planets: from the Greek *planasthai*, to wander, *planetai*, wanderers.

V About God

14. Hence I think it sheer human weakness to investigate the appearance and shape of God. Whoever God is, if he really is distinct [from the world] and in whatever part of it, he consists entirely of sensibility, entirely of sight, entirely of hearing, entirely of soul, entirely of spirit, entirely of himself. Indeed, to believe in gods without number (derived in fact from human failings) such as Chastity, Concord, Intellect, Hope, Honor, Mercy, Loyalty – or, as Democritus liked, merely two, Punishment and Reward – is all the more mindless.

15. Weak and over-burdened humanity, conscious of its own frailty, has separated [God] into parts so that individuals may worship those groups that each needs most. Consequently, we find different names among different peoples and countless deities among the same. Even Underworld deities are divided by type along with illnesses and many diseases too, given our anxious fear and longing to appease them.

16. So there is even a shrine to Fever dedicated by the state on the Palatine [hill in Rome], one to Bereaved Parents at the temple of the Household Gods, and an altar to Misfortune on the Esquiline. For this reason one could reckon that there is an even larger community of heavenly beings than of humans, since individuals of their own accord also make an equal number of their own gods by securing Junos and Genii for themselves. In fact certain peoples have animals for gods – even some disgusting ones, and many it would be quite indecent to describe – and they swear by stinking food and such like.

17. To believe in gods inter-marrying, and that in all this long time no birth has resulted, that some of them are perpetually aged and grey-haired, while others are young and even boys, dark-skinned, winged, lame, hatched from eggs, living and dying on alternate days, is almost childish delusion. But the supreme effrontery is to invent adultery between them, and then quarrels and hatred, and even that there are deities of theft and crime.

18. God is human helping human, and this is the way to everlasting glory. It is the one Rome's leaders have taken, and now – in the direction of heaven – it is being traversed with his children by Vespasian Augustus, the greatest ruler of all time, who is rescuing an exhausted state of things.

19. To enroll such men among the deities is the most ancient means of rendering thanks to benefactors. To be sure, the names both of other gods and of stars that I mentioned above originated from benefactions.

In light of this interpretation of Nature, who would not admit that it is ridiculous for them to be calling each other Jupiter or Mercury or some other names, and that there is a heavenly nomenclature?

20. Should we believe or doubt that the highest being, whatever it is, exercises concern for human affairs, and that it would not be degraded by such a depressing and wide-ranging function? It is hardly worth deciding which is more advantageous to humanity when some have no respect for gods, while others are shamefully respectful.

21. These are slavish adherents of foreign cults, who wear gods on their fingers [i.e. rings]; they also worship monsters, they sentence them and devise their diet, they inflict dire commands on themselves so as to be deprived even of restful sleep. They make no decision about marriages or children or in short anything else except with the help of sacrifices. Others practice deception on the Capitol itself, and commit perjury in the name of Jupiter the lightning-striker – even gaining assistance from their crimes, whereas the previous ones are penalized by their devotion.

22. Yet humanity's speculation about God is further muddled by its invention for itself of a deity midway between these two outlooks: throughout the whole world, in every place and at every hour, everyone is calling out loud by name just to Fortune. She alone is accused, alone put on trial, alone contemplated, alone praised, alone blamed and worshiped with insults, flighty and unstable, indeed by general opinion also considered blind, erratic, changeable, unreliable, and a fickle supporter of the undeserving. All costs are attributed to her as are all gains too, and in humanity's whole ledger she alone features in both columns. And we are so much the victims of chance that Fortune, who demonstrates the uncertainty of God's existence, herself serves as a substitute for God.

23. Another group dismisses her and attributes its fate to their star and the laws of their birth, [with the view] that God, after issuing a single order about what will ever happen for the entire future, then permitted himself to relax in perpetuity. This opinion is beginning to take root, and the path towards it is being taken by masses of both the educated and the ignorant alike.

24. Look out for lightning alerts, predictions of oracles, prophecies of diviners, and even trivial sneezes and missteps taken to be ominous! Deified Augustus recorded that, on the day when he was nearly overthrown by troops mutinying, he had put his left boot on the wrong foot.

25. Short-sighted humanity is overwhelmed by these isolated events, which between them merely show for sure that nothing is certain, and nothing more pathetic or proud than mankind. Indeed, all that other living creatures care about is feeding, which Nature's bounty spontaneously takes care of for them. Their one blessing much superior to all others is that they give no thought to glory, money, ambition or, in particular, death.

26. Even so, life-experience leads to the belief that gods do care about these aspects of human affairs, that punishment for wrongdoing, although sometimes delayed – God has such a burden, after all – never lacks force, and that humans were not created as his close relatives in order to match the beasts in beastliness.

27. But there is comfort to be derived from Nature falling short of perfection when it comes to humans, in particular that not even God can do everything. For, even if he wanted to, he cannot commit suicide, something most advantageous that he gave humans amid life's great penalties. Nor can he give humans eternal life, nor recall the dead, nor arrange to delete the life of someone who has lived, or for someone who has held an office not to have held it. He has no authority over the past except to forget it, nor (as this association with God is also linked with facetious arguments) can he make twice ten not equal 20, let alone accomplish many similar feats. Without doubt the power of Nature is revealed by these means, as well as the fact that this is what we call God. To have digressed onto these topics will not have been irrelevant, because they have gained plenty of attention thanks to the constant inquiries about God.

VI About the nature of the wandering stars

28. From here, let us return to the rest of Nature. Stars, as we have said, are attached to the universe and are not, as common folk suppose, assigned to each of us, distributed among humans to shine according to their fortunes, bright for the rich, less so for the poor, and dim for failures; nor does each human have a star that rises and dies with them, and falling stars do not mean that someone has breathed their last.

29. We are not so closely linked to the sky that the stars' brightness there should also be mortal, like our own fate. When they are thought to fall it is because, after an excessive intake of the fluid they absorb, they give off the surplus with blazing energy, just as we observe happening in our world too when lamps have been lit with a stream of oil.

30. Rather, by nature the heavenly bodies – which bind the universe together, woven into its fabric – are eternal; yet their power makes substantial impact on the earth, and their influence, brightness and size have made it possible to be informed about them in great detail, as we shall show in the proper place [18.210]. Also, it will be more appropriate to refer to the theory of the orbits in the sky when dealing with earth [6.211], because these relate entirely to it; in addition, mention of discoveries made about the zodiac should be postponed no further.

31. Anaximander of Miletus, in the fifty-eighth Olympiad [548–545 BCE], is said to have been the first to explain its slant and thus to have opened the door to such knowledge; he was followed by Cleostratus for the constellations in [the sky], beginning with Aries and Sagittarius; long before, Atlas [had explained] the sphere itself. Moving on from the framework of the universe, we shall now consider what else there is between sky and earth.

32. It is certain that the planet called Saturn is the highest and therefore appears to be the smallest, circling in the largest orbit and returning to its starting-point within a minimum of 30 years. However, the motion of all the wandering stars [planets], including the sun and moon, proceeds in the opposite direction to the world's; in other words they go left, while it always speeds to the right.

33. And although they are lifted away from it by their constant rotation at tremendous speed, being swept westward, even so each travels in the opposite direction along its own path. Consequently the air is not rendered inert when wound into a sluggish mass by the world's unceasing gyration, but is spread around when separated and dispersed by the planets' thrust in the opposite direction.

34. Planet Saturn, however, is naturally cold and frozen, while Jupiter orbits much lower than it does, and so with a faster pace completes a rotation in 12 years. Third, Mars – which some call Hercules – is ablaze with fire because of its closeness to the sun, and orbits in about two years. Because of its excessive heat and Saturn's cold, Jupiter in between the two is balanced by them both and made healthy.

35. Next, the sun's path is indeed divided into 360 degrees but, so that an observation of its shadows may return to the starting-post, five-and-a-quarter days are added to the year. Thus one intercalary day is added every fourth year, so that our time-reckoning may match the sun's journey.

36. Below the sun orbits a huge planet called Venus which wanders in alternating directions and with its additional names rivals the sun and moon. In fact when it comes early and rises before dawn, it is given the name Lucifer as another sun advancing the day; by contrast, when it shines after sunset it is called Vesper for prolonging daylight or taking on the moon's role.

37. Pythagoras the Samian first discovered its nature around the forty-second Olympiad [612–609 BCE], which was 142 years after the city of Rome [was founded]. Now it outdoes all other stars in size, and is actually so bright as to be the only planet with rays that cast a shadow. So there are also many competing names for it: for some call it Juno, others Isis, others Mother of the Gods.

38. Everything on earth is given life by its nature, because at one or another of its risings it sprinkles vitalizing dew which not only fills earth's reproductive organs, but also stimulates those of every living creature. Moreover, its orbit of the zodiac takes 348 days, and according to Timaeus it is never more than 46 degrees distant from the sun.

39. With a similar course but not at all the same size or energy, the star next to it is Mercury, which some call Apollo. A lower orbit makes its circuit nine days speedier; sometimes rising before sunrise, other times shining after sunset, but (as Cidenas and Sosigenes explain) never more than 22 degrees distant from the sun. Consequently the course of these stars is also exceptional and does not match those mentioned above.

40. Those are observed to be either a quarter or a third part of the sky distant from the sun, and often opposite the sun. In addition they all have other, larger circuits of full rotation as specified in the concept of the Great Year.

41. But the last planet, the moon, commands everyone's admiration, it being most familiar on earth and devised by Nature to counteract darkness. The enigma of its many shapes has given mental torture to those who ponder it, resenting how their nearest star is the one they know least about.

42. It is always waxing or waning, at one time looking as if curved into a crescent shape, at another divided in half, at another hollowed out into a circle, as well as pock-marked and instantly shining bright, huge with a full disc and then suddenly nothing, some of the time there all night, at other times [appearing] late and reinforcing the sun's light for part of the day, in eclipse and yet visible during eclipses, and in hiding at the end of the month when not reckoned to be suffering an eclipse;

43. at one stage low, at another high, though not even consistently, but sometimes raised up in the sky and at others touching mountains, now swept up north, now cast down south. Endymion was the first human to detect these individual characteristics [of the moon], and this is the basis of the story told about his love for it. For certain, we are not grateful to those whose work and concern have shed light for us on this source of light, and so amazingly noxious is the human mind that in our historical writing we prefer to feature blood and slaughter – so that those who know nothing about the world itself may be informed about mankind's crimes.

44. Because [the moon] is nearest to the pivot, it has the shortest orbit, covering the same distance in 27 and one-third days that Saturn, the highest star, completes in 30 years as already mentioned [2.32]. Then, after a two-day pause in conjunction with the sun, at the latest on the thirtieth day it departs again on the same course. It has perhaps taught [us] everything that could be known about in the sky:

45. that [first,] the year should be divided into periods of 12 months, since it is for this number of times that [the moon] itself follows the sun as it returns to its starting-point; that [second, the moon], like the rest of

the stars [planets], is controlled by the sun's brilliance insofar as the light with which it shines is all borrowed from there, such as we observe flickering when reflected in water; that hence [third, the moon] only releases moisture with rather mild and partial pressure and even increases it, whereas the sun's rays soak it up; that hence [fourth, the moon's] light visibly varies because it is full only when opposite [the sun], and on all other days displays to earth from itself merely as much [light] as it draws for itself from the sun;

46. that [fifth, the moon] is invisible when in conjunction with [the sun], because when turned in its direction it reflects back there the full quantity of light drawn from there; that [sixth,] stars are undoubtedly sustained by moisture from earth, since a half[-moon] is never visibly blotched and there is evidently not the strength at this stage to absorb more than a normal amount – the blotches being simply dirt from earth lifted with the moisture; that [seventh, the moon's] eclipses and those of the sun – the most amazing events in the entire study of Nature, like a miracle – serve to indicate their size and shadows.

VII About lunar and solar eclipses. About darkness

47. It is quite clear that the sun is hidden when the moon comes in between, and the moon when earth blocks it, also that there is reciprocal impact, with the moon's position in between depriving earth of the sun's rays, and earth likewise depriving the moon. As the latter passes, dusk spreads all of a sudden, and correspondingly the former's shadow dims the moon; the darkness is merely earth's shadow, but the shadow's shape is like a cone and an inverted spinning-top when with just its point it bears down on the moon and does not go beyond its height, since no other star is darkened in the same way and such a figure always tapers off to a point.

48. That shadows are swallowed up by distance is demonstrated by birds flying very high. This, then, is their limit, where the air ends and ether begins. Everything above the moon is clear and full of daylight. The stars are visible to us through the night, however, just as lights remaining are in darkness, and hence the moon's eclipses are at night. Eclipses of both [sun and moon] occur regularly though not monthly, because of the zodiac's slant and the moon's course curving very erratically (as already mentioned [2.43]), and with the stars' motions not always matching to the smallest fraction of a degree.

VIII About the size of stars

49. Such theorizing draws human minds skyward and, as they ponder from there so to speak, reveals the scale of Nature's three greatest parts. To be sure, if earth were larger than the moon, it would be impossible for the whole sun to be blocked from earth while the moon comes in between. From this pair [of findings] will emerge thirdly the immensity of the sun, so there should be no need to investigate its bulk from what our eyes prove and our minds imagine. It is huge

50. because the shadows it casts of trees stretched out along boundaries for any number of miles are equally spaced, just as if it were at the middle for the whole length; and because at an equinox it appears vertically above at the same time to everyone living in the southern region; and because the shadows of those living within the summer-solstice [celestial] circle fall to the north at midday, but to the west at sunrise, which could not happen unless [the sun] was much larger than the earth; and because when rising it surpasses mount Ida in width, broadly encompassing it to right and left, despite in particular being so far away.

51. A lunar eclipse provides secure proof of its size, just as one of its own eclipses does for how small earth is. Now given that there are three shapes for shadows, clearly if the object casting one matches the light-source in size, its shadow is cast in the shape of a column with no end to it; but if the object is larger than the light-source, its shadow has the shape of an upright spinning-top, narrowest at the bottom and likewise of infinite length; if the object is smaller than the light, the shape is that of a cone coming to an end with a tip, and such a shadow is observed in a lunar eclipse.

52. It is plain, with no further room for doubt, that the [sun's] size exceeds earth's, as indeed Nature itself also quietly indicates. For why, when dividing up the seasons of the year, does [the sun] move away during winter or revive earth with dark nights? Without doubt it might burn it up, and actually does burn up a certain part: its size is that great.

IX–X What has been discovered by observing the sky

53. The first Roman-born person to put into circulation an explanation for both [lunar and solar] eclipses was in fact Sulpicius Gallus – consul with Marcus Marcellus [166 BCE] – though at the time [168 BCE] a military tribune who freed the army from fear on the day before King Perseus was defeated by Paulus, when the general led him in front of an assembly to predict an eclipse; and later he wrote his book. However, first of all the Greeks to make an investigation was Thales of Miletus, who predicted a solar eclipse in the fourth year of the forty-eighth Olympiad [584 BCE] during Alyattes' reign, 170 years after the foundation of the city [of Rome]. After them, Hipparchus prophesied both [types of eclipse] for 600 years, including the months [when they would occur] where [various] peoples [live], as well as days, hours, specific locations and visibility among [various] communities. Time attests him to have been none other than a contributor to Nature's plans.

54. What outsized and superhuman men they were who discovered such great deities' rules and have now relieved the mental misery of humans terrified that eclipses [signify] wrongdoing or some kind of death for stars (those sublime-sounding poets Stesichorus and Pindar were clearly frightened by a solar eclipse), or convinced that the dying moon is poisoned, so that they help it by banging and clashing; because of this alarm, in his ignorance the Athenians' general Nicias was too frightened to lead his fleet out of harbor [during the siege of Syracusae, 413 BCE] and weakened their power. All hail to your acumen, you interpreters of the sky and experts on the natural order, finders of proof that has overcome gods and humans!

55. So who, observing all this along with the regularity of the stars' 'labors' [i.e. eclipses] – as has been the fashion to call them – would not leave those born mortal to determine their own fate?
  Now in concise summary form I shall touch upon what is acknowledged about the same matters, explaining them only where necessary and in brief because such exposition does not fit the scheme of this work, and because the impossibility of advancing reasons for every single thing is less remarkable than the agreement reached about some.

X

56. For certain, eclipses repeat their orbits after 223 months. There are solar eclipses only when the moon is in its last or first stage, at what is called the 'conjunction', and lunar ones only when it is full and always just short of their last occurrence. However, every year at specified days and hours there are both solar and lunar eclipses below the earth, but even when they occur above it, they are not visible everywhere, sometimes because of clouds, more often because earth's round mass blocks out the curvature of the universe.

57. Within the last 200 years Hipparchus with his insight has also discovered that a lunar eclipse sometimes occurs five months after the previous one, but seven months in the case of a solar eclipse, and that the latter while above the earth is out of sight twice within 30 days, although the two are not visible in the same places. The most remarkable aspect of this remarkable event is that, at the right moment for earth's shadow to obscure the moon, this is sometimes cast from the west side and at others from the east. [Another of Hipparchus' discoveries is] the reason why, although this obscuring shadow should be under the earth at sunrise, there has before now been an occurrence of the moon eclipsing in the west while both stars [sun and moon] were visible above earth. For eclipses of each of them to occur within 15 days has even happened in our own time during the Vespasians' rule, in the father's third consulship [71 CE] and in his son's second [72 CE].

XI About the moon's movement

58. There is no doubt that the moon always has its crescents turned away from the sun, that it faces east if waxing and west if waning, that it shines for an extra three-quarters plus one-twenty-fourth of an hour [47½ minutes] from the new moon until the full one and for this much less when waning, and that it is always out of sight within 14 degrees of the sun. This fact establishes that the planets are a larger size than the moon, since these occasionally become visible even at seven degrees. But their distance from earth makes them appear smaller, just as the sun's brightness renders the [stars] fixed in the sky invisible by day, even though they are shining just as during the night, something demonstrated by solar eclipses and very deep wells.

XII Movements of planets and patterns of their radiance

59. Those three planets which we have said are positioned above the sun [Mars, Jupiter, Saturn] are out of sight when they move with it, but early in the morning while rising they are separated from it by no more than 11 degrees. After that they are guided by contact with its rays and take up their morning stations, also called their 'first', in a triangle 120 degrees away; then 180 degrees away in the opposite direction they make their evening risings. Again, approaching from the other side, they take up their evening stations – their 'second', 120 degrees away – until at 12 degrees [the sun] catches up and obscures them, in what is called their evening setting.

60. Planet Mars, as it is nearer, feels the rays even from its quadrature, at 90 degrees, from which its motion also gained its name, being called the 'first' or 'second 90' after each rising. The same planet remains in a single station in one sign [of the zodiac] for six months, and at other times spends two months, while the others do not spend as much as four months in each of their stations.

61. In a similar way, the two lower planets [Mercury and Venus] are hidden at their evening conjunction, and after being left behind by the sun they make their morning risings the same number of degrees away; then from the extreme limits of their distance they follow the sun, and having overtaken it they are hidden in their morning setting and go past. Later, at the same distance [from the sun as in the morning], they rise in the evening as far as the limits we have mentioned; from these they make a retrogradation towards the sun and vanish in their evening setting. Planet Venus also occupies two stations – at morning and evening – after each rising, at the furthest extent of its distance. Mercury's stations are for too brief an instant to be detected.

XIII–XIV Why they should appear more distant and at other times closer

62. This scheme of [the planets] becoming brighter as well as less visible is made more baffling by their motion and obscured by many marvels: they change in size and color, they approach the north and depart to the south, and all of a sudden they appear to be closer now to earth, now to the sky. In reporting on such topics, although we shall differ considerably from our predecessors, we acknowledge too the contribution of those who first demonstrated the means to investigate them, so that no-one should give up hope that through the ages advances always occur.

63. All this happens for a great many reasons: first because of circles, which Greeks call *apsides* [loops] in the case of stars (for Greek terminology will have to be used). Each [planet], moreover, has its own [circle], different from those of the universe, since earth with two extremities (called poles)

is the center of the sky and also of the zodiac, which is situated on a slant between them [the poles]. All this can be established beyond any doubt with the use of compasses. So each of these *apsides* extends out from its own different center and consequently the [planets] have different orbits and divergent paths, because the interior *apsides* are necessarily shorter.

64. Therefore, the *apsides* are most distant from earth's center when in Scorpio for Saturn, in Virgo for Jupiter, in Leo for Mars, in Gemini for the sun, in Sagittarius for Venus, in Capricorn for Mercury, at their midpoint in each instance; and at the opposite they are lowest, as well as nearest to earth's center. So it turns out that [the planets] appear to move more slowly while making their most distant circuit, not because they would accelerate or slow down their natural motions – which are fixed and unique to each – but because lines drawn from the very top of an *apsis* have to converge at the center like the spokes of a wheel, and the same motion is felt to be faster here or slower there depending on its proximity to the center.

65. An alternative reason for the [planets'] height is because from their center they have their most distant *apsides* in different constellations – Saturn in the twenty-first degree of Libra, Jupiter in fifteenth of Cancer, Mars in twenty-eighth of Capricorn, the sun in nineteenth of Aries, Venus in twenty-seventh of Pisces, Mercury in fifteenth of Virgo, the moon in third of Taurus. A third reason for their distance is deduced from the sky's dimensions not from orbits, because [our] eyes determine how they rise or fall through the air's expanses.

66. A related matter is the reason for the zodiac's breadth and slant. The stars we have mentioned are drawn through it, and only the part of earth lying below it is habitable, everywhere else towards the poles being barren. Planet Venus alone goes two degrees beyond [the zodiac], which explains why it is feasible for some animals to be born even in the world's deserts. The moon also roams through its entire breadth, but never goes beyond it. Planet Mercury strays very loosely from these [two], but for no more than eight of [the zodiac's] 12 degrees – the latter has this much breadth, after all – and not even equally through these, but two in its middle and four above, two below.

67. Then the sun is drawn unevenly in its middle within two degrees on a winding snake-like path, planet Mars in its middle within four, Jupiter in its middle and two above it [northward], Saturn within two like the sun. This will account for their latitudes when they are falling south or rising north. But this agrees neither with the third reason mentioned [2.65] for the planets rising from earth to the sky – as has been generally, but wrongly, assumed – nor equally [with the notion] that they ascend simultaneously too. To refute those who think this, there is the need to develop a tremendously acute explanation that encompasses all these reasons.

68. It is agreed that at their evening setting the planets are nearest to earth in both altitude and latitude, that their morning risings occur at the start of both, and that their stations are at the midpoints of their latitudes, called 'ecliptics'. It is likewise acknowledged that they accelerate so long as they are in earth's vicinity and decelerate when they move far away. This reasoning is proven most emphatically by the moon's elevations. There is equally no doubt that the three higher [planets] still accelerate in their morning risings, and decelerate when they proceed from their first stations all the way to their second stations.

69. This being the case, it will be obvious that they ascend through the latitudes from their morning rising, because this is the stage at which their movement first begins to slow down; but in their first stations they are also gaining height, since at that point the numbers [i.e. degrees] first begin to go down and the stars to recede.

In each case a separate explanation must be given. [The planets], when struck at the degree we mentioned and by a trinal sun-ray, are prevented from taking a straight course and are elevated up high by blazing force.

70. Our sight is incapable of immediately grasping this, and so [the planets] are assumed to be stationary, hence the term *statio* [station]. Then the same ray's force advances and compels them – struck by its heat – to recede. This occurs much more in their evening rising, when they are pushed out to their topmost *apsides* by the full sun facing them, and they are least visible because of being at their greatest distance away and conveyed at their slowest pace, seeming all the slower given that this happens in the most distant constellations of the *apsides*.

71. After their evening rise they go down in latitude, their motion now decelerating more slowly without, however, accelerating before their second stations; then they lose height too, with the sun-ray overwhelming them from the other side and pressing them down towards earth again with the same force [with which] it raised them skyward from their previous triangle. Whether the rays come from below or above makes that much difference, and all this recurs to a greater extent in the evening setting. This is how the higher stars are to be accounted for; to do so for the rest is more difficult, and no-one before us has explained it.

XIV

72. First, there should be a statement about why planet Venus never goes more than 46 degrees from the sun and Mercury never more than 20, and why they often move back towards the sun within that range. Both have inverted *apsides* situated under the sun, and as much of their orbits is below it as in the case of the previously mentioned planets is above it. And so [Venus and Mercury] cannot be further away, because the curve of their *apsides* lacks greater length there. By the same token, therefore, the edges of their *apsides* impose a limit on both, and the distance of their longitude counterbalances the extension of their latitudes.

73. But why do they not always reach 46 and 20 degrees? In fact they do, but the reason escapes those who compile tables. For it is apparent that their *apsides* too are shifted, because they never cross the sun. So when on one side or the other the edges have fallen to its degree specifically, then the [two] planets also are considered to reach their furthest distances away; when the edges have been within the [planets'] range, then the [two] themselves are also forced to recede promptly by the corresponding number of degrees, since this is always as far as each may go.

74. Here, too, the contrary principle of their motions finds an explanation. For the higher [planets] are raised fastest in the evening setting, [but Venus and Mercury] slowest; the former are furthest from earth when they move slowest, [but] the latter are quickest at this stage because, just as proximity to the center serves to accelerate the former, so too the limit of their orbit does for the latter. The former begin to reduce speed from their morning setting, but the latter to increase it; the former take a receding path from their morning station to their evening one, whereas Venus does this from the evening one to the morning one.

75. Moreover, after its morning rise [Venus] begins to ascend in latitude, and from its morning station its height does too, pursuing the sun (so it is swiftest and highest at its morning setting); but from its evening rising it comes to a lower latitude and slows down, in fact receding and at the same time from its evening station losing altitude. By contrast, planet Mercury begins an ascent in both dimensions after its morning rising, but loses latitude after its evening one and, having followed the sun until it is 15 degrees away, halts and barely moves for four days.

76. Later, [Mercury] loses height and recedes from its evening setting to morning rise, and only it and the moon set in the same number of days as they have risen. Venus' rise is 15 times as many days [as its setting]; Saturn and Jupiter by contrast set in twice as many, Mars too in four times. So great is Nature's variety, but the reason is clear: those that struggle towards the heat of the sun also have difficulty coming down.

XV General principles of planets

77. Much more can be revealed about these secrets of Nature and the laws it obeys – for example, regarding planet Mars, whose course is the most difficult one to observe: it never takes up its station with planet Jupiter at a 120-degree angle, and very rarely with it at a 60-degree one (a number which matches the universe's hexagonal shape), and [these two planets] only rise simultaneously in two constellations, Cancer and Leo. Planet Mercury in fact rarely makes its evening rises in Pisces and most frequently in Virgo, its morning ones in Libra and morning ones also in Aquarius, but most rarely in Leo; it does not make its retrogradation in Taurus and Gemini, and not below the twenty-fifth degree in Cancer;

78. Gemini is the only constellation in which the moon makes a conjunction with the sun twice, and in Sagittarius alone it sometimes makes no conjunction at all; the only constellation in which the old moon and the new are visible in the same day and night is Aries (another occurrence seen by few mortals, hence the fame gained by Lynceus for sighting it); and the planets Saturn and Mars are not to be seen in the sky for 170 days at most, Jupiter for 36 days (or 10 days less at least), Venus for 69 (or 52 at least), Mercury for 13 (or 17 at most).

XVI Reasons why their colors should change

79. The colors [of planets] are determined by the principles of their heights above earth, since they derive their appearance from the atmosphere of whichever [constellation] they have come to in rising, and the orbit of another's path leaves its mark as they approach from either direction, a colder one creating a pale color, a hotter red, a windy drab, and the sun and the intersections of *apsides* and their extreme orbits deep black. Actually each one has its own color: Saturn white, Jupiter plain, Mars fiery, Lucifer bright, Vesper glowing, Mercury radiant, the moon soft, the sun flaming as it rises, radiant thereafter; linked with these causes too is the visibility of those [stars] fixed in the sky.

80. For there are times when a quantity of them is clustered round the half-moon's discs and a calm night is lighting them up gently; at other times they are dispersed, amazing us by how much they have vanished, with the full moon concealing them or when rays from the sun or the [planets] mentioned above have constricted our vision. Yet the moon itself is also undoubtedly aware of variations in the sun-rays impacting it, because the sky's curvature blunts them, bent as they are, except where they deliver right-angle blows. And so, in the sun's quadrature it is split in half, and in its trinal aspect it orbits with its sphere half hollowed out, but this is made full when directly opposite; then again, when waning it assumes the same shapes at equal intervals, following the same principles as those for the three planets above the sun.

XVII The sun's movements. Theory of unequal days

81. But the sun itself has four variations: two in spring and autumn when night and day are the same [length], and it coincides with earth's center at the eighth degree of Aries and Libra; two when its course is changed, in the direction of lengthening the day in winter (in the eighth degree of Capricorn) and likewise the night at the [summer] solstice (in the same degree of Cancer). Unequal [days and nights] are caused by the zodiac's slant, since there is at all times an equal part of the universe above and below the earth, but those constellations that move up directly when rising hold their light for a longer spell, whereas those that move obliquely make the shift in a shorter period.

XVIII Why thunderbolts are assigned to Jupiter

82. Most people are unaware of a discovery made by leading scholars from their major study of the sky, namely that the objects which fall to earth and are called thunderbolts are fires of the three upper planets, although especially of the one of these located in the middle position, perhaps because by this means it discharges the taint of excessive moisture from the upper circle and of heat from the lower; hence Jupiter is said to hurl thunderbolts. Therefore, just as charcoal is spat from a burning log with a crackle, so is heavenly fire from a planet, conveying prophecies with it and still working for the gods even after being discarded [by the planet]. And this occurs along with tremendous atmospheric disturbance, either because accumulated moisture triggers an overflow or because it is stirred up by a planet – pregnant so to speak – somehow giving birth.

XIX Distances between planets

83. There have been many attempts to ascertain the distance of the planets from earth, and it has been claimed that the sun is 19 times further away from the moon than the moon itself is from earth. In fact Pythagoras, a man with a sharp mind, calculated that it is 126,000 stades from earth to the moon, double that from it to the sun, and triple from there to the 12 zodiacal signs. This was also the opinion of our own Gallus Sulpicius.

XX About the planets' music

84. But Pythagoras sometimes invokes music theory as well, saying that the distance from earth to the moon amounts to one tone, from it to Mercury a semitone, and the same from there to Venus, from it to the sun a tone and a half, from the sun to Mars one tone (the same distance as from earth to the moon), from [Mars] to Jupiter a semitone, and the same from there to Saturn, and on to the zodiac a tone and a half. Thus seven tones produce what [Greeks] call *dia pason harmonian*, in other words 'universal harmony'. In this, Saturn moves in the Dorian tone, Jupiter in the Phrygian, and likewise for the rest of the planets – the fine detail being amusing rather than essential.

XXI About the universe's geometry

85. A stade works out to 125 of our [Roman] paces, that is, 625 feet. Posidonius says that mists, winds and clouds attain a height of under 40 stades from earth – from there onwards the air is pure and liquid and the light serene – but that from the turbulent [zone] to the moon it is 2,000,000 stades, and from there to the sun 500,000,000; thanks to this distance even its enormous size does not burn up the earth. Most accounts, however, have stated that clouds rise to a height of 9,000.

These [figures] should be reported because they have been claimed, although they are conjectural and impossible to sort out. However, in these matters there is a method of geometrical argument – one that never misleads and is impossible to reject – open to anyone interested in researching this more deeply, yet not with a mind to establish the distances (because wishing to do this would be a fairly crazy use of time), but merely to make acceptable estimates by intelligent inference.

86. Since it appears from the sun's circuit that the circle along which its globe moves extends nearly 366 degrees, and since the diameter [of a circle] always amounts to the sum of one-third and a little less than three-sevenths of its circumference, it appears that – with half subtracted because of earth coming in the middle – the extent of the sun's distance is nearly one-sixth of this immense range thought to be encompassed by its orbit around earth; but the moon's is one-twelfth, since it travels a circuit so much shorter than the sun's – as it would, situated half-way between the sun and earth.

87. I am astonished by how unscrupulous the human mind may become once tempted by some trivial success, such as when issues like those above offer the opportunity to be shameless. And having dared to guess the distance from the sun to earth, [people] apply the same figure to the sky because of the sun being midway – which also allows them to calculate instantly dimensions for the universe on their fingers. For they reckon the ratio of a diameter to a circumference is seven to twenty-two, as if the sky's dimensions may be calculated straightforwardly from a plumb-line!

88. The Egyptian calculation issued by Petosiris and Nechepso reckons that one degree of the moon's orbit (the smallest, as mentioned above) is just over 33 stades, Saturn's (the widest) double that, and the sun's (midway, as we have said) half of the two. This calculation is utterly shameful, because adding to Saturn's orbit the distance from it to the zodiac itself produces a multiple that is impossible to compute.

XXII–XXIII About shooting stars. About the nature, position and types of comets

89. A few other points about the universe remain, such as stars suddenly being formed even in the sky itself. There are many types of them: Greeks call them comets, we call them 'long-haired' [stars], bristling with blood-red hair and shaggy on top just like dreadlocks. [Greeks] again have the term *pogoniae* for [comets] with a mane streaming from their lower part in the shape of a long beard. There are meteors which vibrate like a dart and are warnings of imminent events. This [type] and its most recent sighting to date are the subject of a celebrated poem by Titus Imperator Caesar written during his fifth consulship [76 CE]. When shorter and tapered to a point, these same ones have been called *xiphiae*, the palest of all, gleaming rather like a sword and without any of the rays which even the *discus*-type – with a shape to be expected from its name, but the color of amber – occasionally shoots from its edges.

90. The *pitheus* is recognizable from its jar-shape, with smoky light in its cavity. The *ceratias* has a horn-shape, as appeared when Graecia fought the battle of Salamis [480 BCE]. The *lampadias* resembles burning torches, the *hippeus* horses' manes, with the speediest motion and spinning round in a circle. A shining comet also exists, with a silvery tail so bright that to observe it is almost outlawed, and displaying in itself a god's face in human form. *Hirci* also exist, circled by something like shaggy hair and a sort of cloud. Once to date, in the 108th Olympiad, 408 years after the founding of the city [346 BCE], a mane-shaped [comet] has changed into a spear-shaped one. It has been noted that the shortest period for which they are seen is seven days, the longest 180.

XXIII

91. Some, moreover, move like planets, others stick to being immobile; nearly all are under the northern [sky] although not in any particular part of it, but mostly in the bright area which has acquired the name Milky Ring [Way]. Aristotle says also that many are seen at the same time – his discovery alone, so far as I know – and that they give warning of severe winds or heat. They also occur in the winter months and at the south pole, but there they lack radiance. A terrible one was found among communities of Aethiopia and Egypt to which Typhon, king at the time, gave his name; it had a burning appearance and was twisted like a spiral, a grim sight too, not so much a star as some sort of burning knot.

92. Occasionally both planets and the rest of the stars have hair that spreads out. While there is never a comet in the western part of the sky, there are altogether frightening stars, not easily appeased, as occurred during the civil war when Octavius was consul [87 BCE], and again during the war between Pompey and Caesar [49–48 BCE], [and] actually in our time more or less when Claudius Caesar was poisoned and passed his rule to Domitius Nero [54 CE]; and then during his Principate there was a ferocious one almost continuously. General opinion is that the parts [of the sky] to which [a comet] shoots itself matters, as does the star from which it gains strength, also what forms it assumes, and in what places it becomes conspicuous:

93. when shaped like a flute it foretells musical skill, when like constellations' private parts disgusting morals, and talent and learning if it makes the shape of an equilateral triangle or a rectangle with the positions of some of the fixed stars; when at the head of the northern or southern Serpent it produces poisonings.

The one place in the whole world where a comet is worshiped is in a temple at Rome; Deified Augustus judged it a very favorable omen to himself, because it appeared when he was a beginner [44 BCE], during games for Venus Genetrix which he was organizing not long after the death of his father Caesar and [as a priest] in the college founded by him.

94. For he expressed his delight publicly with these words: *On the very days of my games a hairy star was visible for seven days in the northern region of the sky. It used to rise about the eleventh hour of the day* [= an hour before sunset] *and was bright and conspicuous in every land. Common folk believed this star to mean that Caesar's soul had been received among the spirits of the immortal gods, and for this reason that emblem was added to the bust of his head that we later consecrated in the forum.* This is what he said in public; privately he was delighted, interpreting [the comet] as born for him and he in it. And, let us be honest, it was to the world's benefit.

There are those who also believe that these stars are everlasting and proceed in their own orbit, but are only visible when abandoned by the sun; but others believe that they are born from random and fiery force, and dissolved as a result.

XXIV About Hipparchus' identification of stars. About celestial prodigies, with historical examples

95. The same Hipparchus is never praised enough; no-one has done more to demonstrate that the stars and mankind are related and that our souls are part of heaven. He also discovered a new star – different [from a comet] – created during his lifetime. Its movement, as it shone, led him to doubt whether this happened often, as well as [to wonder] whether the [stars] we assume to be fixed do move. Consequently he dared something outrageous even for a god – to count the stars for posterity and to list heavenly bodies by name; he invented instruments to mark each one's location and size, so that from then on it would be possible to spot easily not only whether they die and are born, but whether in general some are passing and being shifted, and also whether they are waxing or waning. He left the sky as a legacy to everyone, if anyone might be found to accept this inheritance.

XXV Torches, fireballs [i.e. meteorites]

96. Torches also are conspicuous, but only seen when falling, such as the one that swept over in public view at midday when Germanicus Caesar was organizing a gladiatorial show. There are two types: one called *lampades* or simply 'torches', the other *bolides* like the one seen during Mutina's misfortunes [44–43 BCE]. They differ because 'torches' make long tracks with their front part flaming, whereas a *bolis* is aflame all over and creates a longer wake.

XXVI Celestial beams, opening of the sky

'Beams' also are similarly conspicuous: there are those called *dokoi* [Greek], such as the one when the Lacedaemonians' fleet was defeated and they lost their control of Graecia [394 BCE]. An opening of the sky itself also occurs, called *chasma*.

XXVII About the sky's colors. About celestial blazes

97. A conflagration with the look of blood also occurs – mankind has nothing more dreadful to fear than this – which falls from there to earth, as it did in the third year of the 107th Olympiad [349 BCE], when king Philip was disturbing Graecia.

But I think that these things are generated at set times as in the rest of Nature and not, as is mostly thought, for the various reasons which sharp minds invent. They have certainly served as advance warnings of huge misfortunes, but I do not think that those happened because of these developments [in the sky]. Instead, these developments occurred because the misfortunes were imminent. Their rarity, moreover, has concealed the reasons behind them, and consequently there is not the familiarity with them that there is with the risings and settings treated above and much else.

XXVIII–XXIX About celestial crowns. About suddenly formed circles

98. Stars alongside the sun are also seen for entire days, usually also around the sun's disc like crowns made from ears of corn, as well as multi-colored rings as when Augustus Caesar in his early youth entered the city after his father's death to take his mighty name [44 BCE]. The same crowns occur around the moon and around well-known stars and those fixed in the sky.

XXIX

A bow appeared around the sun in the consulship of Lucius Opimius and Quintus Fabius [121 BCE], a band in the consulship of Gaius Porcius and Manius Acilius [114 BCE], and a circle colored red in the consulship of Lucius Julius and Publius Rutilius [90 BCE].

XXX <Protracted solar eclipses>

Portentous and protracted solar eclipses occur, like the one after the dictator Caesar's murder [44 BCE] and during the Antonine war[3] when there was constant gloom for almost a whole year.

XXXI Numerous suns

99. And again, numerous suns are seen simultaneously, and neither above nor below the actual [sun] itself, but at an angle, never next to earth or opposite, nor at night but at either sunrise or sunset; there are reports that once on the Bosporus [such suns] were seen even at midday, and that they lasted from morning to sunset. Even in antiquity three suns were seen very often, as in the consulships of Spurius Postumius and Quintus Mucius [174 BCE], of Quintus Marcius and Marcus Porcius [118 BCE], of Marcus Antonius and Publius Dolabella [44 BCE], and of Marcus Lepidus and Lucius Plancus [42 BCE]; our generation saw this when Deified Claudius was *princeps*, holding his consulship with Cornelius Orfitus as colleague [51 CE]. To date, there is no report of more than three being seen at the same time.

---

[3] Between Antony and Octavian.

XXXII Numerous moons

Three moons at a time have also appeared, as in the consulship of Gnaeus Domitius and Gaius Fannius [122 BCE].

XXXIII Daylight at night

100. Light from the sky at night – mostly called nocturnal suns – was seen in the consulship of Gaius Caecilius and Gnaeus Papirius [113 BCE] and often at other times, so that what appeared to be daylight shone at night.

XXXIV Burning shields

A burning shield swept across [the sky] sparkling from west to east at sunset in the consulship of Lucius Valerius and Gaius Marius [100 BCE].

XXXV A celestial marvel noted once

It is said that only once, in the consulship of Gnaeus Octavius and Gaius Scribonius [76 BCE], has a spark fallen from a star and grown in size while approaching earth, and that having achieved the size of the moon it shone as if on a cloudy day; and then, when it returned itself to the sky, it became a torch. The proconsul Silanus and his entourage saw it.

XXXVI About stars running this way and that

Another occurrence is that stars seem to run this way and that, although never haphazardly, which is why fierce winds are thought not to originate from that area.

XXXVII About the stars called Castores

101. Stars also appear on sea and land: I have seen star-shaped sparkle adhering to the javelins of soldiers on night-watch in front of the rampart. With some voice-like sound, [stars] also settle upon yard-arms and other parts of ships under sail, and they shift from one spot to another as birds do. When just one comes, that is painful; these cause ships to sink, and they set them alight if they should fall to the bottom of the hull. A pair of them, however, is reassuring and signals a successful voyage, because their appearance is said to drive away the terrible, menacing one called Helena, and hence such divine presence is attributed to Pollux and Castor, which is why those at sea pray to them. When [stars] also sparkle around humans' heads in the evening, that is a major omen. All this lacks clear explanation, concealed as it is within Nature's majesty.

XXXVIII About air

102. So much for the universe itself and for the stars; now on to the rest of what is notable about the sky. For previous generations also called 'sky' what [we] have a different name for – 'air', all appearing empty yet emitting this breath of life. This zone below the moon – indeed far below it, as I note is virtually undisputed – blends unlimited amounts of Nature's upper air with unlimited amounts of earth's exhalations, and is formed by the two types. From here [come] clouds, thunder-claps and other lightning strikes, from here hail, frost, rain, storms, tornados, from here most of mortals' misfortunes as well as strife within the natural order.

103. The stars' force presses down earthly objects straining skyward, and the same draws in its direction whatever does not rise of its own accord. Rain falls, clouds rise, rivers run dry, hail pours down; rays scorch and they strike the earth from all directions at its center, and once weakened they bounce back, taking with them whatever they have been able to. Moisture falls from high up and returns again up high. Empty winds bear down, and they again withdraw with what they have seized. So many creatures, when they breathe, draw breath from [the air] above, but it resists, and the earth streams back breath to the sky as if it were empty.

104. So with Nature in motion this way and that like some sort of catapult, discord is aroused by the speed of the universe. Nor is a halt to the battle permitted, but [the universe] is constantly caught up and spun round, and it displays the causes of things in an immense globe around the earth, repeatedly unfurling another sky through the clouds. This is the winds' kingdom. So here their nature is paramount and virtually encompasses those remaining causes, since bolts of thunder and lightning are generally attributed to the [winds'] force, and indeed it sometimes rains stones because they have been caught up by wind, as well as much else similar. Consequently more must be said at once.

XXXIX About fixed weather conditions

105. It is clear that there are some fixed causes for storms and rainfall, but also some chance ones as well as ones yet to be explained. For who would doubt that summer and winter and whatever changes are recognized in the seasons year by year are caused by the stars' movements? Therefore just as the nature of the sun is understood to regulate the year, so each of the rest of the stars also has its own power that is productive in accordance with its own nature. Some are rich in moisture discharged as liquid, others in [moisture] hardened into frost or condensed into snow or frozen into hail, others in wind, others in warmth or humidity, others in dew, others in cold. But they should not be thought to be the size they appear to the eye since, when their immense distance from earth is taken into account, none is in fact smaller than the moon.

106. Each one, therefore, in its own motion activates its own nature, as the passages of Saturn make very clear with their rainfall. This is a power not only of the stars that move, but also of many that are fixed in the sky, whenever they are pushed by planets approaching or stirred by the impact of rays, as we are aware happens in the case of the Suculae [Little Pigs, Hyades], the Greek name for which comes consequently from rain. But in addition some [move] of their own accord and at fixed times, as with the rising of the Haedi; in fact it is rare for there not to be hailstorms when the constellation Arcturus rises.

XL About the rising of Canicula

107. For who does not know that the sun's heat is intensified when Canicula [Dog Star] rises? The impact of this star is very widely felt on earth: at its rising, seas boil, wine in cellars sways, lagoons stir. Egypt has a wild animal called oryx, which is said to stand facing [Canicula] when it rises and to gaze at it like a worshiper, after first sneezing. There is certainly no doubt that throughout this period dogs are especially liable to contract rabies.

XLI Force of annual seasons

108. For sure, parts of some constellations also have a force of their own, as at the fall equinox and winter solstice when we gather from storms that a star has reached its endpoint, and not just from rainfall and storms, but also from much that both our bodies and the countryside experience. Some people are struck by a star, others periodically suffer upset to their bowels, muscles, head, mind. Olive and white poplar and willows turn their leaves at the solstice. The herb pennyroyal, after being hung up indoors to dry, flowers on the very day of the winter solstice; skins inflated with air burst.

109. This may surprise anyone who fails to notice from everyday experience that one herb, called heliotrope, always gazes at the sinking sun and at every hour turns with it, even when obscured by cloud. In fact more methodical observers have already discovered that the bodyweight of oysters, purple-fish and all shellfish expands and contracts again under the moon's influence, and that shrew-mouse tissue reacts to lunar phases, and that the smallest creature, the ant, is sensitive to stellar forces and always inactive between the old moon and the new.

110. For humans not to know this is all the more disgraceful, especially when they acknowledge that the incidence of eye-diseases in beasts of burden is greater or lesser in line with the moon. Their defense is the vastness of it all, separated at immense distance from earth into 72 constellations comprising the shapes of objects or creatures into which experts have divided the sky. They have actually identified 1,600 stars in them considered remarkable for their impact or appearance, for example the seven called Vergiliae in the tail of Taurus, and in his forehead the Suculae; also Bootes which follows the Septentriones [Great Bear].

XLII About unpredictable weather conditions. About rains and why it should rain stones

111. I would not deny that rain and wind develop from causes other than these, since it is certain that the earth gives off a damp mist, although at other times a smoky one because of heat, and that clouds are produced by liquid ascending high up or out of air condensed into liquid. Their thickness and mass may undoubtedly be inferred from the fact that, when they obscure the sun, it is still clearly visible to those who are diving, no matter how great the depth of the water.

XLIII About thunder and thunderbolts

112. So there is no denying the possibility of fires also falling on [clouds] from the stars above – a common sight in calm conditions – and that the air is really disturbed by their impact, just as weapons whirr when flung;

but when [these fires] reach a cloud, steam that makes a different sound is generated, as when white-hot iron is plunged in water and a plume of smoke swirls up. This is how storms are generated and, if there should be resistance from wind or steam in the cloud, thunder is produced – or, in the case of a fiery eruption, thunderbolts; or, should the struggle be more widespread, lightning-flashes. These split clouds, while [thunderbolts] smash through; thunder is colliding fires hitting one another, which is why there are instantly fiery slits flashing in the clouds.

113. Another possibility is that breath which has left the earth – thwarted when blocked by stars and then halted by cloud – becomes thunder, with Nature strangling the sound for the duration of the brawl, but the noise discharging when the breath bursts out, as happens with a bag filled in this way. Another possibility is that this breath – whatever it is – ignites from the friction when propelled so abruptly. Another possibility is for it to be forcibly ejected when clouds collide – like a couple of stones – with flashes of lightning. But all these are chance happenings. Hence there is no reason or significance to thunderbolts, and hence no rule of Nature for them, although both mountains and seas are hit, and all their other strikes are pointless. But the former [type] are prophetic, coming from on high for distinct reasons and from their own stars.

XLIV–XLVIII Why echoes should occur. Winds: Their types, natures and observations about them

114. Similarly I would not deny that winds or rather gusts can be generated by earth's dry and parched breathing; another possibility is that they are produced by bodies of water exhaling air that is neither condensed into mist nor thickens into clouds; the sun's impact can also cause them, since wind is considered to be merely an airstream which can be generated in numerous other ways as well. For we also see [winds coming] off rivers and snow and the sea even when calm, while others called *altani* rise from land; when these return from the sea, they are called 'turning' ones, but 'land' ones if they continue on.

115. Ceaseless winds are produced by the curves of mountains and their numerous peaks and ridges bent like elbows or broken into shoulders, as well as by the hollow depressions of valleys that because of their ruggedness tear the air which [then] bounces back (hence in many places even causing voices to echo). Caves too [produce winds], such as the one in

Delmatia with a huge mouth yawning open: when a light object is thrown into it, even on a calm day, a violent gale like a whirlwind shoots out; the place is called Senta. There is also said to be some rock in the province of Cyrenaica that is sacred to the south wind and sacrilege for a human hand to touch, because [then] the south wind immediately makes the sand swirl. Even in many houses containers containing liquids and shut up in the dark [still] create their own drafts. There must be a reason for this.

XLV

116. But gusts and winds are very different. The latter are regular and blow steadily, and their impact is global rather than just regional. They are not breezes or storms but winds [specifically], which even have a masculine name, whether born from the world's constant motion and collision with the stars' opposite direction, or whether they are that famous 'breath' which generates everything in Nature and roams here and there as if in some kind of womb, or whether they are air whipped up by the varied impact of planets and their rays' diverse thrusts, or whether they emerge from their own stars (these nearer ones), or drop from ones fixed in the sky. [In any event] it is clear that a law of Nature applies to them, one not unknown, but not yet known in full.

117. Over 20 ancient Greek authors have issued observations about these topics. So I am all the more surprised that, when the world was in turmoil and being divided into kingdoms – into pieces, that is – so many men took the trouble to make such challenging discoveries, especially during wartime when their hosts might not be trustworthy, and when moreover pirates – all mankind's enemies – controlled the transmission of information. As a result, today anyone can learn something about his own region from [these authors'] records – even though they never went there – more reliably than from the locals' knowledge. Nowadays, however, in such blissful peace and with a *princeps* so delighted by productivity in sciences and the arts, nothing more is being learned through new research, and really there is not even a full grasp of the ancients' discoveries.

118. Their rewards [then] were not larger since their success, for all its greatness, was spread among many, and the majority elicited what they did with no reward other than aiding posterity. For mankind's morals have deteriorated, not their profits, and [now] that the sea has been opened up in all directions, with every coast welcoming those who land there, there is a huge mass of voyagers – but for gain, not knowledge; it does not strike them, being short-sighted and fixated solely on greed, that they could make gains more securely through knowledge. Recognizing that there are so many thousands of voyagers, I shall consequently treat winds in more detail than is perhaps appropriate for the project I have undertaken.

XLVI

119. The ancients observed four of them in all – a match for the world having this number of parts (so not even Homer names more) – for reasons later considered to lack insight. Eight were added in the subsequent period, but that was too subtle and detailed; those who came next gladly adopted a compromise which supplemented the shortlist with four from the full one. So there are two in each quarter of the sky: Subsolanus [blows] from the equinoctial sunrise, Volturnus from winter sunrise, with Greeks styling the former Apeliotes, the latter Eurus; Auster from the midday [sun] and Africus from winter sunset, with [Greeks] naming them Notus and Libs; Favonius from equinoctial sunset, Corus from sunset at the solstice, with [Greeks] calling them Zephyrus and Argestes; Septentrio from the north and Aquilo between it and sunrise at the solstice, referred to by [Greeks] as Aparctias and Boreas.

120. Among these, the fuller scheme had inserted four: Thrascias in the area midway between Septentrio and sunset at the solstice; likewise Caecias midway between Aquilo and equinoctial sunrise on the side of sunrise at the solstice; Phoenix midway between winter sunrise and midday; and also between Libs and Notus the combination of both, Libonotus midway between midday and winter sunset.

Nor are we finished; others in fact also added one called Meses between Boreas and Caecias, and Euronotus between Eurus and Notus. There are even certain winds special to particular peoples, ones which do not advance beyond a defined region: so for the Athenians there is Sciron diverging slightly from Argestes and unknown to the rest of Graecia; elsewhere, further up the same one is called Olympias;

121. typically all these names are taken to signify Argestes. Some also call Caecias Hellespontias, and others have other names for the same ones.

Likewise, in the province of Narbonensis the most famous wind is Circius, second to none in ferocity and continuing as far as Ostia by slicing through the Ligustic sea. This wind is not only unknown elsewhere in the sky, but does not even touch Vienna, a city in the same province; great among winds it may be, but when it encounters a modest ridge a few miles ahead [of the city], that halts it! Fabianus states that Auster winds also do not penetrate Egypt. This makes the law of Nature clear that, even for winds, time and limit are prescribed.

XLVII

122. Spring, therefore, opens the sea to voyagers, with west winds softening the winter sky at the start when the sun occupies the twenty-fifth degree of Aquarius; that occurs on the sixth day before the Ides of February [February 8]. This more or less applies also to all [winds] – which I shall situate next – although they come one day earlier during intercalary years, but in the subsequent period again maintain their order. There are those who give the name Chelidonias to Favonius on the eighth day before the Kalends of March [February 22] when they spot swallows, but some call it Ornithias which blows for nine days after the birds' arrival, 71 days after the winter solstice; the opposite to Favonius is the one we have called Subsolanus.

123. The rise of the Vergiliae in the same degree of Taurus on the sixth day before the Ides of May [May 10] signals summer, Auster's time, the opposite wind to Septentrio. But during the hottest part of summer the star Canicula rises with the sun entering the first degree of Leo, on the fifteenth day before the Kalends of August [July 18]. About eight days earlier the Aquilones precede its rise; these are called Prodromi.

124. Two days after its rising, however, these same Aquilones blow more steadily for 40 days; these are called Etesiae. They are thought to be softened by the sun's warmth coupled with the constellation's heat, and among winds their regularity is unmatched. After them, Austri become common again until the constellation Arcturus rises 11 days before the autumn equinox; at this point Corus begins. Corus brings fall; its opposite is Volturnus.

125. Nearly 44 days after this equinox the setting of Vergiliae brings in winter, a date which normally falls on the third day before the Ides of

November [November 11]; now there is the wintry Aquilo, very different from the summer one; its opposite is Africus. However, for seven days prior to the shortest day and as many after, the sea stays calm for halcyons to give birth, which is why these days are so called. For the rest of the season it is wintry. Yet the stormy weather does not close the sea: the pressure to court death and risk winter-seas initially came from the [reduced] threat of death from pirates, but nowadays this pressure comes from greed.

XLVIII

126. The coldest winds are those, as we said, blowing from the north as well as the next one, Corus; these both check the rest and drive clouds away. For Italia, Africus and especially Auster are the damp ones; in Pontus too, Caecias has a reputation for attracting clouds. Corus and Volturnus are dry except when dying down. Aquilo and Septentrio are snowy. Septentrio and Corus bring hail. Auster is hot, Volturnus and Favonius warm; these are drier than Subsolanus, and on the whole all those from the north and west are drier than those from the south and east.

127. Healthiest of all is Aquilo; Auster is harmful, even more so when dry, perhaps because when damp it is colder; animals are thought to have less appetite when it is blowing. The Etesiae almost die down at night and start up from the third hour of the day; in Hispania and Asia they blow from the east, in Pontus from the north, and elsewhere from the south. However, they also blow from the shortest day – when they are called Ornithiae – but more gently and [only] for a few days. Two also change their character in line with their position: in Africa Auster is mild and Aquilo cloudy.

128. All the winds blow when it is their turn, and mostly so that as one dies down its opposite starts up. When those closest to the ones easing rise, they go round from the left side to the right, like the sun. It is the quarter-moon that usually determines their pattern for the month. Yet voyages are made with the same winds in opposite directions by loosening sails, so that at night ships on contrary courses commonly collide. Auster produces larger waves than Aquilo, since the former (being from below) blows from the depths of the sea and the latter from above. In consequence, earthquakes occurring after Auster winds are especially destructive.

129. Auster is stronger at night, Aquilo in the daytime, and [winds] blowing from the east are longer lasting than those blowing from the west. Septentriones normally die down after an odd number [of days], an observation which is also valid in many other areas of Nature; hence odd numbers are thought to be masculine. The sun both inflates and calms winds: it inflates them at sunrise and sunset, and calms them at midday in summertime; and so they are usually lulled at midday or midnight because of being slackened by either undue cold or heat. Winds are also lulled by rain. But they are mostly expected [to come] from where clouds have scattered and made the sky visible.

130. In fact (given a willingness to observe the smallest circuits) Eudoxus thinks that everything recurs identically after four years, not just winds, but also other weather conditions for the most part; and that this cycle always begins in an intercalary year at Canicula's rising. So much for winds in general.

XLIX Cloudburst, typhoon

131. Now for sudden blasts which, as mentioned [2.111], emerge in many forms as they arise from earth's breathing and subside again once cloud-cover has been drawn over. In fact, like running water, random rapid ones produce thunder and lightning (the favored view of some, as we have shown [2.112]). When carried along with greater weight and impetus, if they extensively rupture a dry cloud they produce a storm, one that Greeks call *ecnephias* [cloudburst]; but if with a more restricted gyration they break through in a sunken fold [of cloud] – without fire, that is without lightning – they produce a vortex called a typhoon, that is, a vibrating *ecnephias*.

132. This brings down with it something broken off from an icy cloud, rolling and whirling it round and accelerating its own plunge because of the weight, shifting from place to place with a rapid spin. It is a menace to voyagers above all, smashing not only yard-arms, but also the very ships that it has contorted; the remedy of pouring vinegar (something very cold by nature) as it approaches is feeble. Once repulsed by its own impact, it takes back skyward whatever it has grabbed and sucks it up on high.

L Tornados, *presteres*, vortexes and other monstrous types of storms

133. But if this bursts out of a larger chasm of sinking cloud, yet is not as wide as a storm and makes a crash, they call it a tornado, which flattens everything in its vicinity. When hotter and ignited as it rages, it is called a *prester* which burns as well as crushes what it encounters. However, a typhoon does not occur with Aquilonius, nor an *ecnephias* with snow or when snow is lying; but if it has burst a cloud and burned it at the same time, and was ignited rather than catching fire later, it is a thunderbolt.

134. This differs from a *prester* in the way a flame differs from fire; a blast spreads [a *prester*] far and wide, while [a thunderbolt] forms a mass for its strike. However, a vortex differs from a tornado by its return movement, and as a shrill sound does from a crash; a storm differs from both by its extent, the clouds being scattered really rather than burst. Fog with the look of a monster also develops within cloud, something terrible for voyagers. What is called a *columna* occurs as well, when compacted moisture becomes rigid and stands of its own accord; in the same category too is an *aulon* when – like a pipe – a cloud draws up water.

LI–LVI About thunderbolts: Lands in which they do not fall and why

135. Thunderbolts are rare in winter and summer for opposite reasons, because in wintertime air – after being made dense by a thicker covering of clouds – is compacted, and earth's entire breathing, stiff and icy, extinguishes any fiery vapor it takes in. This is the reason why Scythia and its frozen surroundings are safe from thunderbolts, while by contrast excessive heat [protects] Egypt, since earth's hot, dry breathing very seldom condenses, and [even then only] into slight, feeble clouds.

136. But in spring and autumn thunderbolts come more often, because in both these seasons the causes for them in summer and winter are overturned. For this reason they occur often in Italia, because the air [there] is more mobile, with a milder winter and cloudy summer making it always spring-like and fall-like. Also, in those parts of Italia which slope down from the north towards warmth – the region of the city [Rome] and Campania, for instance – thunderbolts occur equally in winter and summer, something that does not happen elsewhere.

LII Thunderbolt types and marvels

137. Several different kinds of thunderbolts have been reported. Those which arrive dry disintegrate without burning [anything]; damp ones scorch but do not burn; there is a third kind called 'bright', quite remarkable in character, since without leaving any other trace these drain casks with their lids still closed, and gold and bronze and silver are turned to liquid inside pouches which themselves are not singed at all, with not even their wax seal melted. Marcia, a leading Roman lady, was struck while pregnant, and although the fetus was killed, she herself survived without any other injury. Among the prodigies in the Catiline episode [63 BCE], a decurion [council member] from the municipality of Pompeii, Marcus Herennius, was struck by a thunderbolt on a clear day.

LIII Etruscan and Roman observations about them

138. In Etruscan writings the view is that nine gods send thunderbolts, and that there are 11 types, three of them hurled by Jupiter. Romans have retained only two of these, attributing daytime ones to Jupiter and night-time ones to Summanus (rarer naturally because the sky is colder then). Etruria also thinks that ones it calls 'from below' burst out of the ground, and that in wintertime they become particularly fierce and horrible, as are all those [the Etruscans] reckon come from earth, in contrast to the typical type or ones coming from planets instead of from the nearest and less stable element. Clear proof is that all upper ones falling from the sky strike at an angle, whereas those they call 'earthly' stay straight.

139. But it is because these fall from the nearer substance that they are therefore thought to emerge from the earth; for certain, their rebound produces no traces, despite this being the characteristic, not of a strike from below, but of one in the opposite direction. Those who have pursued the topic in more depth think that these ones originate from planet Saturn, just as the burning ones do from Mars, as when Volsinii, the wealthiest Etruscan town, was completely burned up by a thunderbolt. Those that occur when anyone first sets up for himself his own household are also called 'household' ones and predict his entire life. But they think that private ones do not predict beyond ten years – excluding those on the day when a man becomes head of his family or on his birthday – nor public ones beyond 30 years, excluding those [on the day of] a city's foundation.

LIV About calling down thunderbolts

140. According to a tradition in the *Annals*, thunderbolts may be either induced or solicited by certain rites and prayers. There is an old story from Etruria that one was solicited when a monster they called Volta was nearing the city of Volsinii after pillaging its territory and was then lured away by [the city's] king Porsina. Before him – as Lucius Piso, a serious authority, recorded in the first book of his *Annals* – [at Rome king] Numa had made a habit of doing this more often, but [when king] Tullius Hostilius did the same with insufficient concern for the ritual he was struck by a thunderbolt. We also have groves and altars and rites, and among Jupiter's attributes – Stator and Tonans and Feretrius – we recognize Elicius too.

141. People's opinion about this matter varies and everyone has their own feelings. It is the mark of a bold person to believe that Nature takes orders from rituals, just as it is of someone dimwitted to dismiss their beneficial vigor, when knowledge even in the interpretation of thunderbolts has advanced to the point of predicting that others will come on a specific day, and whether they will undo the impact of a previous one or of other earlier ones so far undetected, as in countless instances – public and private – of both types. Consequently, given that it is all a matter of what has appealed to the natural order, some [instances] are definite, others dubious, accepted by some, rejected by others; we shall not overlook what else merits recording in this connection.

LV General principles of thunderbolts

142. It is certain (and unsurprising, since light is faster than sound), that even though they occur simultaneously, lightning is seen before thunder is heard; moreover, that the strike and the sound coincide because Nature so regulates them, but the sound is caused by the thunderbolt starting, not by its strike, and again air is swifter than a bolt. So everything is rocked and blown before being hit, and no-one has been struck prior to seeing a bolt or hearing thunder. [A thunderbolt] on the left is considered lucky, because sunrise occurs on the world's left side. And its approach is noticed less than its rebound, whether because on impact the fire rekindles, or air returns after its task is accomplished or its fire expended.

143. On this view, the Etruscans divided the sky into 16 parts: the first is from the north towards equinoctial sunrise, the second to the south, the third towards equinoctial sunset, the fourth occupies what remains from the west to the north. These parts are then divided into quarters, of which the eight on the east side they called the 'left', and the corresponding number opposite the 'right'. The most dreadful of these are the ones striking north from the west. So the directions from which thunderbolts came and to which they withdrew are matters of the greatest significance. A return towards eastern parts is best.

144. Hence the height of good fortune will be predicted when they come from the first part of the sky and return to that same part, the sign (we understand) given to the dictator Sulla. Ones occurring elsewhere are less favorable or dreadful depending on the part of the universe. Some people consider it improper to make a report about thunderbolts or to listen to any, unless they are conveyed to a host or parent. How really pointless it is to respect this restriction emerged when the temple of Juno in Rome was struck in the consulship of Scaurus [115 BCE], who later was leading citizen.

145. Lightning without thunder occurs more at night than by day. There is one creature it does not always kill – humans – but the rest [die] instantly; no doubt Nature bestows this honor because so many beasts have superior strength. Everything falls down in the opposite direction [from which it was struck]; humans do not recover unless turned round to the direction in which they were struck. Those struck from above collapse. Someone struck while awake is found with eyes closed, but while asleep with them open. It is improper to cremate people who have died in this way; according to traditional religious practice, they should be buried in the earth. No creature is burned by a thunderbolt unless already dead. The wounds of those struck by thunderbolts are colder than the rest of their bodies.

LVI What may never be struck

146. Among the products of the earth, [thunderbolts] do not strike laurel bushes and never penetrate more than five feet underground. People frightened of them consequently consider deeper caves to be safest or tents made from the skins of beasts called seals, because this marine creature uniquely is not struck, just as eagles are the only birds not to be, hence their depiction armed with this weapon. In Italia the construction of towers between Tarracina and the temple of Feronia during a war was abandoned after every one of them had been demolished by a thunderbolt.

LVII Showers of milk, blood, flesh, iron, wool and fired bricks

147. Apart from [all] this in the lower sky, there is documentary record that during the consulship of Manius Acilius and Gaius Porcius [114 BCE] and on many other occasions it rained milk and blood, just as it rained flesh in the consulship of Publius Volumnius and Servius Sulpicius [461 BCE], and nothing of what birds [failed] to snatch rotted. Again [according to the record], it rained iron in Lucania the year [54 BCE] before Marcus Crassus was killed by the Parthians and all the Lucanian soldiers with him (a great number of them in his army); the iron that rained was shaped like sponges, and the *haruspices* prophesied wounds from above. However, during the consulship of Lucius Paulus and Gaius Marcellus [50 BCE] it rained wool around the fortress of Compsa, near which Titus Annius Milo was killed the following year. The year's *Acts* reported that while he was pleading a case it rained fired bricks.

LVIII <Clashing arms and trumpet sounds heard from the sky>

148. We gather that during the Cimbric wars clashing arms and trumpet sounds were heard from the sky, and that both earlier and later this happened often. During Marius' third consulship [103 BCE], the people of Ameria and Tuder viewed armies from east and west clashing in the sky, with the western one driven back. It is not at all surprising, and has been a very frequent sight, that the sky itself should be on fire when clouds have been set alight by a really large blaze.

LIX About stones falling from the sky. Anaxagoras' view of them

149. Greeks draw attention to how, in the second year of the seventy-eighth Olympiad [467 BCE], Anaxagoras of Clazomenae, with his knowledge of writings about the sky, predicted that a rock would fall from the sun in a specific number of days, and this happened during daytime at Aegos river in a part of Thracia, where even nowadays the stone is pointed out (of wagon-load dimensions and with a burnt tinge); a comet was blazing at night-time then too. Anyone who may believe in this prediction must at the same time acknowledge that Anaxagoras' prophetic powers were even more amazing, and that our grasp of the natural order dissolves and everything is thrown into confusion if it be credited that either the sun itself is a stone or there ever was a stone inside it. Nevertheless, it will not be doubted that they have fallen frequently.

150. On this account [a stone] is nowadays worshiped in the gymnasium of Abydos, one of moderate size but (so goes the tale) predicted by Anaxagoras again to fall in the middle of its lands. One is also worshiped at Cassandria, once called Potidaea, where a settlement was established for this reason. I saw one myself that had come down quite recently in a field of the Vocontii.

LX Rainbows in the sky

What we call rainbows occur frequently and are neither a marvel nor a portent. In fact they do not even reliably predict rainy days or calm ones. Quite clearly, a sunbeam plunged into a hollow cloud has its tip thrust back towards the sun and is broken up, with varied colors produced by a mix of clouds, fires and air. For certain, rainbows are only formed opposite the sun and neither in any shape ever except a semicircle, nor at night, although Aristotle does say that one is sometimes seen [then], although he also admits that only on the thirtieth of the lunar [month] is this possible.

151. Moreover in winter they mainly occur when the day becomes shorter after the autumn equinox. When it lengthens again after the spring equinox they do not develop, nor during the longest days around the solstice, but in winter often – during the shortest days, that is. They are high when the sun is low, low when it is high, smaller but extended in width at sunrise or sunset, slim but larger in circumference at midday. In summer, however, they are not seen around midday, but after the autumn equinox [they are] at any hour, although never more than two at the same time.

LXI Nature of hail, snow, frost, mist, dew. Shapes of clouds

152. For many people, as I see it, other occurrences of the same type give no cause for doubt – namely, that hail is produced by frozen rain and snow by a weaker concentration of the same fluid, but frost from cold dew; that snow falls throughout the winter, although hail does not, and hail itself more often falls during daytime than at night, and that it melts much more quickly than snow; that mists do not develop in summer nor under extremely cold conditions, nor dews when it is frosty or hot or windy, but only on calm nights; that liquid shrinks with freezing and the same amount is not found after ice has thawed; that variations in color and shape are seen in clouds corresponding to how far fire mixed into them gains the upper hand or is overcome.

LXII Distinct characteristics of the sky according to location

153. [I see] in addition that for certain locations certain distinct characteristics apply – namely, Africa's summer nights have heavy dew, in Italia rainbows appear daily without fail at Locri and lake Velinus, while at Rhodos and Syracusae the cloud-cover is never thick enough to prevent the sun from being seen at some hour; these [characteristics] will be dealt with more appropriately in their own sections. This is what there is to say about air.

LXIII Earth's character

154. Next comes earth, the one part of the natural order on which we have bestowed an additional name of motherly respect because of her exceptional service. She is thus for humans what the sky is for God, she takes us in at birth, feeds us once born, after our creation always sustains us, and at the very end – after our rejection by the rest of Nature – folds us into her lap, then especially safeguarding us like a mother. There is no service that makes her holier than the one by which she makes us holy too; she even supports our monuments and inscriptions, continues our names and counters time's brevity by extending our remembrance. She is the final divinity which we, when angry, beseech to weigh heavy upon the departed, as if we were unaware that she is the one – uniquely – never angry with humans.

155. Water turns into rain, freezes into hail, swells into waves, hurls down in torrents; air grows thick with clouds, rages with storms: but [earth] is benign, mellow, yielding, and always the servant of mankind's needs. What scents and flavors, what juices, how much that is tangible and colorful she produces under duress or disgorges of her own accord! How honestly she repays loans entrusted to her. How productive she is on our behalf. In the case of noxious living creatures, the breath of life is to blame; [earth] must accommodate their seed and sustain what is born. The harm lies in the wickedness of what gives birth. She rejects a snake once it has bitten a human, and even exacts penalties on behalf of the powerless. She supplies medicinal herbs in plenty and is always delivering for mankind's benefit.

156. Indeed it can even be credited that, out of pity for us, she created poisons in order to prevent starvation – a death quite out of line with earth's benefits – from consuming us with prolonged decline when tired of life, or to stop precipices from scattering our mangled bodies, or our being tormented by the punishment of the noose that perversely confines breath when aiming to expel it, or – should we aim for death in the depths – our burial just making us [fish-]food, or torture in irons tearing our bodies apart. Hence out of pity [earth] created something that would kill us with the easiest gulp, our body still intact and without the least bloodshed – the effortless act of thirsty people – and such that after death we would be untouched by bird or beast, and those who had taken their own lives would be preserved for the earth.

157. Let us admit the truth: what earth created as a remedy for our afflictions is what we make into life-threatening poison. For do we not use iron, which is indispensable to us, in a similar way too? We would not be justified in complaining either, even if [earth] had intentionally produced poison for wrongdoing. In fact we are ungrateful to this one element in Nature. What pleasures and outrages are there that she is not at humans' disposal for? We toss her into the sea or dig her up to make way for channels. At all hours she is tortured by water, iron, fire, wood, stone, crops and much more so that she may serve our pleasures instead of our sustenance.

158. Indeed what she suffers on her top, outermost layer may appear tolerable [by comparison with how] we delve into her innards, mining veins of gold and silver and quarrying copper and lead. We even hunt for gems and certain tiny stones by driving shafts deep down, and we extract her innards to look for a gem to be worn on someone's finger. How many hands are worn out so that a single knuckle may sparkle! Should any Underworld beings exist, the tunnels for greed and luxury would definitely have dug them up by now. Yet it surprises us that earth has produced doers of harm.

159. I believe that in fact wildlife protects her and keeps off profane hands; we do not dig where there are snakes and work on veins of gold amidst poisonous roots. Yet in that connection, this divinity is still more merciful, because all these opportunities for wealth lead in the direction of crime and slaughter and war. We soak her with our blood and cover her with unburied bones. She nonetheless, as if reprimanding our craziness, eventually offers us her cover and even conceals the crimes of mankind. Not to know her character in this respect too I would rate among the offenses of our ungrateful minds.

LXIV About her form

160. But it is her shape first on which opinion is unanimous; for certain, we speak of the earth as a sphere, and we acknowledge the globe's enclosure within poles. For her form is not that of a perfect sphere, given the great height of the mountains and width of the plains, but her circumference would make the shape of a perfect sphere if lines were to extend around everywhere; this occurs by the very force of Nature, although not for the same reasons that we raised in the case of the sky. For there the convex hollow turns in upon itself and leans from every direction upon its pivot; this, the earth, rises up solid and dense as if swelling, and projects outward; the universe is centripetal, but earth is centrifugal, and the continuous revolution of the universe around her forces her immense globe into the form of a sphere.

LXV Whether there may be antipodes

161. Here there is huge dispute between scholarship and folklore over whether humans are distributed everywhere around earth and stand with their feet pointed at each other, whether the pole is the same for all, and whether at the center [the impact of] their tread is the same whatever their location. [Common folk] ask why those situated on the opposite side should not fall off, regardless of there being equally good reason for those people to wonder why *we* do not fall off! A compromise opinion – plausible even to the ignorant masses – is that earth with her uneven globe form (comparable to a pinecone's shape) does nonetheless have inhabitants all the way around.

162. But how much does this matter when another marvel arises, namely that earth herself hangs and does not fall (nor we with her)? This might call into question the force behind breath, especially when confined within the universe, or the feasibility of a fall when Nature resists that and fails to provide anywhere to fall to; for just as fire dwells nowhere but in fire, water but in water, air but in air, so earth – hemmed in by everything else – has no place but in herself. The surprise, however, is the formation of a globe from so much flatness (seas and plains). This insight is endorsed by that outstandingly learned man Dicaearchus, who was commissioned by kings to measure mountains and declared Pelium – with an altitude of 1,250 paces – the highest, inferring that this does not count towards [earth's] total roundness. This strikes me as a dubious conjecture, as I am aware that some peaks in the Alps rise to a great height, 50,000 paces and more.

163. But what common folk debate most of all is this: whether they should be required to believe that water also has a shape that draws it to a point. And yet nothing in the natural order is more openly visible. Everywhere, hanging drips are rounded into small balls, and when they land on dust or are placed on a leaf's soft surface they are seen to be perfectly spherical; when a cup is filled, a swelling occurs (in the middle especially) that subsides because of the liquid's fineness and fluidity, and so is more easily recognized in theory than by observation. And it is even more remarkable that, when a minimal amount of liquid is added to a full cup, the extra overflows; but the opposite occurs when heavy items are added – often as many as 20 *denarii* – evidently because after their absorption into the liquid these create a peaked [surface], whereas what is poured onto a swollen one runs off.

164. It is for the same reason that land may not be seen from a ship but is plainly visible from a ship's mast, and why, as a vessel proceeds into the distance, if something shiny be fastened to the masthead, it appears to sink gradually and eventually disappears. Lastly, what other shape would hold together the ocean – which we regard as the ends [of the earth] – and stop it from falling off when there is no border beyond to confine it?

Just how the outermost sea, despite its globe shape, should not fall off is altogether amazing. On the other hand, to their own great delight and great credit Greek researchers demonstrate with sophisticated mathematics that it cannot happen, even were the seas to be flat and to have the shape that they seem to.

165. For while water is drawn from a high level to a lower one and this is its acknowledged characteristic, no-one doubts that on any shore it has reached as far as the slopes there permit; and it is undoubtedly evident that anywhere lower is closer to earth's center, and all lines extending from there to the nearest water are shorter than those drawn to the furthermost sea from the water's starting-point. All water anywhere, therefore, is centripetal and hence does not fall off, since it presses inwards.

LXVI How water is bound up with earth. The reason for rivers

166. Thus we should believe this to be the shape skillfully made by Nature so that earth, arid and dry, could not congeal of her own accord without moisture, nor could water stay in place without earth's support; hence they would be joined in a mutual grip, the former spreading out her folds, and the latter penetrating throughout, inside and out, above and below, through veins running all over like chains, even bursting out on the tops of ridges. After being driven there by breath and ejected by earth's weight, [water] spurts as if from a tube and is at such minimal risk of falling down that it vaults to the greatest height possible. This clearly explains why seas do not rise despite so many rivers draining into them daily. So across her entire globe earth is encompassed by sea flowing around, a fact not in need of investigation to prove it, but already known through experience.

LXVII Whether ocean circles earth

167. Today the entire west is open to navigation from Gades and the pillars of Hercules all around Hispania and the Galliae. Most of the northern ocean was opened at the instigation of Deified Augustus when a fleet sailed around Germania to the Cimbrians' cape, and from there – after sighting an immense sea or learning of it from reports – to the Scythian steppe and [others] frozen with excessive moisture. Hence for there to be no sea here – where moisture is such a dominant force – is extremely unlikely. Next, from the east, the whole part under the same star sloping from the Indian sea to Caspian sea was fully opened to navigation by Macedonian forces during the reigns of Seleucus and Antiochus, who wanted the [two] named Seleucid and Antiochid after themselves.

168. Much of the ocean shore around the Caspian has been explored as well, and almost the entire north has been rowed over in one direction or other, furnishing emphatic proof – yet with no room left for speculation – of the Maeotic marsh, which is either a bay of that ocean (as I recognize many have believed), or an overflow separated from it by a narrow strait.

On the other side of Gades, from the same western point, a great part of the southern bay around Mauretania is today open to navigation. In fact Alexander the Great's victorious [campaigns] traversed the greater part of it and of the east as far as Arabian bay, where ships' figureheads from Hispaniensian shipwrecks are said to have been identified during the period when Augustus' son Gaius Caesar was in charge there.

169. When Carthago was a leading power, Hanno too circumnavigated from Gades to the Arabian border, a voyage about which he issued a written account, as did Himilco when dispatched at the same time to investigate Europa's outer limits. In addition, according to Nepos Cornelius, some contemporary of his, Eudoxus, when escaping from king Lathyrus went out from Arabian bay and was conveyed all the way to Gades; and long before him, [says Nepos,] Caelius Antipater personally had seen a man who made a trading voyage from Hispania to Aethiopia.

170. With reference to the northern way round, Nepos again says that Quintus Metellus Celer, Afranius' consular colleague [60 BCE] but at the time proconsul of Gallia, was presented by the king of the Suebi with a gift of Indians who on a trading voyage from India had been swept to Germania by storms. So seas that are all-encompassing in every direction split the globe and deprive us of part of the world, with a viable route neither from there to here nor from here to there. This is a realization that serves to expose mortal vanity, and appears to insist that I should feature this entire entity – whatever it is, and where people are never satisfied – as if it were available to be viewed.

LXVIII What part of earth may be habitable

171. In the first instance it is evidently reckoned that [this entity] is half [of the globe, situated] so that no part of it should give way to the ocean itself, which is located around the entire middle discharging and reabsorbing all further water, and supplying whatever rises into clouds and the very stars (so big and so many of them!): so, then, how extensive a space should it be thought to occupy? A domain of such vast heft should be far-flung and boundless.

172. Add the sky which removes more from the remainder. For even though [the globe] has five parts called zones, whatever lies below the remotest two surrounding both poles is all weighed down by pernicious cold and eternal frost (one of these two is called Triones Septem, the other opposite is named Austrinus); in both there is perpetual murkiness along with light made dim by an obstructed view of the milder stars and only made clear by frost. But the middle part of earth, where the sun orbits, is scorched by flames, burned by it being so close, and made torrid. Only the two [zones] in between the scorched one and the cold ones are temperate, and the stars' heat makes each inaccessible to the other. So the sky has cut off three-quarters of the earth; how much of the ocean it has seized is unclear.

173. But the one portion left to us is also to some extent at even greater risk since, as we shall explain, the ocean again by discharging into many bays comes so threateningly close to the inner seas that Arabian bay is [only] 115 miles from Egyptian sea, and the Caspian 375 from the Pontic. So when this same widespread [ocean] comes in through so many seas which separate Africa, Europa, Asia, how much land does it take over?

174. Moreover, let the dimensions of so many rivers and such extensive marshes be calculated, add in lakes and lagoons as well, and then too the ridges that rise to the sky and are hard to see, and then the woodlands and rugged valleys, and wilderness as well as what is desert for any number of reasons; let all these parts be subtracted from earth, or more frankly (as has often been said) from a dot of a world – since in the universe earth is merely that. This is the object of our glory, this our home; here we hold office, here exercise authority, here covet riches, here create turmoil for mankind, here we instigate even civil wars and by slaughtering one another make earth less crowded.

175. And this – if I may pass over the collective insanities of peoples – is where we drive off those next to us and secretly dig up a neighbor's soil for our own land. He, then, who has marked out an estate that extends furthest and who has incontrovertibly ejected his neighbors, how large a part of earth might he be so proud of? Or, after extending it to match his greed, what segment of her may he eventually [expect to] retain once dead?

LXIX That earth is in the center of the universe

176. Unshakable arguments establish her to be in the center of the entire universe, the clearest of them being the equal hours of the equinox. For were she not in the center, the days and nights could not be equal; surveying instruments have also detected what they emphatically establish, namely that at the time of an equinox sunrise and sunset may be seen on the same line, just as sunrise at the summer solstice and sunset at the winter one have their own line; only with [earth] located in the center could this happen.

LXX About the zones' slant

177. However, it is three circles interlaced with the zones mentioned above [2.172] that create seasonal differences: the summer solstice one [Tropic of Cancer] on the side of the zodiac that for us is highest in the direction of northern regions, and opposite towards the other pole the winter solstice one [Tropic of Capricorn], and also the equator that runs in the middle circuit of the zodiac.

LXXI About variations in earth's tilt

The cause of what else amazes us is to be found in the shape of earth herself, which by the same arguments is understood (along with her waters) to resemble a globe. Thus there is no doubt that for us the stars of the northern regions never set, while their southern counterparts never rise, while again the northern ones are out of sight in the south because the globe, elevating itself, blocks sightlines across the middle lands.

178. Trogodytice and its neighbor Egypt do not see the Septentriones [Great Bear], nor does Italia see Canopus or the so-called Berenice's Hair, or also the one which during Deified Augustus' rule was named Caesar's Throne, conspicuous stars there. The rising high ground is so clearly curved that to observers there – in Alexandria – Canopus seems elevated nearly a quarter of a sign above earth, whereas from Rhodos it is somehow brushing the very ground, and absolutely invisible in Pontus (where Septentrio is at its highest). The latter is concealed from Rhodos and even more from Alexandria, while in Arabia during the month of November it is hidden during the first watch and shows itself during the second, [although] in Meroe it appears for a short time in the evening during the summer solstice as well as being visible at daybreak for a few days before Arcturus rises.

179. It is sailors' voyages that mostly make these discoveries, because the sea bends towards some stars and away from others, and because ones that had been hidden by the curvature of the sphere are suddenly visible as though rising from the deep. Contrary to some accounts, it is not the case that at this upper pole the universe makes itself higher; in that event [these stars] would be visible everywhere. Rather, the same ones are thought to be higher by anyone nearest and submerged by those far away, just as right now this pole appears to be high up to [people] situated below it; likewise, to those who have made the crossing to earth's downward-sloping [side], these [stars] rise of their own accord, while the ones that had been high up here sink [there] – something only possible with a spherical shape.

LXXII Where eclipses may not be seen and why

180. So those living in the east are not aware of solar or lunar eclipses in the evening, nor are western residents of morning ones, and in fact for them midday ones occur later than for us [in Rome]. When Alexander the Great was victorious at Arbela [331 BCE], a lunar eclipse was reported at the second hour of the night, and the same one occurred in Sicilia as [the moon] was rising. The occurrence of a solar eclipse a few years ago on the day before the Kalends of May [April 30] in the consulship of Vipstanus and Fonteius [59 CE] was observed between the seventh and eighth hour of the day in Campania, while the commander Corbulo in Armenia reported it being sighted between the tenth and eleventh hour of the day, with the globe's curvature making something alternately visible in one area, hidden in another. But were earth flat, everything would be visible to everyone at the same time and nights would not vary in length, because equal periods of 12 hours would be observed by those at locations other than the equator – a uniform match everywhere which does not occur currently.

LXXIII The reason for daylight on earth

181. Consequently, although they are alike [everywhere], it is not night and day at the same time all over the world, with the blockage created by the globe bringing night and the space surrounding it day. Many experiments have led to this discovery, such as at Hannibal's towers in

Africa and Hispania, and indeed in Asia when panic about piracy prompted comparable defensive measures in the form of lookouts: their warning fires lit at the sixth hour of the day [noon] were often discovered to have been seen by those furthest to the rear at the third hour of the night. Alexander (mentioned above) had a courier, Philonides, who covered the 1,200 stades from Sicyon to Elis in nine daylight hours and then, even though the route is downhill, very often made his return by the third hour of the night. The explanation is that outbound he was traveling with the sun, but on the return he was going past the sun coming from the opposite direction; this is why those sailing west, even on the shortest day, exceed the distance they sail during the night because they are traveling with the sun itself.

LXXIV Relevant gnomonics

182. Also, instruments for telling the time are not usable everywhere, because the sun's shadows change of their own accord every 300 stades, or 500 at most. So in Egypt, at noon on the day of the equinox, the shadow of the shadow caster (what is called the gnomon) measures just over half the gnomon; in the city of Rome the shadow is one-ninth shorter than the gnomon; in the town Ancona it is one thirty-fifth longer; at the same hours in the region of Italia called Venetia shadow and gnomon match.

LXXV–LXXVI Where and when there are no shadows; where twice a year. Where shadows may be cast in opposite directions

183. Reports state that likewise in the town Syene, situated 5,000 stades up[river] from Alexandria, no shadow is cast at noon on the day of the solstice, and that a shaft made to test this was fully lit up. As a result it is evident that the sun is directly above that place then, and Onesicritus writes that this is the case at the same time in India too, up Hypasis river. It is also established that in Berenice, city of the Trogodytae, as well as 4,820 stades away in the same people's town Ptolemais (founded at the edge of Red sea for the first elephant-hunts), the same happens for 45 days before the solstice and for as many days after it, and that for those 90 days shadows are thrown south.

184. Back in Meroe – an inhabited island and capital of the Aethiopian people 5,000 stades from Syene on Nile river – the shadows disappear twice a year, when the sun is in the eighteenth degree of Taurus and then in the fourteenth of Leo. In India there is a mountain of the Oretes people named Maleus, next to which shadows are thrown south in summer and north in winter; Septentrio is visible there for only 15 nights. In the same [part of] India, at the very well-known harbor Patalis, the sun rises on the right, and shadows fall south.

185. When Alexander was staying there [325 BCE], it was noted that Septentrio was seen only during the first part of the night. Onesicritus, his guide, wrote that Septentrio is not visible in Indian locations where there are no shadows; these places are called 'shadeless', and hours are not measured there.

LXXVI

And Eratosthenes reported that throughout Trogodytice shadows fall in the opposite direction for two 45-day periods annually.

LXXVII Where days are longest, where shortest

186. So it is that, because of variation in how long light continues, the longest day in Meroe is 12 and eight-ninths equinoctial hours, but 14 hours at Alexandria, 15 in Italy, 17 in Britannia where the bright summer nights undoubtedly confirm what theory urges us to believe: namely that, when on the days of the solstice the sun comes closer to the top of the world with a narrow spread of light, the parts of earth below have continuous days for six months and continuous nights when the sun has shifted in the opposite direction towards winter.

187. Pytheas of Massilia writes that this happens on Thule island, a six-day voyage north away from Britannia, and in fact some also make the same claim about Mona, which is about 200 [miles] away from the town Camalodunum in Britannia.

LXXVIII About the first sundial

Anaximenes of Miletus, a student of Anaximander (already mentioned [2.31]), developed this theory of shadows (also called gnomonics), and he was the first to display the timepiece – called sciothericon [shadow-hunter] – at Lacedaemon.

LXXIX How days may be defined

188. Different peoples have observed the actual day differently: Babylonians between successive sunrises, Athenians between successive sunsets, Umbrians from midday to midday, all common folk from dawn till dusk, Roman priests and those who fixed the civil day (as well as Egyptians and Hipparchus) from midnight to midnight. It is in any case evident that the breaks in daylight between sunrises are briefer near the solstices than at the equinoxes, because the position of the zodiac slants more about its midpoints, but is more upright near the solstice.

LXXX Differences between peoples according to the world's principles

189. Connections should be established with matters linked to these celestial causes. For without doubt Aethiopians are scorched by the heat of a nearby star and are born with a burnt appearance, their beards and hair frizzy, while in the opposite region of the world people have pale, icy skin, with straight fair hair. These latter people are truly fierce because of the harshness of their climate, whereas the instability of the others' [climate] makes them shrewd, with their legs demonstrating that the nature of the heat causes their sap to be drawn upwards, while in the case of the latter people it is driven down to their lower parts by falling moisture. Their environment produces hefty beasts, that of the others [produces] creatures of various shapes, especially birds of many types made swift by force of heat;

190. in both regions [people's] bodies are tall, in the one instance from force of heat, in the other from supporting moisture. In earth's middle region, because of a healthy mix from both sides, there are areas productive in all respects, bodies are medium-build with considerable blend of complexions, manners polite, senses supple, intellects fertile and able to comprehend Nature whole. Here are people with systems of rule, which those on the margins have never had, nor too have they even submitted to the others, because of being cut off and solitary in line with the vastness of Nature that encroaches on them.

LXXXI–LXXXIII About earthquakes. About chasms in the earth. Signs of an impending quake

191. Babylonian theory holds that both earthquakes and chasms, along with everything else, are caused by the force of the three planets to which they attribute thunderbolts, and moreover that these occur while [those planets] are moving with the sun or in conjunction with it and especially when at right-angles with earth. In this connection, if we believe it, a certain remarkable and unforgettable prophetic act is credited to the scientist Anaximander of Miletus, who is said to have warned the Lacedaemonians to protect their city and buildings because an earthquake was imminent. Their whole city did then collapse, as well as a large piece of mount Taygetus jutting out in the shape of a ship's stern; this broke off and brought down further damage on this disaster. Another prediction, also heaven-sent, is credited to Pythagoras' teacher Pherecydes, who was alerted by drawing water from a well and gave his fellow citizens advance warning of an earthquake.

192. If these accounts are true, then just how different from God can such men appear to be in their lifetime? Really, let each individual's judgment be left to weigh these issues: I have no doubts in thinking that winds are the cause. For earth tremors never occur unless the sea is calm and the sky so still that there is no support for birds to fly, because all the breath that propels them has been removed. [Tremors] also only ever occur after there has been wind, because there are surely blasts secreted in cavities and [earth's] hidden hollows. On earth a tremor is nothing other than a thunderclap in a cloud, nor is a chasm anything other than when lightning bursts out as confined air struggles and strains to free itself and escape.

LXXXII

193. So tremors vary in character, and the effects produced are amazing: in one place walls are flattened, in another they are drawn into a deep chasm, in another there is a pile-up of rocks, in another rivers well up and sometimes even fires or hot springs, and in another river courses are altered. Before and during [tremors] there is terrible noise, sometimes like rumbling, at others like bellowing or human shouting, or the clash of weapons coming to blows. It depends on what sort of substance is absorbing [the tremor] and the shape of either the caverns or channel through which it passes; its volume is more subdued in narrow confines, loud where there are bends, echoing in firm surroundings, agitated in wet ones, quivering in flooding, frenzied against solid surfaces.

194. So even with no quake sound is often produced. And the shaking is never straightforward, but there is tremor and vibration. Sometimes a chasm stays open displaying what it has ingested, and sometimes it hides this by closing its mouth and bringing soil back again so that no traces protrude; typically, cities are engulfed and agricultural land swallowed up. But coastal areas are especially tremor-prone, though mountain ones are by no means free of such misfortune; a discovery of mine is that tremors have occurred rather often in the Alpes and Appenninus.

195. Earthquakes are more frequent in autumn and spring, as is lightning; consequently, the Galliae and Egypt have very few tremors, since summer in the latter acts to prevent them, as does winter in the former. Similarly, they happen more often at night than in the daytime; however, the largest earthquakes occur in the morning and evening, but they are frequent near dawn and around midday. They also happen during solar and lunar eclipses (because there is a lull in storms then), but especially when heat follows rain or rain heat.

LXXXIII

196. Also sailors confidently sense one as waves suddenly swell up when there is no breeze or a shock shakes [the ship]; in fact even on ships doorposts shudder just as they do in buildings and their creaking is prophetic; indeed even birds, their courage gone, stay perched. There is even a sign in the sky that precedes an impending quake either in the daytime or a little after sunset in fine weather: cloud like a thin line strung out over a wide span.

LXXXIV Precautions against future quakes

197. [Another sign is when] water in wells is muddier and smells bad. In wells there is a [possible] precaution too, one that caves also frequently provide: for they exhale breath that has been trapped. This is noticed in entire towns: those where plenty of channels have been bored for drainage are shaken less, and [buildings] suspended above them are much safer, as may be realized from Neapolis in Italia, where its thick part is vulnerable to such falls. Arches of buildings, as well as the corners of walls and doorposts, are the safest parts because they rebound at every other thrust; also tremors do less damage to walls built with clay brick.

198. There is also a great difference in the actual type of quake, given the several ways in which they shake: one that quivers is safest, with buildings audibly vibrating and a quake alternately rising to swell and then sinking; no harm is caused either when colliding roofs strike one another from opposite directions, since every blow is matched by a rebound; danger comes with a billowing roll and wave-like rocking, or when the entire quake propels itself in one direction. However, tremors cease once the wind has found release; but in the event of their continuing, it is 40 days before they stop and usually even later, as has happened with some continuing for as long as a year or two.

LXXXV Unique predictions of earthquakes

199. I find – actually in books of Etruscan science – that a mighty foreshadowing of an earthquake once occurred in the Mutina region during the consulship of Lucius Marcius and Sextus Julius [91 BCE]. For two mountains collided with one another, dashing forward and [then] recoiling with the greatest din, and between them during the day flames and smoke rose skyward, while a large crowd of Roman *equites* and their families and travelers looked on from Via Aemilia. All the farm-buildings [there] were crushed in the collision and a great many animals inside them killed, in the year [moreover] before the Social War, which perhaps did more harm to the actual territory of Italia than the civil wars. In our own day too, no less remarkable a sign was recognized in the last year of Nero's principate (as related in our account of its events) [68 CE], when meadows and olive trees with a public road running between them switched to opposite sides in territory of the Marrucini, on an estate belonging to Vettius Marcellus, a Roman *eques* and agent for Nero's holdings.

LXXXVI Miraculous earthquakes

200. When earthquakes occur the sea surges at the same time, evidently poured in by the same breath or withdrawn from recesses in the earth as it caves in. The greatest earthquake in human memory occurred during the principate of Tiberius Caesar, when 12 cities of Asia were flattened in a single night [17 CE]; the one with the most prolonged series of shocks occurred during [the Second] Punic War, when 57 were reported to Rome all in the same year, the year [217 BCE] in fact when neither the Carthaginians nor the Romans fighting at lake Trasimenus felt this greatest of quakes. Really, such a disaster is no straightforward matter, nor does danger lie just in the quake itself, but as much or more in the premonition; never has the city of Rome shaken without this being advance warning of some future occurrence.

LXXXVII Locations where the sea has receded

201. What causes the creation of lands is the same, when that same breath powerful enough to raise the ground-level has lacked the strength to burst through. For [lands] are not only created by influx of rivers – as the Echinades islands have been built up by Achelous river and the greater part of Egypt by Nile, to which, if we believe Homer, the journey from Pharos island took a night and a day – nor by the sea receding, as happened at Cercei according to [Homer] again. There are reports that this happened both in the harbor of Ambracia for a distance of 10,000 paces and for 5,000 at the Athenians' Piraeus; also at Ephesus where [the sea] once used to lap against the temple of Diana. In fact, if we believe Herodotus, up from Memphis there was sea as far as the mountains of Aethiopia and also from the plains of Arabia, [and] around Ilium and all Teuthrania, and where the Maeander built up plains.

LXXXVIII The creation of islands explained

202. Lands are also created in another way and suddenly emerge in some sea, as if Nature is balancing itself and replacing what a chasm had swallowed up elsewhere.

LXXXIX Which ones have been created and when

This is the tradition preserved about Delos and Rhodos, islands that have long been famous; lesser ones were created later, Anaphe beyond Melos, Neae between Lemnus and the Hellespont, Halone between Lebedos and Teos, Thera and Therasia among the Cyclades in the fourth year of the 135th Olympiad [237 BCE], and between these same [two] 40 years later Hiera (same as Automate), and Thia two stades from it 242 years later – in our time on the eighth day before the Ides of July [July 8] in the consulship of Junius Silanus and Laelius Balbus [19 CE].

203. Before our time, among the Aeolian islands next to Italia one also appeared, likewise one next to Creta 2,500 paces long with hot springs, and another in the third year of the 163rd Olympiad [126 BCE] in Tuscan bay, this one ablaze with violent wind; tradition records that great quantities of fish were floating round it, and that those who ate them died instantly. So too the Pithecussae are said to have risen in Campanian bay, and on them later mount Epopos, when suddenly a flame shot from it and it was reduced to the level of the plain; here also a town was swallowed up into the depths, and a lagoon emerged from another earthquake, while yet another produced Prochyta island after mountains had been toppled.

XC Lands broken up by sea

204. The natural order also formed islands in this way: it ripped Sicilia from Italia, Cyprus from Syria, Euboea from Boeotia, Atalante and Macria from Euboea, Besbicum from Bithynia, Leucosia from Sirens' cape.

XCI Islands joined to mainland

Again, it deprived islands of sea and joined them to land, Antissa to Lesbos, Zephyrius to Halicarnasus, Aethusa to Myndus, Dromiscos and Perne to Miletus, Narthecusa to cape Parthenius. Hybanda, once an Ionian island, is nowadays 200 stades from the sea, Ephesus has Syrie landlocked, as has its neighbor Magnesia the Derasidae and Sapphonia. Epidaurum and Oricum have ceased to be islands.

XCII Lands changed completely by sea

205. First of all, [sea] – if we believe Plato – has entirely removed land over a vast area where the Atlantic sea is, then in the interior [i.e. Mediterranean] where today we see Acarnania submerged by the Ambracian bay, Achaia by the Corinthian, Europa and Asia by the Propontis and Pontus. In addition, the sea has forced its way through Leucas, Antirrhium, Hellespont, the two Bospori.

XCIII Lands that have shrunk spontaneously

Let me move on from bays and lagoons [to state that] the very earth eats itself up. It has devoured the very high mount Cibotus along with Carice town; [mount] Sipylus in [the territory of] Magnesia and the once very famous city there called Tantalis; the territories Galene and Gamale of cities in Phoenice along with them also; Phegium the highest ridge in Aethiopia. It is not as if shorelines should fail to advance treacherously too!

XCIV Cities swallowed up by the sea

206. Pontus has removed Pyrrha and Antissa around Maeotis, as has the Corinthian bay Helice and Bura, there [still] being traces of these visible in the depths. [Sea] seized – along with most of the population – over 30,000 paces suddenly broken off from Cea island, as well as half Tyndaris city in Sicilia and whatever is missing from Italia, likewise Eleusis in Boeotia.

XCV About airholes

This should be enough said about earthquakes and about any occurrences where at least cities' ruins survive. So at this point let us talk about earth's wonders rather than Nature's crimes. And, by Hercules, the heavenly bodies did not prove harder to describe!

207. [I shall talk about] the wealth of mines in all their variety, wealth and productivity, originating in so many centuries, even though every day so much worldwide is damaged by fires, collapses, shipwrecks, wars, fraud, and so much wasted by luxury and masses of people; the very varied palette of gems, diverse markings in stones and among these the brightness of some blocking everything except daylight; the power of medicinal springs; flames of fire perpetually shooting up from so many places for so many centuries; elsewhere deadly fumes either discharged from holes or made lethal by the site's actual location, in certain places only for birds (as in the Soracte region close to the city [Rome]), in others for further creatures except humans, and sometimes for humans too as in the territories of Sinuessa and Puteoli.

208. These are called 'spiracula' [airholes] or 'charonea' by others, holes discharging deadly fumes, like the place near the Mephitis temple at Ampsanctus in the territory of the Hirpini, where anyone who has entered dies; similarly a site at Hierapolis in Asia is harmless only to the priest of the Great Mother. Elsewhere there are prophetic caves where those intoxicated by the fumes may predict the future, as at the very famous oracle of Delphi. What other explanation could some human give to account for all this, except that Nature's divine presence has spread everywhere and is repeatedly bursting out at this location or that?

XCVI About land that never stops shaking. About islands always in motion

209. There are in fact places where the land shakes when walked on, such as [an area of] nearly 200 *iugera* which riders traverse in the territory of Gabii not far from the city of Rome; the same occurs in the territory of Reate. Certain islands are always floating about, as in the territories of Caecubus and Reate again, also of Mutina and Statonia, on lake Vadimon, in the dense woodland near Aquae Cutiliae (never seen in the same place by night as by day), and in Lydia ones called Calaminae shifted not just by breezes but also possible to punt in any direction desired, a refuge of many citizens during the war with Mithridates. There are also small ones at Nymphaeum, called Saliares because they are moved with foot-beats keeping time with a musical group's singing. On Italia's great Tarquiniensian lake two islands carry around groves, with the wind giving their outline now a triangular shape, now circular, but never square.

XCVII Locations where there is no rain

210. Paphos has a famous shrine of Venus with a particular area in it where rain does not fall, the same applies around an image of Minerva in Nea, a town in the Troas: also, leftover sacrifices here do not rot.

XCVIII Collection of earth's marvels

Next to Harpasa town in Asia stands a formidable rock that can be moved with one finger but remains stationary if pushed with full body-weight. On the peninsula of the Tauri in the state Parasinus there is earth which heals all wounds. But around Assos in the Troas originates a stone that eats away all bodies; it is called sarcophagus.

211. There are two mountains next to Indus river: the nature of one is to grip everything iron, while the other repels it, so on one people wearing nailed shoes find it impossible to uproot their soles, but on the other to put them down. It has been observed that a plague has never occurred in Locri and Croton, nor an earthquake in Ilium, while in Lycia there are in fact always 40 days of calm after an earthquake. In Arpi's territory grain that is sown fails to grow; and at the Mucian altars in [the territory] of Veii, as well as in that of Tusculum and in the Ciminian woodland, there are places where [it is impossible] to pull out something driven into the ground. Hay produced in [the territory] of Crustumerium is harmful there, but wholesome elsewhere.

XCIX On what principles the sea's tides should rise and fall

212. Plenty has also been said about the nature of water, but it is especially remarkable that the sea's tides rise and ebb, and indeed in very many ways, but the cause [lies] in the sun and moon. Between the moon's two risings there are two high and two low [tides] every 24 hours; swells occur first when the world elevates itself with [the moon], then from midday they recede as it sinks from the peak of the sky towards sunset, and after sunset they flood again from below as it approaches the sky's lowest part and that opposite the meridian.

213. After this, until [the world] rises again, [the tides] hold themselves back, with their return flow never at the same time as the day before (as if panting because of a greedy star that voraciously draws the sea with it and regularly rises elsewhere than the day before), yet at equal intervals they always go in and out every six hours – not hours of each day or night or place, but equinoctial ones; so, reckoned in ordinary hours, they are of unequal duration whenever these hours prove to be longer than equinoctial ones during either the day or night, only being equal everywhere at an equinox.

214. This is solid proof – completely transparent and even expressible in everyday language – of how dimwitted it is to deny that the same stars move to a lower position and rise again, and that they present a similar appearance to earth (or rather to the whole of Nature) in the same actions of rising and setting, with the star's course or other impact being just as plain under the earth as when it is conveyed past our own eyes.

215. Moreover, there is multiple variation due to the moon – seven days in the first instance: in fact tides are moderate from the new moon to the half-moon, from there they rise higher and surge highest at full moon; then they become lower, on the seventh day matching their level on the first, and increasing again with the half-moon on the other side; at the conjunction of sun and moon they match the level at full [moon]; with the same [moon] withdrawing north and further away from earth, they are lower than when it has passed south and exerts its force with nearer pressure. Every eight years the moon's hundredth circuit brings them back to the starting-point of their shifts and to the corresponding advances. The sun's annual agency boosts all these increases; they swell most at the two equinoxes and more fully at the autumn one than the spring, but are feeble at the winter solstice and even more so at the summer one.

216. Nonetheless [these increases] do not occur at the exact points of time I have stated but a few days later, just as not at the full or newest moon but afterwards, and not at the very instant the universe displays or conceals the moon or deflects it from the central zone, but about two equinoctial hours later, which is the effect of everything happening in the sky always falling to earth slower than the sight of them, as with lightning, thunder and thunderbolts.

217. But all tides cover and expose larger expanses in the ocean than in the rest of the sea, whether because [the sea] in its entirety is more energetic than in its parts, or because its open vastness is impacted more effectively by a star's force when operating unrestrained, whereas confined spaces act to curb it; this is why neither lakes nor rivers experience comparable impact. According to Pytheas of Massilia, tides above Britannia rise 80 cubits.

218. But in addition interior seas are enclosed by land as in a harbor; yet in some locations more expansive extent produces submission to [star-] power, as seen from the numerous examples of three-day crossings from Italia to Utica over a calm sea, with no drive from sails [but] a surging tide. Such a pace, however, is detected more around shorelines than in the open sea; in bodies too, the extremities have more intense feeling of the veins' pulse – that is, of breath. Yet in most estuaries, because of stars' different risings in each region, the timing of tides varies even though they occur for the same reason, as in the Syrtes.

C Where tides may equally occur according to no principle

219. There are also some of a peculiar nature, for instance a more frequent rise and fall in Tauromenitanian strait, and in the Euboean one seven times each day and night; the same tide halts three times per month, on the moon's seventh, eighth and ninth [days]. At Gades the spring – enclosed like a well – nearest to the shrine of Hercules sometimes rises and falls with the ocean, [but] sometimes does both at a different time; the other spring in the same place keeps in step with the ocean's movements. There is a town on the banks of Baetis where the wells fall when the tide rises and ebb when it falls, and meantime do not change level; in Hispalis town one well is like that and the rest normal. Also, the Pontus always passes out into Propontis, with sea never flowing into Pontus in reverse.

CI–CV Sea marvels

220. All seas are cleansed at the full moon, some even at a fixed time; around Messana and Mylae dung-like filth is disgorged onto the shore – hence the story about the sun's oxen being stabled there. To this (so that I should omit nothing I have learned) Aristotle adds that animals only die when the tide is ebbing; this observation has often been made about the Gallic ocean, and is a finding applicable to humans too.

CII The moon's power over land and sea

221. Hence arises the sound presumption that the moon is truly to be considered the star of breath; it is the one that saturates earth and fills bodies as it approaches, [then] empties them as it departs; so, as it waxes, shells are enlarged and its breath is felt especially by whatever is bloodless, but blood too (even of humans) is enhanced and reduced with its light, and leaves and fodder also feel it (as will be explained at the appropriate point [18.321*]), since the same force penetrates everything.

CIII The sun's [power]

222. Thus the sun's heat makes liquid evaporate, and we have taken it for granted that this is [the action of] a male star, scorching and absorbing everything.

CIV Why the sea is salty

So the tang of salt is baked into the sea's wide expanses. This is either because, once what is fresh and tender has been drained from it (this extraction being very easy for fiery force), everything more bitter and coarse would be left (hence surface seawater is fresher than that deep down) – a sounder explanation for its bitter taste than [claiming that] the sea is earth's never-ending perspiration; or because the heat mixed with it is mostly a product of arid conditions; or because earth's own nature soaks in, as it does into healing waters [at spas]. As one instance, once Dionysius, tyrant of Sicilia, had been removed from power, there occurred the prodigy that for one day the seawater in the harbor became fresh.

223. The moon, on the other hand, is regarded as female, a charming star that exudes night-time moisture and attracts it without removing it; this is evident from how its glance causes wild animals' dead bodies to putrefy, and how when [creatures] are asleep she retrieves the sleeper's drowsiness – concentrating it in the head – [then] thaws ice and relaxes everything with moistening breath. So Nature's give-and-take is balanced and always sufficient, with some of the stars drawing elements together and others in fact discharging them. But the moon finds sustenance in fresh waters, just as the sun does in the sea.

CV Where the sea is deepest

224. Fabianus says that the deepest sea measures 15 stades. Others say that in the Pontus opposite the Coraxi people, about 300 stades from the mainland, there is an immense sea-depth – they call it Bathea Ponti – where the bottom has never been found.

CVI Marvels from springs and rivers

What makes this all the more remarkable is fresh water gushing out, as if from pipes, right by the sea; for even the nature of water is not without its marvels. Fresh water floats on the sea, no doubt because it is lighter; so too seawater, heavier by nature, gives floating objects more support. In fact there are even some instances of freshwater streams flowing intermixed one above another, like the river floating on lake Fucinus, Addua on Larius, Ticinus on Verbannus, Mincius on Benacus, Ollius on Sebinnus, Rhodanus on Lemannus (this last one is across the Alps, the previous ones are in Italia) – guest-passages of many miles, removing no more than just their own water which they brought in. This has also been reported in Syria's Orontes river and many others.

225. But some, with loathing for the sea, flow under the bottom of it, like Arethusa, the spring at Syracusae where items are returned after being thrown into the Alpheus which flows through Olympia and empties out on the Peloponnesian coast. Ones that flow underground and are brought back up again include the Lycus in Asia, Erasinus in Argolica, Tigris in Mesopotamia, while items dropped in Aesculapius' spring at Athens are returned in the one at Phalerum; also a river submerged in Atina's plain comes out after 20 miles, as does the Timavus in [the territory] of Aquileia.

226. Nothing can be sunk in Judaea's lake Asphaltites which produces bitumen, nor in [lake] Aretissa in Greater Armenia; it in fact being alkaline nurtures fish. In the Sallentine region near the town Manduria is a lake full to the brim, which neither falls when water is emptied from it nor rises when some is added. Wood thrown into the Cicones' river and into lake Velinus in Picenum is covered with a stone coating; in the Colchian river Surius the coating continues to harden to the point of creating a mostly stone surface. Similarly in Silerus river beyond Surrentum not only brushwood immersed there but also leaves petrify, [though] otherwise its water is safe to drink; rock forms at the outlet of the marsh at Reate, <and olive trees and green bushes grow in Red sea>.[4]

227. Yet it is remarkable how hot many springs are by nature, as occurs even up in the Alpine range and in the sea itself between Italia and Aenaria in Baian bay, as well as in Liris river and many others. Fresh water is drawn from the sea in plenty of places, such as at the Chelidoniae islands and at Arados and in the Gaditan ocean. Green grass grows in the hot springs of the Patavini, frogs in those of the Pisani, and fish in those of the Vetulonii in Etruria not far from the sea. A river in [the territory] of Casinum called Scatebra is cold and fuller in summer; in it, as in [lake] Stymphalis in Arcadia, water-mice are born.

228. Cold though it may be and prone to douse torches dipped in it, Jupiter's spring at Dodone sets alight extinguished ones when they are moved near it; this same spring always falters at midday – hence it is called *Anapauomenon* [intermittent in Greek] – [but] later revives and gushes till midnight, and from then gradually falters again. Clothes laid out above a cold spring in Illyrian territory catch fire. Jupiter Hammon's

---

[4] This final clause may be a later addition.

lagoon is cold in the daytime and hot at night. A spring called 'Sun's' in Trogodytic [territory] is fresh and especially cold around midday; later it gradually becomes warmer till by midnight it is spoiled by its heat and brackishness.

229. The source of the Padus always dries up at midday in summer as if taking a siesta. A spring on Tenedus island always overflows after the summer solstice from the third hour of the night till the sixth, while on Delos island Inopus spring falls or rises as the Nile does and in step with it. Opposite Timavus river there is a small island in the sea with hot springs that crest and dwindle in step with the sea tide. In Pitinum region across Appenninus is Novanus river, in full spate at every [summer] solstice [but] dried up in winter.

230. In Faliscan territory all the drinking water makes oxen white, but in Boeotia Melas river makes sheep black, Cephisus flowing from the same lake makes them white, Penius again makes them black, and Xanthus near Ilium makes them red, hence the river's name. In Pontus the Asiaces river soaks fields where mares pasturing nurse their young with black milk. In the territory of Reate a spring called Neminie shoots up now in one place and now in another, indicating variations in the harvest. A spring in the harbor at Brundisium offers voyagers untainted water. What is called water of Lyncestis has the sourness of wine and intoxicates people; similar water is found in Paphlagonia and in the Cales region.

231. Mucianus – three times consul [64, 70, 72 CE] – believes that on Andrus island at the temple of Father Liber is a spring that always flows tasting like wine on the Nones of January [January 5]; this day is called *Theodosia* [god's gift in Greek]. A drink from the Styx near Nonacris in Arcadia kills instantly, despite there being nothing distinctive about its smell or color; three springs on Liberosus hill in the territory of the Tauri do the same without antidote or fatal pain. In Carrinensian territory in Hispania two springs flow side by side, the one spouting everything out, the other gulping it down; another among the same people makes all fish look golden, although out of its water there is nothing distinctive about their color.

232. In the territory of Comum by lake Larius a large spring rises and falls hourly. On Cydonea island off Lesbos is a hot spring active only in springtime. Lake Sannaus in Asia is colored by the wormwood growing around it. In the cave of Apollo Clarius at Colophon there is a pool from which those who drink receive amazing prophecies, but such drinking reduces their lifespan. Our own time, too, has seen rivers flow backwards during the final stages of Nero's principate, as we have recorded in our account of it.

233. That all springs are colder in summer than in winter aren't we surely now aware of, as well as of the following quite amazing feats of Nature? Namely, that bronze and lead sink in lump form, but float when flattened out, that some objects of the same weight sink while others float, and that heavy bodies are more easily moved in water; stone from Syrus, regardless of its bulk, floats [but] sinks after it has been broken up. Fresh corpses drift to the bottom, but are raised up as they begin to swell; it is not easier to beach an empty vessel than a loaded one; rainwater is fresher than other types for the production of salt, and salt only forms when fresh water is mixed in;

234. sea water freezes slower and heats up faster; the sea is warmer in winter, and saltier in autumn; all its water is calmed by oil, and for this reason divers spit it from their mouths because it alleviates the [sea's] natural harshness and carries down light; snow does not fall on high seas; although all water is conveyed downward, springs shoot upward even at the base of Aetna, which burns enough to belch out sand in a ball of flames for 50, [even] 100 miles.

CVII–CX Related marvels of fire and water

We should also record some marvels of fire, Nature's fourth element, but first ones [originating] from water.

CVIII About fossil pitch

235. In Commagene's city Samosata there is a marsh exuding a flammable mud called *maltha* [fossil pitch]. When it touches anything solid, it sticks to it; besides, after being touched, it also sticks to those trying to escape it. This is how [the city's people] defended their walls against Lucullus' attack [69 BCE]: soldiers were burned by their own weapons; water also inflames it; experiments have shown that it is only extinguished by earth.

CIX About naphtha

The nature of naphtha is similar; this is the name for the liquid-like bitumen that flows in the Babylon area and that of the Austaceni in Parthia; it is closely related to fire which, having seen it from no matter where, instantly leaps across to it. The story goes that Medea's rival was burnt up this way, after she had approached an altar to sacrifice and her garland caught fire.

CX Places which may always be burning

236. However, among mountain marvels Aetna always burns at night and provides enough fuel for blazing so long, even with the snow on it in wintertime and the frost with which it covers the ash produced. It is not only there that Nature rages with threats to burn up the world: mount Chimaera in Phaselis blazes, and indeed does so day and night with perpetual flames; Ctesias of Cnidus says that its fire is further inflamed by water, but extinguished by earth or dung. Also in Lycia the mountains of Hephaestus blaze when touched with a flaming torch, and to such an extent that rocks on the banks of rivers and sand in their very water burn too, and rain feeds this fire; there are stories that if somebody takes a stick set alight by it and draws a furrow, streams of fire then follow. In Bactrians' territory the peak of Cophantus burns at night;

237. in that of the Medes so does flat country, as also in Sittacene on the border of Persis, likewise in Susa at Turris Alba [white tower] from 15 openings, the largest of which burns by day too. In Babylon [fires] blaze from some kind of *iugerum*-wide fishpond, so too – like stars – on the plains near mount Hesperu among the Aethiopians; something similar occurs in the Megalopolitans' territory. By Theopompus' account, if there has been a break in the flow from the lovely Nymphaeum basin – which does not burn the foliage of the thick woodland above it, and is always blazing next to a cold spring – that is a terrible omen for the local Apolloniatae; it is boosted by rainfall and discharges bitumen to blend with the spring's unpalatable water, which otherwise is less substantial than any bitumen.

238. But who should be surprised by all this? In mid-sea the Aeolian island Hiera close to Italia, along with the sea itself, burned for several days during the Social War, until placated by an embassy from the senate.

However, the range called Theon Ochema in the Aethiopians' territory burns with the greatest blaze, discharging waves of flames as hot as the sun. In so many places, with so many blazes does the natural order burn earth!

CXI Marvels of fire itself

239. Besides, considering how prolific this one element's mode of operation is and how it reproduces itself and grows from the smallest sparks, what should be expected to happen when there are so many pyres on earth? What nature does it have that feeds the greediest appetite in the whole world without harming itself? To be added [to the pyres] are countless stars and the immense sun, manmade fires and those ingrained naturally in rocks and when sticks are rubbed together, as well as those from clouds and the sources of thunderbolts: for certain, it will be beyond all miracles that there has been a day on which everything was not in flames, especially when hollow mirrors facing the sun's rays set off a blaze even more easily than any other form of fire.

240. What about the countless small, but natural, ones that spring up? In the Nymphaeum [2.237] a flame set alight by rain comes out of a rock; at Aquae Scantiae too one comes out, weak as it passes and not long-lasting in any substance (an ash-tree that spreads over this blazing spring keeps on thriving); one comes out in the Mutina region on days dedicated to Vulcan. There are references in the sources to the ground burning in fields belonging to Aricia if charcoal is dropped there; to stones with oil on them blazing up in the territory of Sabinum and Sidicinum; to a flame at once appearing if wood is placed on some sacred rock in the Sallentine town Gnatia; to ashes placed on the altar of Juno Lacinia in the open air remaining motionless no matter in what direction gales blow.

241. [Sources also say that] fires suddenly appear both in waters and in bodies, even human ones: the whole of lake Trasimenus burned; when as a boy Servius Tullius was asleep, a flame shot from his head; Valerius Antias tells of Lucius Marcius having a similar experience in Hispania when he was speaking after the death of the Scipios and encouraging soldiers to take revenge. More on this later, and in fuller detail;[5] for

---

[5] If Pliny kept this promise, the passage does not survive.

now the current presentation amounts to a variety of marvels across the entire range. And moving on from explaining Nature, [my] mind is keen to steer readers' attention by hand on a tour of the whole world.

CXII Size of the whole world

242. Our part of earth, to which I am referring, in effect floats in the ocean that surrounds it (as was said [2.167–173]), and its widest extent from east to west – that is, from India to the pillars dedicated to Hercules at Gades – is 8,578 miles on the authority of Artemidorus, but 9,818 on that of Isidorus. Artemidorus adds 991½ more beyond Gades around cape Sacrum to cape Artabrum, which is the longest foreland projecting from Hispania.

243. The calculation is made from a pair of routes: [first,] from Ganges river and its mouth, where it flows into the eastern ocean, through India and Parthyene to the Syrian city Myriandrus situated on Issic bay 5,215; from there by the shortest voyage to Cyprus island, Patara in Lycia, Rhodos, Astypalaea (island in Carpathian sea), Taenarum in Laconica, Lilybaeum in Sicilia, Caralis in Sardinia 2,113, then Gades 1,250; this makes the total distance from the eastern sea 8,578.

244. Another better known route, mainly by land, extends from the Ganges to Euphrates river 5,169, then Mazaca in Cappadocia 244, then through Phrygia, Caria and Ephesus 499, from Ephesus across Aegean sea to Delos 200, Isthmos 212½, then by land and Laconian sea and Corinthian bay to Patrae in Peloponnesus 90, Leucas 87½, the same to Corcyra, Acroceraunian [mountains] 82½, Brundisium 87½, Rome 360, Alpes as far as Scingomagus village 519, through Gallia to Illiberis in Pyrenaei mountains 468, to the ocean and shore of Hispania 831, with crossing to Gades 7½; by Artemidorus' calculation this makes a distance of 8,945.

245. But earth's breadth from its southern point to the north amounts to just under half as much [as its width], 5,462 according to Isidorus, a clear demonstration of how much has been consumed by heat on one side and cold on the other; for I do not think earth has [some part] missing or that it is not globe-shaped; rather, there are uninhabitable [parts still] undiscovered on both sides. This measurement runs from the shore of the

Aethiopic ocean (where it is at least inhabited) to Meroe 625, then Alexandria 1,250, Rhodos 584, Cnidus 87½, Cous 25, Samus 100, Chius 94, Mytilene 65, Tenedos 119, cape Sigeum 12½, Pontus mouth 312½, cape Carambis 350, Maeotis mouth 312½, Tanais mouth 275, a journey that can be made 79 less if short-cuts by sea are taken.

246. After the Tanais mouth, the most thorough authorities have produced no precise figures; Artemidorus considered what lies beyond to be undiscovered, although he acknowledged that Sarmatian peoples live around Tanais over towards the north; Isidorus added 1,250 to as far as Thule, an estimate worthy of a clairvoyant. So far as I am aware, the Sarmatians' territory is known to extend for at least the distance just indicated. In any case, how big should this region be to accommodate countless peoples who constantly shift where they settle? Hence, I think it is an uninhabitable region of much greater size that lies further beyond; I am informed too of huge, previously unknown islands beyond Germania.

247. This is what I would consider worth recording about [earth's] length and breadth. However, Eratosthenes – who has a detailed appreciation of all branches of learning and is the pre-eminent expert in this one, approved (I see) by everybody – makes its total circumference 252,000 stades, which in Roman units comes to 31,500 miles: a shameless assertion, but one put forward with such shrewd reasoning that there is no embarrassment in accepting it. Hipparchus – who is amazing for confuting him and for all the rest of his painstaking efforts – adds just under 26,000 stades.

248. Dionysodorus' conviction is different, and I would not omit this supreme instance of Greek fantasy. He was a Melian famous for his knowledge of geometry; he died an old man in his native land; his female relatives who were his heirs arranged his funeral. During the following days, when performing the appropriate rituals, they are said to have found in his tomb a letter in Dionysodorus' name written to his survivors: [it said that] he had passed from his tomb to earth's lowest point 42,000 stades away. There have been experts in geometry who interpret this to mean that the letter was sent from the center of the earth, which is the longest distance down from the top and still the sphere's midpoint. Hence there followed the calculation that its circumference is stated to be 252,000 stades.

CXIII Harmony theory of the universe

The harmony theory – which holds that the natural order makes sense on its own terms – adds 12,000 stades to this calculation, and makes earth a ninety-sixth piece of the entire universe.

Totals: Facts and inquiries and observations, 417

[Roman] Sources

Marcus Varro. Sulpicius Gallus. Emperor Titus Caesar. Quintus Tubero. Tullius Tiro. Lucius Piso. Titus Livius. Cornelius Nepos. Sebosus. Caelius Antipater. Fabianus. Antias. Mucianus. Caecina who [wrote] about the Etruscan discipline. Tarquitius who [wrote] about the same. Julius Aquila who [wrote] about the same. Sergius Plautus.

Foreign Sources

Hipparchus. Timaeus. Sosigenes. Petosiris. Nechepso. Pythagoreans. Posidonius. Anaximander. Epigenes. Eudoxus. Democritus. Critodemus. Thrasyllus. Serapion [who wrote about] gnomonics. Euclid. Coeranus the philosopher. Dicaearchus. Archimedes. Onesicritus. Eratosthenes. Pytheas. Herodotus. Aristotle. Ctesias. Artemidorus of Ephesus. Isidorus of Charax. Theopompus.

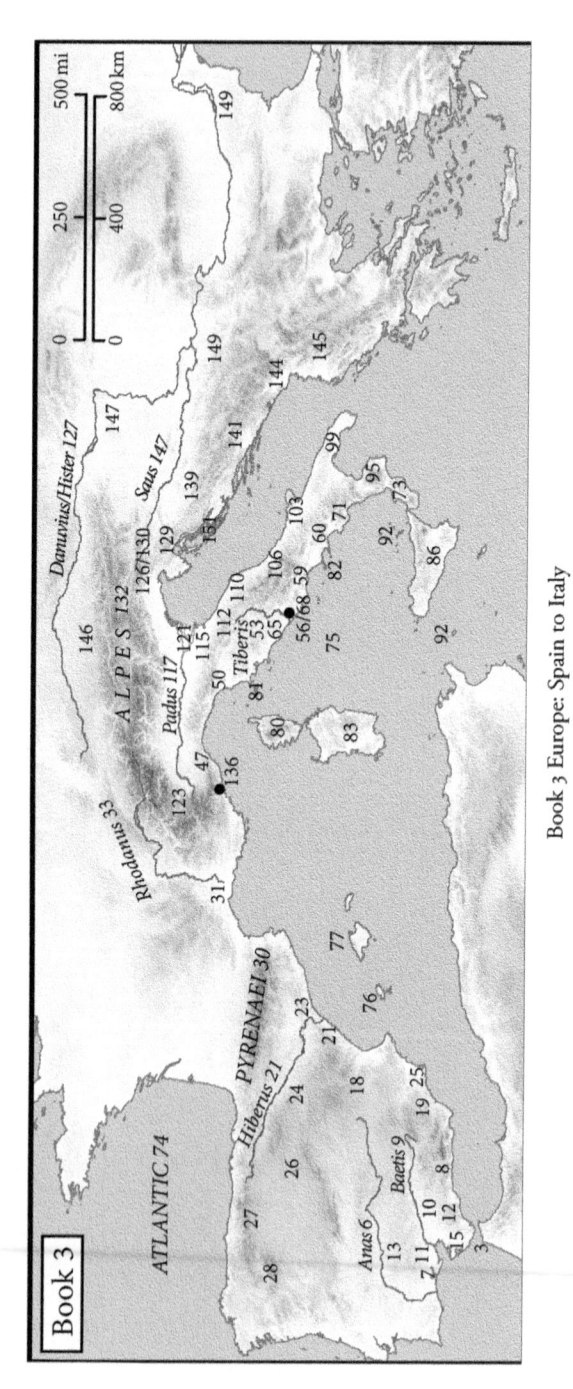

Book 3 Europe: Spain to Italy

BOOK 3

Europe: Spain to Italy

[Division of the world into parts]

1. So much for the position and wonders of the earth, waters and stars, as well as the character and dimensions of the universe. Now on to its parts, although this too is considered a never-ending task, not undertaken lightly or without incurring some criticism. In this field above all others it is fairer to make allowances, since it is far from surprising that someone born human should not know all things human. So I shall not follow any single writer, but rather whichever one I shall judge to be the most truthful for each region, because with few exceptions each takes the greatest care to describe locations in which he was active.

2. So I shall not blame or refute anyone. The bare names of places will be set down and will be given with all possible brevity, deferring their fame and the reasons for it to the appropriate sections; for at this point the treatment is comprehensive. This is why I should like it to be understood that names will be mentioned without reference to their fame, just as they were originally before any record of achievement; they form a kind of catalog, although one of the world and of Nature.

3. The whole world is divided into three parts, Europa, Asia, Africa. It starts where the sun sets at the Gaditan strait, where the Atlantic ocean breaks into the interior seas and spreads out. Entering from here, Africa is on the right, Europa on the left, Asia between them. The Tanais and Nile rivers form boundaries. According to Turranius Gracilis, who was born nearby, the ocean mouth just mentioned stretches to 15 miles long and five wide from Mellaria, a village in Hispania, to cape Album in Africa.

4. Titus Livius and Nepos Cornelius have stated the width to be seven miles at its narrowest point but ten at its greatest: so slight a mouth opens up to such an immense expanse of water. If it were very deep this might be less remarkable, but it is not. In fact whitewater streaks constantly scare ships' keels, and for this reason the area has often been called the threshold of the internal sea [Mediterranean]. Mountains situated either side of the narrowest part of the passage enclose the strait; Abila in Africa and Calpe in Europa mark the limits of Hercules' labors. Hence locals call them that god's pillars and they believe that, after excavating the passage, he let in the sea that had previously been shut out, and so changed the face of Nature.

I [Europa]

5. Therefore I shall start with Europa. It reared the community which has conquered all peoples, and is by far the loveliest part of the earth, of which it has mostly and rightly been considered half rather than a third, when the entire globe is split in two [along a dividing-line] from Tanais river to the Gaditan strait. Ocean pours the Atlantic sea through the space described and with its eager progress overwhelms the lands that dreaded its arrival, even lapping at those that resist with a winding, broken coastline. It especially hollows out Europa with repeated coves, although with four principal bays: the first is curved in an immense arc from mount Calpe at the tip of Hispania, as mentioned, to Locri and cape Bruttium.

II [Hispania Further]

6. The first land there is the Hispania called Further or Baetica; then from the border at Murgi there is Nearer or Tarraconensis up to the Pyrenaei range. Further [Hispania] is divided lengthwise into two provinces, since Lusitania in fact extends beyond Baetica's north side and is separated from it by Anas river. This rises in Laminitanian territory in Nearer Hispania, initially pours itself into bogs, then contracts back to a narrow course or completely submerges in underground channels; time and again it takes pleasure in resurfacing and is poured out into Atlantic ocean. Tarraconensis, meanwhile, is attached to the Pyrenaei and slopes down one whole side of it, and at the same time stretches diagonally from Hiberic sea across to Gallic ocean. It is separated from Baetica and Lusitania by mount Solorius, and by the Oretana, Carpetana and Astures ranges.

III Baetica

7. Baetica, named after the river cutting it in half, surpasses all these provinces with its rich culture and a touch of thriving and distinctive brilliance. It has four judicial *conventus* at Gades, Corduba, Astigi, Hispalis. In all, its towns number 175, 9 of them colonies, 10 municipalities of Roman citizens, 27 granted ancient Latin rights, 6 free, 3 allied, 120 tribute-paying. Those of them worth mentioning or easily pronounced in Latin, starting from the ocean shore at Anas river: town Ossonoba called Aestuaria at the confluence of the Luxia and Urius, Hareni mountains, Baetis river, Curensian shore with a winding bay; opposite it Gades which will be described among the islands, Juno's cape, Baesippo harbor, Baelo town, Mellaria, a strait exiting the Atlantic sea, Carteia called Tartesos by Greeks, mount Calpe.

8. Then on the inner [Mediterranean] shore: towns Barbesula and Salduba with their rivers, Suel town, Malaca one of the allied towns with its river. Then Maenuba with its river, Firmum Julium also called Sexi, Sel, Abdara, Murgi where Baetica ends. Marcus Agrippa thought that [the inhabitants of] this coast were entirely of Carthaginian origin; but from the Anas and facing towards Atlantic ocean is Bastulan and Turdulan. Marcus Varro relates that Hiberians and Persians and Phoenicians and Celts and Carthaginians penetrated the whole of Hispania.[1] [According to him,] the *lusus* [Latin = playfulness] of Father Liber or the *lyssa* [Greek = frenzy] of his fellow revelers gave Lusitania its name, and the whole [peninsula] was named after his deputy Pan [His*pan*ia]. But I consider stories about Hercules and Pyrene or Saturn as wholly fictitious.

9. The Baetis rises in Tarraconensis province, not (as in some accounts) at Mentesa town, but in the Tugiensian forest, near which Tader river provides Carthaginian territory [Carthago Nova] with water. [The Baetis] avoids Scipio's tomb at Ilorci, turns west and heads for Atlantic ocean, while giving its name to the province. Although unremarkable initially, it then absorbs many rivers, depriving them of fame and water. It first enters Baetica from Ossigitania and flows smoothly through a pleasant channel past several towns situated on its right and left banks.

---

[1] Here and throughout, the Latin *Poeni* has been rendered as Carthaginians.

10. Between it and ocean's coast the best-known inland [towns and peoples]: Segida also called Augurina, Ulia also called Fidentia, Urgao also called Alba, Ebora also called Cerialis, Iliberri also called Florentini, Ilipula also called Laus, Artigi also called Julienses, Vesci also called Faventia, Singili, Ategua, Arialdunum, Agla Minor, Baedro, Castra Vinaria, Cisimbrium, Hippo Nova, Ilurco, Osca, Oscua, Sucaelo, Unditanum, Tucci Vetus; all of these in the part of Bastitania facing the sea. In the Cordubensian *conventus* directly by the river: Ossigi also called Latonium, Iliturgi also called Forum Julium, Ipra, Isturgi also called Triumphales, Ucia and 14 miles further inland Obulco also called Pontificense, then Ripa, Epora one of the allied towns, Sacili Martialium, Onuba and, on the right, the colony Corduba called Patricia where Baetis first becomes navigable, towns Carbula and Detumo, Singilis river which empties into Baetis on the same side.

11. Towns of the Hispalensian *conventus*: Celti, Axati, Arva, Canama, Naeva, Ilipa also called Ilpa, Italica with colony Hispal also called Romulensis to its left, and on the other side town Osset also called Julia Constantia, Lucurgentum also called Juli Genius, Orippo, Caura, Siarum, Maenuba river a tributary of the Baetis on its righthand side. In the Baetis estuary: towns Nabrissa (also called Veneria) and Colobana; colonies Hasta also called Regia, and inland Asido also called Caesarina.

12. Singilis river plunges into Baetis at the place in the list already mentioned [3.10] and flows by colony Astigi also called Augusta Firma, from which point it is navigable. The remaining colonies exempt from tribute in this *conventus* are Tucci also called Augusta Gemella, Ituci also called Virtus Julia, Ucubi also called Claritas Julia, Urso also called Genetiva Urbanorum; Munda used to be one until seized along with a son of Pompey [the Great]; free towns Astigi Vetus, Ostippo; tribute-payers Callet, Callicula, Castra Gemina, Ilipula Minor, Marruca Sacrana, Obulcula, Oningi, Sabora, Ventippo. Not far away on Maenuba river, which is itself navigable, lie Olontigi, Laelia, Lastigi.

13. The region that stretches from Baetis to Anas river beyond the places already mentioned is called Baeturia, divided into two parts and as many peoples: Celtici (bordering Lusitania) in the Hispalensian *conventus*;

Turduli, living close to Lusitania and Tarraconensis, bring their cases to Corduba. That Celtici are descended from Celtiberi of Lusitania is plain from their religion, language and the names of their towns, which in Baetica are distinguished by additional names:

14. [the name] Fama Julia has been attached to Seria, Concordia Julia to Nertobriga, Restituta Julia to Segida, Contributa Julia to Ugultunia (although nowadays it is also called Curiga), Constantia Julia to Lacimurga, Fortunales to the Stereses and Aeneanici to the Callenses. Besides these in Celtica: Acinippo, Arunda, Arunci, Turobriga, Lastigi, Salpesa, Saepone, Serippo. The rest of Baeturia (which, as mentioned, is Turdulan and in the Cordubensian *conventus*) includes the far from undistinguished towns Arsa, Mellaria, Mirobriga, Regina, Sosintigi, Sisapo.

15. In the Gaditan *conventus* Regina has Roman citizens. Latin rights: Laepia Regia, Carisa also called Aurelia, Urgia also called Castrum Julium as well as Caesaris Salutariensis; tribute-payers: Besaro, Belippo, Barbesula, Blacippo, Baesippo, Callet, Cappa along with Oleastro, Iptuci, Ibrona, Lascuta, Saguntia, Saudo, Usaepo.

16. Marcus Agrippa recorded [Baetica's] total length as 475 miles, its width 258, but that was when its borders reached as far as Carthago [Nova]; for this reason great errors in the computation of the distance very often crop up, with the size of provinces changed in some instances and routes in others, the mileages becoming higher or lower. Over a great length of time the seas have encroached, or elsewhere shorelines have advanced, while rivers have become curved or have straightened out their bends. In addition, measurements are variously made from various different starting-points and following different routes. So the result is that no two [sources] agree.

17. Nowadays Baetica's length from the border town Castulo to Gades is 250, and 25 miles more along the sea-coast from Murgi; its width along the coast from Carteia to the Anas is 234. Who would believe that Agrippa – a man so very painstaking, and over this task especially, when his aim was to present the world for the world to see – had calculated wrongly, and Deified Augustus as well? For the latter completed the portico which housed it, following Marcus Agrippa's plan and notes after his sister had made a start.

IV Hispania Nearer

18. Like several provinces, the old shape of Hispania Nearer has been changed considerably, bearing in mind the claim by Pompey the Great on his trophies erected in the Pyrenaei that he had secured control of 866 towns between the Alpes and the borders of Hispania Further. Nowadays the whole province is divided into seven *conventus*: those of Carthago, Tarraco, Caesaraugusta, Clunia, the Astures, Lucus, the Bracares. There are also islands, which will be treated separately [3.76–78]. Otherwise, besides the 293 states that are incorporated with others, the province contains 179 towns, 12 of them colonies, 13 towns with Roman citizens, 18 with ancient Latin rights, 1 allied, 135 tribute-paying.

19. The first people on the coast are Bastuli; after them (reported in order moving back inland) Oretani of Mentesa and Carpetani on the Tagus; next to them Vaccaei, Vettones and Celtiberian Arevaci. Towns closest to the coast: Urci and Baria (which is assigned to Baetica), Bastitania region, next Deitania, then Contestania, colony Carthago Nova from whose cape (named after Saturn) the crossing to Caesarea, a city in Mauretania, is 197 miles. Back on the coast Tader river, tax-exempt colony Ilici, hence Ilicitan bay. Icositani are incorporated into it.

20. Next Lucentum with Latin rights, tribute-paying Dianium, Sucro river, and at one time a town, the border with Contestania. Edetania region extends back towards Celtiberian territory and has a charming lake spreading out in front of it. Colony Valentia three miles from the sea, Turium river, and the same number of miles from the sea Saguntum a town of Roman citizens famous for its loyalty, Udiva river.

21. Ilergaones' region; Hiberus river with its profitable seaborne commerce rises among the Cantabri not far from town Juliobrica, flows for over 450 miles, and can handle ships for 260 after town Vareia; it is why Greeks have given the name Hiberia to the whole of Hispania. Cessetania region, Subi river, colony Tarraco a creation of the Scipios just as Carthago [Nova] was of Carthaginians. Ilergetes' region, town Subur, Rubricatum river after which the Laeetani and Indigetes.

22. After them, reported in order moving back inland, the Ausetani and Lacetani at the foot of Pyrenaei, and across Pyrenaei the Ceretani, then Vascones. Meanwhile on the coast: colony Barcino also called Faventia,

Baetulo and Iluro towns with Roman citizens, Arnum river, Blandae, Alba river, Emporiae a twin city of ancient indigenous people and Greeks descended from Phocaeenses, Ticer river. [The shrine of] Pyrenaea Venus is 40 from it on the other side of the cape.

23. To add to what has just been said, remarkable facts about individual *conventus* will now be mentioned. At Tarraco 42 communities plead their cases. Among those with Roman citizens: the best-known are Dertosani and Bisgargitani. Among those with Latin rights: Ausetani, Ceretani (some also called Juliani, others Augustani), Edetani, Gerundenses, Iessonienses, Teari also called Julienses; among tribute-payers: Aquicaldenses, Aesonenses, Baeculonenses.

24. In Edetania region Hiberus river flows by Caesaraugusta, a tax-exempt colony where the town was previously called Salduba and serves 55 communities. Those with Roman citizens: Bilbilitani, Celsenses (previously a colony), Calagurritani also called Nasici, Ilerdenses of the same stock as Surdaones near Sicoris river, Oscenses in Suessetania region, Turiassonenses. With ancient Latin rights: Cascantenses, Ergavicenses, Graccurritani, Leonicenses, Osicerdenses. Tarracenses are allies. Tribute-payers: Arcobrigenses, Andelonenses, Aracelitani, Bursaonenses, Calagurritani also called Fibularenses, Conplutenses, Carenses, Cincienses, Cortonenses, Damanitani, Ispallenses, Ilursenses, Iluberitani, Lacetani, Libienses, Pompelonenses, Segienses.

25. Island residents aside, 65 communities convene at Carthago [Nova]: the colonies Acci Gemellensis and Libisosa also called Foroaugustana (both of which have been given Italian rights), and colony Salaria. Townsfolk with ancient Latin rights: Castulonenses also called Caesarii Iuvenales, Saetabitani also called Augustani, Valerienses. Best-known tribute-payers: Alabanenses, Bastitani, Consaburrenses, Dianenses, Egelestani, Ilorcitani, Laminitani, Mentesani also called Oretani, Mentesani also called Bastuli, Oretani also called Germani, Segobrigenses (from the capital of Celtiberia), Toletani (from the capital of Carpetania) settled on Tagus river, then Viatienses and Virgilienses.

26. Varduli lead 14 communities to the Cluniensian *conventus*, of whom only the Alabenenses are worth mentioning. Turmogidi lead four, including Segisamonenses and Segisamajulienses. Carietes and Vennenses go to the same *conventus* with five states, including the Velienses.

Celtiberian Pelondones come too with four communities, of whom the Numantini used to be famous, just like the Intercatienses, Palantini, Lacobrigenses, Caucenses among the 17 states of the Vaccaei.

27. Among the nine Cantabric communities only Juliobriga merits mention, and among the ten states of the Autrigones [only] Tritium and Virovesca. Areva river gave its name to the Arevaci. They have six towns: Segontia and Uxama whose names often crop up elsewhere, and in addition Segovia and Nova Augusta, Termes and Clunia itself at the border of Celtiberia. The rest looks towards the ocean, with the Varduli (among those already mentioned [3.26]) and Cantabri.

28. Joining them are 22 communities of Astures divided into Augustani and Transmontani, and the magnificent city Asturica. Among them are Gigurri, Paesici, Lancienses, Zoelae. The entire population amounts to 240,000 free individuals. The Lucensian *conventus* has 16 communities, all – Celtici and Lemavi excepted – insignificant and with barbarous names, but amounting to about 166,000 free individuals. Similarly, the 24 states of the Bracares amount to 285,000 people of whom, apart from the Bracares themselves, it is not absurd to name Biballi, Coelerni, Callaeci, Equaesi, Limici, Querquerni.

29. Nearer Hispania's length is 607 miles from the Pyrenaei to the border at Castulo, and a little more along the coast; its width 307 from Tarraco to the shore at Oiarso, beginning from the foot of Pyrenaei, where it is narrowed into a wedge between the two seas; then it gradually expands and more than doubles its width once it touches Further Hispania.

30. Nearly all of Hispania has mines of lead, iron, copper, silver, gold in abundance, while Nearer has *selenite* mines, and Baetica cinnabar ones. There are also marble quarries. Emperor Vespasian Augustus conferred Latin rights on the whole of Hispania at a time when it was shaken by state-wide turmoil. Pyrenaei mountains separate Hispania and Gallia with capes extending into the two seas either side.

V Narbonensis province

31. The part of the Galliae lapped by the internal sea is called Narbonensis province. Previously called Bracata, it is separated from Italia by the Varus river and Alpes range (a great protection for Roman rule), and from the rest of Gallia to the north by the Cebenna and Jures mountains. Among provinces it is second to none in the quality of its agriculture, the reputation of its people and their conduct, and their high level of wealth; it is, in short, more truly Italia than a province.

32. On the coast the Sordones' region and inland that of the Consuarani, Tecum and Vernodubrum rivers, towns Illiberis (mere vestige of a once-great city), Ruscino Latinorum, Atax river flowing from Pyrenaei through lake Rubrensis, Narbo Martius a colony of the Tenth legion 12 miles from the sea, Arar and Liria rivers. Otherwise few towns because of coastal marshes:

33. Agatha once the possession of Massilienses, and the region of the Volcae Tectosages, and where the Rhodians' Rhoda used to be, hence many call the most productive river of the Galliae Rhodanus; it hurls itself down from the Alpes through lake Lemannus and carries along the sluggish Arar, as well as Isara and Druantia which are no less rapid than it. Its two unexceptional mouths are called Libica, one Hispaniensian, the other Metapinan; a third, the largest, is called Massaliotic. Some sources say that there was also a town, Heraclea, at the Rhodanus mouth.

34. Beyond are canals dug from Rhodanus, famous as Gaius Marius' work and bearing his name, Mastromela marsh, Maritima Avaticorum town and (above it) Campi Lapidei remembered for Hercules' battles; the region of the Anatilii and inland those of the Dexivates and Cavares; towards the sea again, those of the Tricores and inland of the Tritolli, Vocontii and Segovellauni, then Allobroges. But on the coast: the Phocaeensian Greeks' Massilia an allied [city], cape Zao, Citharista harbor, region of the Camactulici, then Suelteri and above them Verucini.

35. On the coast, moreover, Athenopolis founded by Massilienses, Forum Juli a colony of the Eighth legion [also] called Pacensis and Classica, a river named Argenteus, region of the Oxubii and Ligauni, and above them Suebri, Quariates, Adunicates. But on the coast Latin town Antipolis, region of the Deciates, Varus river flowing down from mount Caenia in the Alpes.

36. Inland, colonies Arelate of the Sixth legion, Baeterrae of the Seventh, Arausio of the Second, Valentia in the territory of the Cavares, Vienna in that of the Allobroges. Latin towns: Aquae Sextiae in the territory of the

Salluvii, Avennio in that of the Cavares, Apta Julia in that of the Vulgientes, Alebaece in that of the Reii Apollinares, Alba in that of the Helvi, Augusta in that of the Tricastini, Anatilia, Aerea, the Bormani, the Comani, Cabellio, Carcasum in the territory of the Volcae Tectosages, Cessero, Carbantorate in the territory of the Memini, Caenicenses, Cambolectri also called Atlantici,

37. Forum Voconi, Glanum, the Libii, the Lutevani also called Foroneronienses, Nemausum in the territory of the Arecomici, the Piscinae, Ruteni, Samnagenses, Tolosani in the territory of the Tectosages on the border with Aquitania, Tasgoduni, Tarusconienses, Umbranici, Vasio and Lucus Augusti (the two capitals of the allied state of the Vocontii), and in fact 19 insignificant towns, as well as 24 incorporated with the Nemausienses. From those Alpine dwellers emperor Galba added to the register Avantici and Bodiontici, whose town is Dinia. Agrippa says that Narbonensis province is 370 miles long, 248 wide.

VI–X Italia to Locri

38. Italia next, with its Ligures first, then Etruria, Umbria, Latium, and there the mouths of the Tiberis and Rome, capital of the world, 16 miles distant from the sea. Thereafter the shore of the Volsci and of Campania, then that of Picenum and Lucania and Bruttium, the southernmost point to which Italia extends into the sea from the nearly crescent-shaped Alpes range. After it the coast of [Magna] Graecia, then the Sallentini, Poediculi, Apuli, Paeligni, Frentani, Marrucini, Vestini, Sabini, Picentes, Galli, Umbri, Tusci, Veneti, Carni, Iapudes, Histri, Liburni.

39. I am aware that I can justifiably be thought ungrateful and slack if I should speak about this land in a casual or cursory way – one that is both child and parent of all lands, chosen by divine will to make heaven itself more brilliant, to link dispersed dominions, to moderate their habits, and through the medium of language to bring into dialogue the clashing, crude tongues of so many peoples, to give humans civilization and, in short, to become the single fatherland of all the world's peoples.

40. But what am I to do? Who could only touch upon the great distinction of all its places and not be gripped by the great brilliance of each of their features and peoples? The city of Rome is in a class of its own here, with its face deserving such a joyful neck: how much effort should be devoted to describing it? How should just Campania's coast be described, somewhere of such bliss and divine charm as to be quite clearly a unique creation of Nature at its happiest?

41. Consider too its vigorous, year-round healthiness, such mild climate, fertile fields, sunny hills, safe pastures, shady groves, generous variety of woodlands, mountain breezes, high yields of crops, vines and olives, splendid wool from its flocks, its bulls' fat necks, so many lakes, and its rivers and springs streaming so freely all over, so many seas, harbors, and the lap of the land everywhere accommodating trade, in fact eagerly projecting into the sea as if to aid mankind.

42. I omit mention of its talents and customs, its men, and its peoples who were overpowered by [both] negotiation and action. Greeks themselves – people most prone to promote their own glory – have delivered their verdict on it by naming a limited part Magna Graecia! Without doubt I should proceed in this section as I did in describing the sky [in Book 2], namely by touching upon just certain points and a few stars. I merely ask my readers to bear in mind that I am rushing in order to describe the whole in a single work.

43. [Italia], therefore, is very similar to an oak leaf, although much longer than it is broad; it curves itself to the left at the top and ends in the shape of an Amazon's shield, where the projection from the center is called Cocynthos with its crescent-shaped bays throwing out two wings, Leucopetra on the right, Lacinium on the left. Its length extends 1,020 miles from the foot of the Alps at Augusta Praetoria through the city [Rome] and Capua, then zigzags to the town Regium situated on its shoulder; here the curve of its neck, so to speak, begins. If calculated all the way to Lacinium the figure would be much higher, except that such a bend here would seem to create a sideways shift.

44. Its width varies, being 410 miles between the two seas Lower [Tuscan] and Upper [Hadriatic], and the Varus and Arsia rivers; but near the middle around the city of Rome, from the mouth of Aternus river flowing into the Hadriatic sea to Tiberis' mouths, it is 136, and a little less from Castrum Novum on the Hadriatic sea to Alsium on the Tuscan sea. Nowhere, however, is it more than 200 wide. The whole perimeter from the Varus to the Arsia is 2,049 miles.

45. From neighboring lands the distances are: 100 miles from Histria and Liburnia in some places, 50 from Epirus and Illyricum, less than 200 from Africa according to Marcus Varro, 120 miles from Sardinia, 1½ from Sicilia, less than 80 from Corcyra, 50 from Issa. It certainly stretches out to sea in the southward part of the sky, but a methodical quest for accuracy would place it between the sixth hour and the first at midwinter.

46. In now itemizing its perimeter and cities, I must first state that the source I follow for the purpose is Deified Augustus and his division of all Italia into 11 regions, but in an order determined by the shorelines' course. In a very concise account it really is impossible to maintain [an order] just from city to neighboring city, so inland I shall follow his alphabetical arrangement, noting colonies just as he did in his enumeration. To describe their locations and origins is not easy: the Ligurian Ingauni, for example – to say nothing of others – have been made 30 land grants.

VII

47. So after Varus river Nicaea founded by Massilienses, Palo river, the Alpes and Alpine peoples with their many names, but especially Capillati, with the town Cemenelum belonging to the state of the Vediantii, harbor of Hercules Monoecus, Ligustinian coast. The best-known of the Ligures beyond the Alpes: Sallui, Deciates, Oxubi, and on this [Italian] side Veneni, Turri, Soti, Bagienni, Statielli, Binbelli, Maielli, Caburriates, Casmonates, Velleiates and the towns of those on the coast to be mentioned next.

48. Rutuba river, Album Intimilium town, Merula river, Album Ingaunum town, Vada Sabatia harbor, Porcifera river, Genua town, Fertor river, Delphinus harbor, Tigulia inland and Segesta Tigulliorum, Macra river, the boundary of Liguria. Behind all the places mentioned above are the Appenninus mountains, Italia's highest, extending in an unbroken chain from the Alpes to the Siculan strait.

49. On its other side along Padus river, Italia's richest, everywhere shines with famous towns: Libarna, colony Dertona, Iria, Vardacate, Industria, Pollentia, Carrea also called Potentia, Forum Fulvi also called Valentinum, Augusta Bagiennorum, Alba Pompeia, Hasta, Aquae Statiellorum. In Augustus' division this is region nine. Between the Varus and Macra rivers the coast of Liguria extends 211 miles.

VIII

50. Next to it is [region] seven, where Etruria begins at Macra river. Its name has changed often; here in ancient times Pelasgi expelled Umbri and themselves [were expelled] by Lydi, who were called Tyrreni after their king and later in the Greek language Tusci because of how they sacrificed. The first town in Etruria is Luna, with its famous harbor. The colony Luca is located back from the sea and nearer to Pisae between the Auser and Arnus rivers; it was founded by Pelopidae or Teutani, a Greek people. Vada Volaterrana, Caecina river, Populonium once the only Etruscan town on this coast.

51. After this the Prile and then Umbro rivers, the latter navigable, and from then on the region of Umbria and Telamo harbor, Cosa founded by the Roman people in the territory of the Volcientes, Graviscae, Castrum Novum, Pyrgi, Caeretanus river, and seven miles inland Caere itself, called Agylla by the Pelasgi who founded it, Alsium, Fregenae, Tiberis river 284 miles from the Macra. Inland, colonies Falisca founded by Argos according to Cato and called 'of Etruscans', Lucus Feroniae, the Rusellanian, Seniensian, Sutrinian.

52. Otherwise the Arretini Veteres, Arretini Fidentiores, Arretini Julienses, Amitinenses, Aquenses called Taurini, Blerani, Cortonenses, Capenates, Clusini Novi, Clusini Veteres, Florentini settled where Arnus flows past, Faesulae, Ferentinum, Fescennia, Hortanum, Herbanum, Nepet, Novem Pagi, Praefectura Claudia of Foroclodium, Pistorium, Perusia, the Suanenses, Saturnini who used to be called Aurini, Subertani, Statonienses, Tarquinienses, Tuscanienses, Vetulonienses, Veientani, Vesentini, Volaterrani, Volcentani also called Etrusci, Volsinienses. In the same area the Crustuminan and Caletran territories preserve the names of their old towns.

IX Tiberis, Rome

53. Tiberis, which in the past was called Thybris and before that Albula, flows down from about the mid-point of the Appenninus' length in the territory of the Arretini. At first it is narrow and not navigable unless its flow has been collected in reservoirs and then released, just as is done in the case of the tributaries Tinia and Clanis with a nine days' build-up unless help comes from rainfall. But even then, because Tiberis is so rough and rocky, it is impossible to proceed far except with rafts rather than boats. In the course of 150 miles it separates Etruria from the Umbri and Sabini not far from Tifernum, Perusia and Ocriculum, [and] later less than 16 miles outside the city [Rome] Veii's territory from the Crustuminan, then that of Fidenae and Latium from the Vaticanus.

54. But below the Clanis of Arretium it is reinforced by 42 rivers, principally Nar and Anio, the latter itself also navigable and enclosing Latium from behind. With aqueducts and so many springs also supplying the city, it can carry any and all the great vessels from the Italian sea. No merchant handles the whole world's products more smoothly, and of rivers everywhere almost none can have as many villas built on its banks or overlooking it.

55. And no river is more restricted, hemmed in as it is on both sides. Even so, it does not resist, despite often flooding suddenly, with the water cresting higher in the city itself than anywhere else. On the contrary, it is regarded as offering prophetic warnings because really its action is always a religious concern, not just havoc.

56. Ancient Latium has been preserved from the Tiberis to Cercei, a length of 50 miles: so slender were the empire's original roots. With frequent changes, it has been settled by different peoples at different times: Aborigines, Pelasgi, Arcades, Siculi, Aurunci, Rutuli, and (beyond Cercei) Volsci, Osci, Ausones, so that the name of Latium extended to Liris river. At the outset there was Ostia, a colony founded by a Roman king, Laurentum town, Sol Indiges' spot, Numicius river, Ardea founded by Danae, Perseus' mother.

57. Then Aphrodisium (only in the past), colony Antium, Astura river and island, Nymphaeus river, Clostra Romana, Cercei once in fact an island surrounded by an expanse of sea (to credit Homer) and nowadays by a plain. What we can pass on about this matter for people's information is remarkable. Theophrastus was the first foreigner to go to the trouble of writing anything about the Romans; before Theopompus, nobody had mentioned them. He merely said that the city was captured by Galli. Clitarchus, the next after him, said only that an embassy was sent to Alexander.

58. But [Theophrastus, offering] more than rumor, recorded a measurement of Cercei's island as 80 stades in that book he wrote when Nicodorus was an Athenian magistrate in the 440th year of our city [314 BCE]. So whatever land is attached to the island beyond a ten-mile radius was added to Italia after that year.

59. Another marvel near Cercei is the Pomptine marsh, a place where the three-times consul Mucianus reported that there had been 24 cities. Then Aufentum river, above which is the town Tarracina called Anxur in the language of the Volsci, and the place where Amyclae or Amynclae used to be until it was destroyed by serpents; then the site of Speluncae, lake Fundanus, Caieta harbor, town Formiae also called Hormiae, assumed to be where the Laestrygones lived in the past. Beyond it used to be a town Pirae, [nowadays] there is the colony Minturnae divided by Liris river once called Clanis, and Sinuessa the last town in Extended Latium, formerly named Sinope according to some.

60. From here the famous, fortunate Campania; in this bay begin hills covered in vines and the glorious intoxication from [their] juice that is famous worldwide, as well as (say the ancients) the object of intense rivalry between Father Liber and Ceres. From here extend the Setian and Caecuban territories, with the Falernian and Calenian ones joined to them. Then rise the Massici, Gaurani and Surrentini mountains. There the Leborinian plains spread out and, by means of hulling, the harvest produces delicious spelt. Hot springs drench the coastline here, and its reputation for shellfish and fine fish is unmatched throughout the seas. Nowhere else is the olive oil superior, it too being an object of rivalry to satisfy mankind's pleasure. Osci, Greeks, Umbri, Tusci, Campani have occupied [Campania].

61. On its coast, Savo river, Volturnum town with its river, Liternum, Cumae [founded by] Chalcidenses, Misenum, Baiae harbor, Bauli, lakes Lucrinus and Avernus; next to the latter was once a town Cimmerium, then colony Puteoli once called Dicaearchea, and after it the Phlegraean plains, Acherusia marsh next to Cumae.

62. Back on the shore Neapolis, also [founded by] Chalcidenses and called Parthenope after a Siren's tomb, Herculaneum, Pompeii with mount Vesuvius in view nearby and watered by Sarnus river, Nucerian territory and nine miles from the sea Nuceria itself; Surrentum with cape Minerva, where Sirens once lived. The sailing distance from Cercei is two short of 80. In Augustus' division of Italia this region from the Tiberis is reckoned as the first.

63. Inland, colonies Capua named after its 40 [miles] of plains, Aquinum, Suessa, Venafrum, Sora, Teanum also called Sidicinum, Nola; towns Abellinum, Aricia, Alba Longa, [and those] of the Acerrani, Allifani, Atinates, Aletrinates, Anagnini, Atellani, Aefulani, Arpinates, Auximates, Abellani, Alfaterni (those named after Latin territory as well as Hernician and Labican), Bovillae, Caiatia, Casinum, Calenum, Capitulum Hernicum, Cereatini also called Mariani, Corani descended from the Trojan Dardanus, Cubulterini, Castrimoenienses, Cingulani,

64. Cabienses on mount Albanus, Foropopulienses in Falernian [territory], Frusinates, Ferentinates, Freginates, Fabraterni Veteres, Fabraterni Novi, Ficolenses, Fregellani, Forum Appi, Forentani, Gabini, Interamnates Sucasini also called Lirenates, Ilionenses, Lanivini, Norbani, Nomentani, Praenestini whose city was once called Stephane, Privernates, Setini, Signini, Suessulani, Telesini, Trebulani also called Ballienses, Trebani, Tusculani, Verulani, Veliterni, Ulubrenses, Urbanates,

65. and above all Rome itself, whose other name it is considered sacrilege to utter because of the secrets of religious rites. When Valerius Soranus divulged what had been suppressed with the finest and most responsible devotion, he soon paid the penalty. It seems relevant to interject here an instance of an ancient rite established specifically for maintaining this silence: the statue of the goddess Angerona, to whom sacrifices are offered on the twelfth day before the Kalends of January [December 21], represents her with mouth covered and sealed.

66. Romulus left the city with three gates or – if we accept sources recording the highest number – four. When Vespasian was emperor and censor [with Titus] 826 years after the city's foundation [73 CE], its walls measured 13 miles and 200 paces around, encompassing seven hills. It is divided into 14 regions, with 265 shrines of Lares at crossroads. Measurement of the distance in a straight line running from the milestone erected at the head of the Roman Forum to the individual gates (37 of them today, counting each of the 12 [double] gates only once, and ignoring seven ancient ones which no longer exist) produces [a total figure of] 20 miles and 765 paces.

67. But if all the roads to the far edge of the built-up area, including the Praetorian camp, are measured from the same milestone through the neighborhoods, the total is just over 60 miles. If the height of the buildings were to be factored in, this would undoubtedly produce an appropriate estimate and establish that no other city in the whole world could match this one for size. To the east it is closed off by Tarquinius Superbus' rampart, a remarkable construction of the first order; for he raised it level with the walls where the city was most exposed to attack from the plain. Elsewhere it was fortified by towering walls or steep slopes, although [now] spreading construction has added many cities.

68. Otherwise in the first region the notable towns in Latium used to be Satricum, Pometia, Scaptia, Politorium, Tellena, Tifata, Caenina, Ficana, Crustumerium, Ameriola, Medullum, Corniculum, Saturnia where Rome is now, Antipolis which nowadays is the Janiculum and part of Rome, Antemna, Camerium, Collatia, Amitinum, Norbe, Sulmo,

69. and with these the Albensian communities which regularly sacrificed together on mount Albanus: the Albani, Aesolani, Accienses, Abolani, Bubetani, Bolani, Cusuetani, Coriolani, Fidenates, Foreti, Hortenses, Latinienses, Longani, Manates, Macrales, Munienses, Numinienses, Olliculani, Octulani, Pedani, Poletaurini, Querquetulani, Sicani, Sisolenses, Tolerienses, Tutienses, Vimitellari, Velienses, Venetulani, Vitellenses.

70. Thus 53 peoples from ancient Latium have perished without a trace. In Campanian territory a town Stabiae existed right up until the day before the Kalends of May [April 30] in the consulship of Gnaeus Pompeius and Lucius Cato [89 BCE], when Lucius Sulla, a legate in the Social War, destroyed it; nowadays it is reduced to a farmstead. Here Taurania perished too. The remains of Casilinum are disappearing. In addition, according to Antias, after Apiole, a town with Latin rights, was captured by king Lucius Tarquinius, he used the spoils to begin construction of the Capitol. Picentian territory for 30 miles from Surrentinum to Silerus river belonged to the Tusci and was famous for a temple to Argive Juno founded by Jason. Inland is Salernum's town Picentia.

X

71. The third region, Lucanian and Bruttian territory, begins at the Silerus; here too the inhabitants have changed frequently. Its occupiers have been Pelasgi, Oenotri, Itali, Morgetes, Siculi, [then] most notably Greek communities and most recently Lucani, Samnite descendants under their leader Lucius. The town Paestum called Posidonia by Greeks, Paestanus bay, town Elea called Velia nowadays, and cape Palinurum from which it is a 100-mile crossing to Columna Regia across the receding bay.

72. Next Melpes river, town Buxentum [or] Pyxus in Greek, Laus river. There also used to be a town of the same name. After this the Bruttian shore, town Blanda, Baletum river, Parthenius harbor founded by Phocians and Vibo bay, site of Clampetia, town Tempsa called Temese by Greeks, and Terina founded by Crotonienses with the huge Terinaeus bay. Inland the town Consentia.

73. On the peninsula Acheron river from which the people of the town are called Aceruntini. Hippo which nowadays we call Vibo Valentia, Hercules' harbor, Metaurus river, Tauroentum town, Orestes' harbor and Medma. Town Scyllaeum, Crateis river, mother (we are told) of Scylla. Then Columna Regia, the Siculan strait and two facing capes, Caenus in Italia and Pelorum in Sicilia, 12 stades apart, and 93 from there to Regium.

74. After that Sila an Appennine forest 15 miles from cape Leucopetra, and from it 51 to Locri named after cape Zephyrium and 303 from the Silerus.

So ends Europa's first gulf. The seas in it are named [as follows]: from where it rushes in, Atlantic, [which] others call Magnum; the strait by which it enters is called Porthmos by Greeks, but Gaditan by us. After the entrance, and so long as it flows by the provinces of Hispania, it is called Hispanian, but by some Hiberic or Baliaric; then Gallic in front of Narbonensis province, and after that Ligustic;

75. from there to Sicilia island it is called Tuscan, although some Greeks call it Notian, others Tyrrenan, but we mostly call it Lower. Beyond Sicilia (as far as the Sallentini, that is), Polybius calls it Ausonian, but Eratosthenes calls all of it between the ocean mouth and Sardinia Sardoan, then Tyrrenan from there to Sicilia, then from there to Creta Siculan, thereafter Cretic.

XI–XIV Sixty-four islands, including:

XI Baliares

76. The first of all the islands throughout these seas were called Pityussae by Greeks because of their pine bushes; nowadays each of the two is called Ebusus, and they are allied states with a narrow channel flowing between them. They extend for 46 [miles] and are 700 stades from Dianium, the same [distance] as from Dianium to Carthago Nova by land; it is just as far from the Pityussae by sea to the two Baliares, as well as to Colubraria in the direction of the Sucro.

77. Greeks called the Baliares – noted for expert slingers – Gymnasiae. The larger is 100 miles long, 475 round. Its towns with Roman citizens are Palma and Pollentia; with Latin rights, Guium and Tucis; and there used to be an allied one of the Bocchori. The smaller [island] is 30 away, 40 long, 150 round. It has the states Iamo, Sanisera, Mago.

78. Twelve [miles] out to sea from the larger [island] is Capraria, where shipwreck threatens, and beyond the region of the city Palma are the Menariae and Tiquadra [islands] and a small one of Hannibal. The soil of Ebusus repels snakes, while that of Colubraria nurtures them, and so it is unsafe for all except those who bring soil from Ebusus; Greeks called it Ophiusa [snake island]. Ebusus does not produce rabbits, which devastate the harvests of the Baliares.

79. There are about 20 other small [islands] in the shallow sea: Metina along the coast of Gallia at Rhodanus mouth, and then one called Blascorum, and three called Stoechades by the neighboring Massilienses because they lie in a row. The names of each are Prote, Mese also called Pomponiana, and the third Hypaea; after them Sturium, Phoenice, Phila, Lero and, opposite Antipolis, Lerina on which (according to tradition) used to be the town Berconum.

XII Corsica

80. In the Ligustic sea is Corsica which Greeks called Cyrnos, but it is nearer the Tuscan [sea]. Projecting from north to south, 150 miles long, 50 at its widest point, 325 round. It is 62 from Vada Volaterrana [in Italia], has 32 states and the colonies Mariana founded by Gaius Marius and Aleria founded by the dictator Sulla. On its near side is Oglasa [island]; in between, and 60 miles from Corsica, Planasia named after its appearance, since it is at sea-level and so deceives ships.

81. Urgo, a larger island, and Capraria which Greeks called Aigilion, as well as Igilium and Dianium (also called Artemisia), both lying opposite Cosa's shore; also Barpana, Menaria, Columbaria, Venaria, Ilva called Aethalia by Greeks, with its iron mines, 100 round and ten from Populonium. [From Ilva] 28 to Planasia. After these, and beyond Tiberis' mouths in the territory of Antium, Astura, then Palmaria, Sinonia and, opposite Formiae, Pontiae.

82. In the Puteolan bay Pandateria, Prochyta (named not after Aeneas' nurse, but because it was a discharge from Aenaria), Aenaria (named after the anchorage for Aeneas' ships which Homer called Inarime), Pithecusa (named not after its numerous monkeys, as some have thought, but from its pottery workshops). Between Pausilypum and Neapolis is Megaris; then eight from Surrentum, Capreae well-known as *princeps* Tiberius' stronghold and 11 miles round;

XIII Sardinia

83. Leucothea; and out of sight on the edge of the African sea Sardinia, less than eight miles from the tip of Corsica, with the channel indeed being narrowed by small islands called Cuniculariae, as well as by Phintonis and Fossae, from which the strait itself is called Taphros.

84. Sardinia – extending 188 along the east, 175 west, 77 south and 125 north – is 565 round. Its cape Caralis is 200 from Africa, 1,250 from Gades. It also has two islands off cape Gorditanum named after Hercules, Enosis off Sulcensis, Ficaria off Caralis.

85. Some also locate the Leberidae and Callode, and one they call Heras Lutra close to [Sardinia]. Its best-known communities: Ilienses, Balari, Corsi; among its 18 towns those of the Sulcitani, Valentini, Neapolitani, Bitienses, Caralitani (who have Roman citizens), and Norenses; one colony, called Ad Turrem Libisonis. Timaeus called Sardinia itself Sandaliotis because its shape is like a sole, and Myrsilus called it Ichnusa because it resembles a footprint. Opposite Paestanus bay is Leucosia, named after the Siren buried there; opposite Velia are Pontia and Isacia, the two sharing the single name Oenotrides, proof that Italia was once held by Oenotri; opposite Vibo, small [islands] called Ithacesiae after Ulysses' look-out.

XIV Sicilia

86. More famous than all, however, Sicilia, called Sicania by Thucydides and Trinacria or Trinacia by more [authors], because of its triangular shape. According to Agrippa, its perimeter extends 618 and it was once attached to Bruttian territory, but was then torn away by the encroaching sea [which made] a strait 15 long and 1½ wide by Columna Regia. It was in view of this undoubted separation that Greeks gave the name Regium to the town situated on the edge of Italia.

87. In this strait is the rock Scylla, as well as the whirlpool Charybdis, both famously treacherous. In the triangle itself (as so called above), the cape called Pelorum opposite Scylla faces Italia, Pachynum (440 from the Peloponnesus) faces Graecia, and Lilybaeum faces Africa, 180 from cape Mercurii [there] and 190 from [cape] Caralis in Sardinia. So the distances between these capes and the lengths of coasts are: by land route from Pelorum to Pachynum 186, then to Lilybaeum 200, then to Pelorum 242.

88. Five colonies here and 63 cities or states. Starting from Pelorum on the coast facing the Ionian sea the town Messana, whose inhabitants are Roman citizens and are called Mamertini, cape Drepanum, colony Tauromenium (in the past Naxos), Asines river, mount Aetna with its amazing night-time fires. Its crater measures 20 stades round; its hot ash reaches as far as Tauromenium and Catina, and its roar to [mount] Maroneus and Gemelli hills.

89. Three Rocks of the Cyclopes, Ulysses' harbor, colony Catina, Symaethum and Terias rivers. The Laestrygonian plains inland. Towns Leontini and Megaris, Pantagies river, colony Syracusae with its Arethusa spring (although the springs of Temenitis and Archidemia and Magea and Cyane and Milichie also soak Syracusan territory), Naustathmus harbor, Elorum river, cape Pachynum. After it, on the front of Sicilia, Hyrminum river, Camarina town, Gelas river, Agragas town which we have called Agrigentum,

90. colony Thermae, the Achates, Mazara, Hypsa, Selinuus rivers, town Lilybaeum after which the cape is named, Drepana, mount Eryx, towns

Panhormum, Soluus, Himera with its river, Cephaloedis, Aluntium, Agathyrnum, colony Tyndaris, town Mylae and, back to where I began, Pelorias.

91. Inland and possessing Latin rights: Centuripini, Netini, Segestani; tribute-payers: Assorini, Aetnenses, Agyrini, Acestaei, Acrenses, Bidini, Citarini, Drepanitani, Ergetini, Echetlienses, Erycini, Entellini, Enini, Egguini, Gelani, Galacteni, Halaesini, Hennenses, Hyblenses, Herbitenses, Herbessenses, Herbulenses, Halicuenses, Hadranitani, Imacarenses, Ichanenses, Ietenses, Mutustratini, Magellini, Murgentini, Mutycenses, Menaini, Naxi, Noini, Petrini, Paropini, Phintienses, Semelitani, Scherini, Selinunti, Symaethii, Talarenses, Tissienses, Triocalini, Tyracinenses. Messenian Zanclaei are on the Siculan strait.

92. Facing towards Africa the islands Gaulos, Melita (87 from Camarina, 113 from Lilybaeum), Cossyra, Hieronnesos, Caene, Galata, Lepadusa, Aethusa which others have spelled Aegusa, Bucinna, Osteodes 75 from Soluus, and Ustica across from the Paropini. But on this [Italian] side of Sicilia, opposite Metaurus river about 25 miles from Italia, seven islands called Aeoliae; the same ones also called Liparaei, or Hephaestiades by Greeks and Volcaniae by us; Aeoliae because Aeolus was king there in Trojan times.

93. In the past Lipara, with its town of Roman citizens and named after Aeolus' successor king Liparus, was called Milogonis or Meligunis. It is 25 from Italia, and a little under five round. Between it and Sicilia another [island], once called Therasia, nowadays Hiera, because it is sacred to Volcanus with a hill on it that spews flames at night.

94. A third one Strongyle six miles from Lipara facing east, where Aeolus was king. Only its more liquid flames distinguish it from Lipara, and we are told that from the smoke its inhabitants forecast which winds will blow three days ahead; hence the assumption that the winds obeyed Aeolus. A fourth [island] Didyme, smaller than Lipara. Fifth Ericusa, sixth Phoenicusa are left as pasture for the nearby islands. Last and likewise smallest Euonymos. So much for Europa's first gulf.

XV–XX Italia from Locri to Ravenna

95. After Locri begins the side of Italia called Magna Graecia, dented by three bays of the Ausonian sea (because Ausones were the first to hold it).

According to Varro, it extends 86 [miles], although most sources say 75. Countless rivers on this coast, but worth mentioning after Locri are the Sagra and ruins of a town Caulon and of Mustia, camp Consilinum, cape Cocynthum which some consider the longest in Italia. Then the bay and town Scolacium, Scylaceum also called Scylletium by its Athenian founders, a location that the reach of Terinaeus bay makes into a peninsula, on which is a harbor called Castra Hannibalis, where Italia is at its narrowest: 40 miles wide. So this is the location where Dionysius the elder wanted a channel [cut] and [the land] added to Sicilia.

96. The navigable rivers here: Carcinus, Crotalus, Semirus, Arogas, Thagines. Inland, town Petilia, mount Clibanus, cape Lacinium. Off its coast ten [miles] from land the island of the Dioscores, another called Calypsus which is thought to have been called Ogygia by Homer, as well as Tyris, Eranusa, Meloessa. Agrippa reports that [the cape] is 70 from Caulon.

97. Europa's second bay begins at cape Lacinium; it curves in a great ring and ends 175 away at Acroceraunium, a cape of Epirus. [On the bay] town Croton, Neaethus river, and Thuri a town between two rivers, Crathis and Sybaris, where there used to be a city of the same name. Heraclea, once called Siris, similarly situated between the Siris and Aciris. The Talandrum, Casuentum rivers, town Metapontum where Italia's third region ends.

98. The only inland Bruttii are the Aprustani, but of the Lucani there are Atinates, Bantini, Eburini, Grumentini, Potentini, Sontini, Sirini, Tergilani, Ursentini, Volcentani to whom Numestrani are linked. In addition, Thebae Lucanae no longer exists according to Cato, and Theopompus says there used to be a city of the Lucani, Pandosia, where Alexander of Epirus died.

XVI

99. The second region, adjacent [to the third], encompasses the Hirpini, Calabria, Apulia, the Sallentini, with a 250[-mile] bay called Tarentinus after the town of the Lacones located at the [bay's] innermost point and incorporating a maritime colony which had been there, 136 from cape Lacinium; it propels Calabria into [forming] a peninsula opposite [the cape]. Greeks called [the peninsula] Messapia after a chief [Messapos], and before that Peucetia after Peucetius, Oenotrus' brother, in the territory of the Sallentini. The capes [Sallentinum and Lacinium] are 100 apart. By land route the peninsula from Tarentum to Brundisium extends 45 in width, and much less from Sasine harbor.

100. After Tarentum the mainland towns Uria, called 'Messapian' to distinguish it from the one in Apulia, Sarmadium; but on the coast Senum, Callipolis which nowadays is Anxa, 75 from Tarentum. After another 33 a cape called Acra Iapygia, where Italia extends furthest into the sea. From there, after 19 miles, the town Basta, and Hydruntum at the boundary between the Ionian and Hadriatic seas; the shortest crossing to Graecia is here, opposite the Apolloniates' town, with a strait 50 [miles] wide at most running in between.

101. King Pyrrhus of Epirus was the first to consider bridging this span to make an unbroken crossing by foot, and after him Marcus Varro when he was in command of Pompey's fleets during the pirate war [67 BCE]; but other concerns hindered both. After Hydruntum the deserted Soletum, then Fratuentium, Tarentinus harbor, Miltopes anchorage, Lupia, Balesium, Caelia, Brundisium 50 miles from Hydruntum and outstanding in Italia for its fine harbor, as well as for its more secure, albeit longer, crossing of 225 to the Illyrian city Dyrrachium.

102. Bordering Brundisium the territory of the Poediculi: nine young men and the same number of young women from the Illyrians propagated their 12 communities. Towns of the Poediculi: Rudiae, Gnatia, Barium; rivers Iapyx named after a son of king Daedalus (Iapygia Acra is also named for him), Pactius, Aufidus which flows down from the Hirpini mountains at Canusium.

103. From here Apulia of the Daunii, so called after their leader the father-in-law of Diomedes. Here town Salapia, famous as the place where Hannibal fell in love with a prostitute, Sipuntum, Uria, Cerbalus river, the end of Daunian territory, Aggasus harbor, cape of mount Garganus (234 round Garganus from [cape] Sallentinum or Iapygium), Garnae harbor, lake Pantanus, Fertor river with its harbors. Teanum Apulorum as well as Larinum, Cliternia, Tifernus river from which Frentana region begins.

104. Thus there are three kinds of Apulians: Teani named after their Graian leader; Lucani subjugated by Calchas, whose territory Atinates nowadays occupy; Daunii who, in addition to the places already mentioned, occupy the colonies Luceria, Venusia, towns Canusium, Arpi which at some time in the past was called Argos Hippium when founded by Diomedes, and later Argyripa. Here Diomedes destroyed the Monadi and Dardi peoples as well as two cities which have been made proverbial jokes, Apina and Trica.

105. Otherwise, inland in the second region a single colony [in the territory] of the Hirpini, once called Maleventum but with a change of name more auspiciously called Beneventum; the Aeculani, Aquiloni, Abellinates also called Protropi, Compsani, Caudini, Ligures (some also called Corneliani, others Baebiani), Vescellani. The Ausculani, Aletrini, Abellinates also called Marsi, Atrani, Aecani, Alfellani, Atinates, Arpani, Borcani, Collatini, Corinenses and Cannenses famous for Rome's disaster [there], Dirini, Forentani, Genusini, Herdonienses, Irini, Larinates also called Frentani, Metinates of Garganus, Mateolani, Neretini, Natini, Rubustini, Silvini, Strapellini, Turnantini, Vibinates, Venusini, Ulurtini. Inland Calabrian peoples: Aezetini, Apamestini, Argetini, Butuntinenses, Deciani, Grumbestini, Norbanenses, Palionenses, Stulnini, Tutini. Sallentini peoples: Aletini, Basterbini, Neretini, Uzentini, Veretini.

XVII

106. The fourth region, which includes quite the bravest peoples of Italia, comes next. On the coast, in the territory of the Frentani starting from Tifernum: Trinium river with its harbors, towns Histonium, Buca, Hortona, Aternus river. Inland, Anxani also called Frentani, Upper and Lower Carecini, Juanenses. Teatini in the territory of the Marrucini. In that of the Paeligni: Corfinienses, Superaequani, Sulmonenses. In that of the Marsi: Anxatini, Antinates, Fucentes Lucenses, Marruvini. In that of the Albenses: Alba by lake Fucinus. In that of the Aequiculani: Cliternini, Carseolani.

107. In that of the Vestini: Angulani, Pennienses, Peltuinates to whom are joined Aufinates Cismontani. In that of the Samnites who used to be called Sabelli and (by Greeks) Saunitae: colonies Bovianum (Vetus and another named after the Eleventh legion), the Aufidenates, Aesernini, Fagifulani, Ficolenses, Saepinates, Tereventinates. Sabine people: Amiternini, Curenses, Forum Deci, Forum Novum, the Fidenates,

Interamnates, Nursini, Nomentani, Reatini, Trebulani (also called Mutuesci and Suffenates), Tiburtes, Tarinates.

108. In this area the Comini, Tadiates, Caedici and Alfaterni have disappeared from the territory of the Aequicoli. According to Gellianus, the Marsic town Archippe, founded by the Lydian leader Marsyas, was submerged by lake Fucinus; likewise, according to Valerianus, a town of the Vidicini in Picenum was destroyed by Romans. Sabini – who some have thought were called Sebini because of their religion and divine rites – live on the damp hills around the Velini lakes.

109. Nar river draws from them and conveys the sulphurous water of the [hills] to the Tiberis, while the Avens, disappearing into [these lakes], refills them from mount Fiscellus near the woods of Vacuna and Reate. But from a different direction the Anio, after rising on Treba's mountain, brings to Tiberis [water from] three famously lovely lakes which have given their name to Sublaqueum. Marcus Varro says that lake Cutilia (with an island floating in it), in the territory of Reate, is the *umbilicus* [center-point] of Italia. Below the Sabini is Latium, with Picenum to the side, Umbria behind; the Appenninus range buttresses the Sabini on both sides.

XVIII

110. The fifth region is Picenum, which was once very densely populated. 360,000 Picentines came under the protection of the Roman Republic. They are descendants of Sabini, after a 'sacred spring' had been vowed. They occupied from Aternus river, where the territory of Hadria is now, and colony Hadria six [miles] from the sea. Vomanum river, Praetutian and Palmensian territories, also Castrum Novum, Batinum river, Truentum with its river (the only community of Liburni left in Italia), the rivers Albula, Tessuinum, Helvinum where the Praetutian region ends and the Picentine begins.

111. Town Cupra, Castellum Firmanorum, and above it colony Asculum the best-known in the interior of Picenum, Novana. On the coast Cluana, Potentia, Numana founded by Siculi, colony Ancona also founded by them and located on cape Cunerum just where the coast bends like an elbow, 183 from Garganus. Inland, the Auximates, Beregrani, Cingulani, Cuprenses also called Montani, Falerienses, Pausulani, Planinenses, Ricinenses, Septempedani, Tolentinates, Traienses, Pollentini at Urbs Salvia.

XIX

112. The sixth region, joined to these, encompasses Umbria and the Gallic territory this side of Ariminum. The Gallic coast also called Togata Gallia begins at Ancona. Siculi and Liburni held most of this region, especially the Palmensian, Praetutian and Hadrian territories. They were expelled by Umbri, who were expelled by Etruria, and it by Galli. Umbrian people are considered the most ancient in Italia and are thought to have been called Ombrii by Greeks because they had survived the downpours that flooded the earth. Records state that the Tusci conquered 300 of their towns.

113. Nowadays on the coast: Aesis river, Senagallia, Metaurus river, colonies Fanum Fortunae, Pisaurum with its river, and inland Hispellum, Tuder. Otherwise Amerini, Attidiates, Asisinates, Arnates, Aesinates, Camertes, Casuentillani, Carsulani, Dolates also called Sallentini, Fulginiates, Foroflaminienses, Forojulienses also called Concupienses, Forobrentani, Forosempronienses, Iguini, Interamnates also called Nartes, Mevanates, Mevaniolenses, Matilicates, Narnienses whose town was previously called Nequinum,

114. Nucerini also called Favonienses and Camellani, Ocriculani, Ostrani, Pitinates (some also called Pisuertes, others Mergentini), Plestini, Sentinates, Sarsinates, Spoletini, Suasani, Sestinates, Suillates, Tadinates, Trebiates, Tuficani, Tifernates (some also called Tiberini, others Metaurenses), Vesinicates, Urvinates (some also called Metaurenses, others Hortenses), Vettonenses, Vindinates, Visuentani. In this area there have disappeared the Feliginates and whoever possessed Clusiolum above Interamna, as well as the Sarranates with the towns Acerrae (also called Vafriae) and Turocaelum (also called Vettiolum), together with the Solinates, Curiates, Falinates, Sapinates. Also disappeared are the Arinates with Crinivolum, and the Usidicani as well as Plangenses, Paesinates, Caelestini. Cato reports that Ameria, mentioned above [3.113], was founded 963 years before the war with Perseus [king of Macedon, 171–168 BCE].

XX About the Padus

115. The eighth region is bounded by Ariminum, Padus [river], Appenninus [mountains]. On the coast: Crustumius river, colony Ariminum with Ariminum and Aprusa rivers, Rubico river once the border of Italia. After it the Sapis and Utes and Anemo [rivers], Ravenna a Sabine town with Bedesis river 105 [miles] from Ancona, and not far from the sea Butrium an Umbrian town. Inland, colonies Bononia (called Felsina when it was Etruria's leading city), Brixillum, Mutina, Parma, Placentia;

116. towns Caesena, Claterna, Forums of Clodius, Livius, Popilius, the Druentini, Cornelius, Licinius; the Faventini, Fidentini, Otesini, Padinates, Regienses founded by Lepidus, Solonates, and Saltus Galliani also called Aquinates, Tannetani, Veleiates also called Vettiregiates, Urbanates. In this region the Boi have disappeared (Cato claims there were 112 tribes of them), also the Senones who captured Rome [in the early fourth century BCE].

117. The source of the Padus is worth visiting. It flows from the recesses of mount Vesulus in the territory of the Ligures Bagienni, which rises to be the highest Alpine peak, then buries itself in a subterranean channel and emerges again in the territory of the Forovibienses. No river is more famous; Greeks called it Eridanus, and Phaethon's punishment gave it distinction. It swells when the snows melt at the rising of Canis, overwhelming fields rather than boats; yet it claims for itself nothing of what it has swept up, and it generously fertilizes where it deposits.

118. From its source it is 300 miles long plus two less than 90 more because of its meandering. It gathers up not only navigable Appennine and Alpine rivers, but also vast lakes that discharge themselves into it. It conveys to the Hadriatic sea 30 rivers in total. Best-known of these from the Appennine side: Iactus, Tanarus, Placentia's Trebia, Tarus, Incia, Gabellus, Scultenna, Rhenus; from the Alpes side: Stura, Orgus, two Duriae, Sesites, Ticinus, Lambrus, Addua, Ollius, Mincius.

119. No other river grows more in so short a distance. In fact, it is pressed by the mass of water and propelled to the sea, a weight on the land, even though split into streams and canals for the 120 [miles] between Ravenna and Altinum. However, where it discharges more widely, it creates what are called the Seven Seas. It is carried by the Augustan canal to Ravenna, where it is called Padusa and was once called Messanicus. The mouth nearest to here has the size of a harbor and is called Vatrenus. This is where, when triumphing over Britannia, Claudius Caesar launched into the Hadria[tic sea] in something really more like an enormous house than a ship!

120. This mouth used to be called Eridanum, and by others Spineticum after the city Spina which was next to it – a very powerful place (as can be believed from its treasuries at Delphi) founded by Diomedes. Here, from Forocorneliensian territory, Vatrenus river swells Padus. Then the next mouth is that of Caprasia, then that of Sagis, then Volane once called Olane – all [called] the Flavian canal, which the Tusci first made from the Sagis so that the river's flow drained crosswise into the marshes of the Atriani called Seven Seas. The Tuscan town Atria has a fine harbor on the sea once called Atriatic and nowadays Hadriatic.

121. Next the deep Carbonaria, Fossiones and Philistina (which some call Tartarus) mouths, all created by overflow from the Philistina canal supplemented by the Atesis from the Tridentine Alps and the Togisonus from the territory of the Patavini. These also partly create the nearby harbor Brundulum, just as the two Meduacus [rivers] and Clodia canal make Aedro [harbor]. Padus merges with these [watercourses] and flows out through them; by most accounts, it makes the shape of a triangle 2,000 stades round between the Alps and the sea-coast, as does the Nile in Egypt with what is called the Delta.

122. It is a disgrace to borrow an account of Italia from Greeks. Nonetheless, Metrodorus of Scepsis says that [Padus] is so called because around its source there are many pine trees of a type called *padus* in Gallic; in fact in the Ligurian language the river itself is called Bodincus, which means 'bottomless'. Reinforcing this argument, a town near Industria, where the river begins to be especially deep, had the old name Bodincomagum.

XXI–XXII Italia across the Padus

123. The eleventh region is called Transpadana after this [river]. Although entirely inland, it brings in imports from every sea along the river's productive channel. Towns Vibi Forum, Segusio, colonies Augusta Taurinorum at the foot of the Alps with its ancient Ligurian stock (Padus is navigable from here), then Augusta Praetoria of the Salassi next to the twin entrances to the Graian and Poenine Alps (the Carthaginians are said to have come through the latter, and Hercules through the Graian), town Eporedia founded by the Roman people as instructed by Sibylline books. Galli call men who are good breakers of horses 'Eporedians'.

124. Vercellae of Libicii descended from Sallui, and Novaria founded by Vertamocori, today a village of Vocontii and not (as Cato thinks) of Ligures. It was [Ligurian] Laevi and Marici who founded Ticinum not far from the Padus, just as Boi who crossed the Alpes founded Laus Pompeia, and Insubres [founded] Mediolanum. According to Cato, Comum and Bergomum and Licini Forum and some surrounding communities are descended from Oromobii, but he admits that he does not know the origin of the people who Cornelius Alexander informs [us] sprang from Graecia, given that their name means 'mountain-dwellers'.

125. In this area an Oromobian town Parra – Cato said the Bergomates came from here – has disappeared, although still today the location reveals itself to be more elevated than advantageous. Also gone are the Caturiges (Insubrian exiles) and Spina mentioned above [3.120], as well as Melpum outstanding for its wealth, which Nepos Cornelius said was destroyed by Insubres and Boi and Senones on the same day that Camillus captured Veii.

XXII

126. Next comes Italia's tenth region, bordering on the Hadriatic sea. It includes Venetia, Silis river coming from the mountains of Tarvisium, Altinum town, Liquentia river coming from the mountains of Opitergium, and a harbor of the same name, colony Concordia, rivers and harbor of Reatinum, Greater and Lesser Tiliaventum, Anaxum to which the Varamus flows, the Alsa, Natiso with [tributary] Turrus which flow past the colony Aquileia, 15 miles from the sea.

127. This is the region of the Carni, and next to them the Iapudes, Timavus river, fortress Pucinum famous for its wine, Tergestine bay, colony Tergeste 33 from Aquileia. Six miles further Formio river 189 miles from Ravenna, the ancient limit of an enlarged Italia, but nowadays [just] of Histria. Most authors, and even Nepos who lived along Padus, have stated wrongly that [this region] is named after Hister river which flows out into the Hadria from Danuvius river (also called Hister) opposite the Padus mouths, so that the inflows from opposite directions make the sea between them fresh water.

128. But there is no river flowing from Danuvius into Hadriatic sea. I believe that [authors] have been deceived because the ship Argo came down a river not far from Tergeste into Hadriatic sea, but there is still no agreement on which river. More careful writers report that [the ship] crossed the Alpes by portage, and that it proceeded along Hister, then Savus, then Nauportus which rises between Emona and the Alpes and gains its name from this event.

XXIII Histria

129. Histria extends out like a peninsula. Some sources have said that it is 40 [miles] wide, 125 round, with the same distances for neighboring Liburnia and Flanatic bay. Others say 225, yet others 180 for Liburnia. Some have extended Iapudia 130 as far as the Flanatic bay to the rear of Histria, and have then made Liburnia 150 [round]. Tuditanus, who conquered the Histri, inscribed on his statue there: *From Aquileia to Titius river 2,000 stades.* Histria's towns with Roman citizens: Agida, Parentium, colony Pola nowadays called Pietas Julia, founded long ago by Colchians, 105 from Tergeste. Next Nesactium town and Arsia river, nowadays the border of Italia. The crossing from Ancona to Pola is 120 miles.

130. Colonies of the tenth region inland: Cremona, Brixia in the territory of the Cenomani, Ateste in the territory of the Veneti; towns Acelum, Patavium, Opitergium, Velunum, Vicetia. Tuscan Mantua is the one remaining across the Padus. According to Cato, the Veneti are descended from Trojan stock, and the Cenomani lived among Volci next to Massilia. The towns of the Feltrini and Tridentini and Beruenses are Raetic, Verona is Raetic and Euganean, and that of the Julienses is Carnian. Then there are those that need not be mentioned in more detail: Alutrenses, Asseriates, Flamonienses (both Vanienses and others called Curici), Forojulienses called Transpadani, Foretani, Nedinates, Quarqueni, Tarvisani, Togienses, Varvari.

131. Along the coast in this area have disappeared Irmene, Pellaon, Palsicium, Atina and Caelina belonging to Veneti, Segesta and Ocra belonging to Carni, Noreia belonging to Taurisci. Also, according to Lucius Piso, a town at the twelfth milestone from Aquileia was destroyed by Marcus Claudius Marcellus against the senate's wishes.

In this region too there are 11 famous lakes, as well as rivers propagated by them or that swell when they take in a river and then let it go, as does [lake] Larius the Addua [river], [lake] Verbannus the Ticinus, [lake]

Benacus the Mincius, [lake] Sebinnus the Ollius, [lake] Eupilis the Lambrus, all tributaries of Padus.

132. Caelius says that the Alpes extend 1,000 miles in length from the Upper to Lower sea, while Timagenes subtracts 25 miles. However, Cornelius Nepos' figure for their width is 100 and Livius says 3,000 stades, but the two measure from different points. Certainly, they sometimes extend over 100 miles where they separate Germania from Italia, but elsewhere not more than 70, a contraction attributable to Nature's foresight. The width of Italia at their base comes to 745 miles measured from the Varus through Vada Sabatia, the Taurini, Comum, Brixia, Verona, Vicetia, Opitergium, Aquileia, Tergeste, Pola, the Arsia.

XXIV Alpes and Alpine peoples

133. Many communities live in the Alpes, but the noteworthy ones between Pola and the Tergeste region are the Fecusses, Subocrini, Catali, Menoncaleni, and next to the Carni those once called Taurisci but nowadays Norici. Bordering them the Raeti and Vindelici, all split into many states. Raeti are thought to be descendants of Tusci who, under their leader Raetus, were driven out by Galli. Then on the side of the Alpes facing Italia, the Euganean peoples have Latin rights and, by Cato's count, 34 towns.

134. Among them Trumpilini, a community who sold themselves and their territory, then Camunni and several similar peoples associated with neighboring municipalities. Cato again thinks that Lepontii and Salassi are Tauriscan peoples; in the case of the Lepontii, nearly everyone else accepts a Greek explanation for their name, according to which they had been in Hercules' entourage but were left behind after they had suffered bodily injury from frostbite when traversing the Alpes. So, too, Graii living in the Graian Alpes were in the same army, as were Euganei with their outstanding ancestry, hence their name. Their leading group are Stoeni.

135. Vennonenses and Sarunetes, who belong to the Raeti, live where Rhenus river rises, and those Lepontii called Uberi live at the source of Rhodanus in the same Alpine region. In addition, some who live here have received Latin rights, such as the Octodurenses and neighboring Ceutrones, the Cottian states and Turi descended from Ligures, Bagienni

Ligures and those called Montani, and several branches of Capillati bordering Ligustic sea.

136. It does not seem irrelevant to insert here the inscription from an Alpine trophy, as follows:

> To Imperator Caesar Augustus, son of a god, pontifex maximus, imperator for the fourteenth time, in the seventeenth year of his tribunician power, the Senate and People of Rome [offer this trophy] because under his command and auspices all the peoples of the Alpes extending from Upper sea to Lower were brought under the rule of the Roman people. Alpine peoples conquered: Trumpilini, Camunni, Venostes, Vennonetes,
>
> **137.** Isarci, Breuni, Genaunes, Focunates, four Vindelici peoples, Cosuanetes, Rucinates, Licates, Catenates, Ambisontes, Rugusci, Suanetes, Calucones, Brixenetes, Leponti, Uberi, Nantuates, Seduni, Varagri, Salassi, Acitavones, Medulli, Ucenni, Caturiges, Brigiani, Sogionti, Brodionti, Nemaloni, Edenates, Esubiani, Veamini, Gallitae, Triullati, Ecdini, Vergunni, Egui, Turi, Nematuri, Oratelli, Nerusi, Velauni, Suetri.

138. Not included are the 15 Cottian states, which had not been our enemies, nor those states attributed to municipal towns by Pompey's law.

[So] this is Italia sacred to the gods, these its peoples, these the communities' towns. Above all, it was this Italia which, in the consulship of Lucius Aemilius Paulus and Gaius Atilius Regulus [225 BCE], after a Gallic rising was reported, alone and without any outside help (and even at that time without the Transpadani), armed 80,000 cavalry and 700,000 infantry. Among lands, it is also second to none for the output of all its mines. But an ancient senatorial decree bans this activity, requiring Italia to be spared in this respect.

XXV Illyricum, Liburnia

139. Liburnian people spread from the Arsia all the way up to Titius river. Among them used to be Mentores, Himani, Encheleae, Bulini and those whom Callimachus called Peucetii, but nowadays all are generally called by the single name Illyricum. Few of the communities deserve mention or have easily pronounceable names. Cases are brought to the Scardonitan *conventus* by the Iapudes and 14 Liburnian states, among which there should be no objection to naming the Lacinienses, Stulpini, Burnistae, Olbonenses. Within this *conventus* and in possession of Italian rights are the Alutae, Flanates (from whom a bay is named), Lopsi,

Varuarini, Asseriates who are tax-exempt and, on islands, Fertinates, Curictae.

140. Otherwise, along the coast and starting from Nesactium: towns Alvona, Flanona, Tarsatica, Senia, Lopsica, Ortoplinia, Vegium, Argyruntum, Corinium, Aenona, the state of Pasini, Telavium river marking the boundary of Iapudia. Besides those noted above, islands with their towns in its bay: Absortium, Arba, Crexi, Gissa, Portunata. Back on the mainland the colony Iader is 160 from Pola, then 30 further the island Colentum, 43 [further] the mouth of Titius river.

XXVI Delmatia

141. Liburnia ends and Delmatia begins at Scardona on that river, 12 miles from the sea. Then the ancient region of the Tariotae and fortress Tariona, Diomedes' cape or (according to some) peninsula of Hyllis 100 round, Tragurium which has Roman citizens and is noted for its marble, Siculi where Deified Claudius sent veterans, colony Salona 112 from Iader.

142. This is where the Delmatae seek their rights, with their strength split into 342 *decuriae* [divisions], the Deuri into 25, Ditiones into 239, Maezei into 269, Sardeates into 52. In this region are Burnum, Andetrium, Tribulium, fortresses celebrated for battles there. Island peoples, the Issaei, Solentini, Separi, Epetini, also seek [their rights in Salona]. After these the fortresses Petuntium, Nareste, Oneum. Colony Narona, the third *conventus*, 85 miles from Salona, is located on a river with its name 20 miles from the sea. According to Marcus Varro, 89 states used to come here;

143. nowadays almost the only ones known are the Cerauni with 24 *decuriae*, Daversi with 17, Desitiates with 103, Docleatae with 33, Deretini with 14, Deraemistae with 30, Dindari with 33, Glinditiones with 44, Melcumani with 24, Naresi with 102, Scirtari with 72, Siculotae with 24 and Vardaei, once plunderers of Italia, with no more than 20 *decuriae*. Besides these, Ozuaei, Partheni, Cavi, Haemasi, Masthitae, Arinistae have occupied this region.

144. The colony Epidaurum is 100 miles from Naro river. Starting from Epidaurum are the Roman citizen towns Rhizinium, Acruium, Butuanum, Olcinium which used to be called Colchinium because it was founded by Colchians, Drino river and above it the Roman citizen town Scodra 18 from the sea, as well as the fading memory of many towns in Graecia and of powerful states too. For in this region were the Labeatae, Senedi, Rudini, Sasaei, Grabaei; although they should strictly speaking be called Illyrians, both the Taulanti and Pyraei retain their name. On the coast, cape Nymphaeum. The Roman citizen town Lissum 100 miles from Epidaurum.

145. The province of Macedonia begins after Lissum, with the Partheni peoples and behind them Dassaretae, Candaviae mountains 78 miles from Dyrrachium, right on the coast Denda which has Roman citizens, colony Epidamnum which because of its inauspicious name Romans called Dyrrachium, Aous river which some call Aeas, Apollonia once a Corinthian colony and four miles inland; within its territory a famous Nymphaeum, and barbarian Amantes and Buliones live nearby. Back on the coast the town Oricum founded by Colchians. From there Epirus begins and the Acroceraunia mountains, which we have made the end of this bay of Europa. Oricum is 80 from cape Sallentinum in Italia.

XXVII Norici

146. Behind the Carni and Iapudes, where the great Hister runs, Raeti adjoin Norici. Their towns: Virunum, Celeia, Teurnia, Aguntum, Iuvaum, all Claudian [settlements], and the Solvensian [settlement] which is Flavian. Next to the Norici are lake Pelso and deserted territory of the Boi; occupied there now, however, are Sabaria a colony of Deified Claudius, and town Scarabantia Julia.

XXVIII Pannonia

147. After that, acorn-bearing lands of Pannonia where the Alpine chain smooths out, turning from north to south through central Illyricum and tapering off with gentle slopes to right and left. The part facing Hadriatic sea is called Delmatia and Illyricum as already mentioned; the north-facing part is Pannonia, bounded by the Danuvius. In it, colonies Emona, Siscia. Famous and navigable rivers flowing into Danuvius are Draus a notably violent one from the territory of the Norici, and Saus a more placid one from the Alpes Carnicae, 120 apart. Draus flows through the Serretes, Serapilli, Iasi, Andizetes, while Saus flows through the Colapiani and Breuci.

148. These are the leading peoples; in addition, the Arviates, Azali, Amantini, Belgites, Catari, Cornacates, Eravisci, Hercuniates, Latovici, Oseriates, Varciani. Mount Claudius with Scordisci in front, Taurisci behind. In the Saus an island, Metubarbis, the largest formed by a river. Besides them other memorable rivers: Colapis flows into Saus by Siscia, making an island there called Segestica with its pair of channels; another river, Bacuntius, flows into Saus <by the> town Sirmium, where the state consists of Sirmienses and Amantini. From here 45 to Taurunum, where Saus merges with Danuvius. Further up, Valdasus and Urpanus, respectable rivers in their own right, flow into [Saus].

XXIX Moesia

149. Adjoining Pannonia is the province called Moesia, which advances with Danuvius all the way up to Pontus [Black Sea]. It begins at the confluence just mentioned [Danuvius and Saus]. Within it the Dardani, Celegeri, Triballi, Timachi, Moesi, Thracians and Scythians next to Pontus. Its famous rivers rising from territory of the Dardani: Margus, Pingus, Timachus, Oescus from [mount] Rhodope, Utus from [mount] Haemus, Asamus, Ieterus.

150. At its fullest, the width of Illyricum amounts to 325 miles, with a length of 530 from Arsia river to Drinius river. Agrippa reports that it is 175 from the Drinius to cape Acroceraunium, and that the entire bay [encompassed by] Italia and Illyricum is 1,700 round. With the borders we have marked out [3.100], there are two seas in it, the Ionian in the first part and, further in, the Hadriatic, which is called Upper.

XXX Islands of the Ionian and Hadriatic [seas]

151. There are no islands in the Ausonian sea worth mentioning besides those already described; in the Ionian along the Calabrian shore off Brundisium there are a few whose obstructive location enables a harbor to be formed, as well as Diomedia off the Apulian shore famous for its monument to Diomedes, and another island of the same name which some call Teutria. Off the coast of Illyricum cluster over 1,000 islands: here it is the nature of the sea to be shallow, and sounds with narrow channels break through. Off the mouth of Timavus are famous [islands] that have hot springs which rise with the tide; [also] Cissa by the territory of the Histri, Pullariae, and ones called Absyrtides by Greeks after Medea's brother who was killed there.

152. Next to these those called Electrides, [some of them] a source of amber which Greeks call electrum – a most definite example of their silliness, as it still remains to be established [just] which ones they mean! Opposite Iader is Lissa and [the island] already mentioned; opposite the Liburni several Crateae [islands] and quite a number of Liburnicae, and Celadussae; opposite Tragurium are Bova and Brattia praised for its goats, Issa which has Roman citizens, and Pharia with its town. Corcyra also called Melaena is 25 from Issa with its town founded by Cnidii, and between it and Illyricum Melite from which Melitaean puppies gain their name, according to Callimachus. Fifteen from it, the seven Elaphites, while in the Ionian sea two miles off Oricum is Sasonis, a well-known pirate base.

****

Totals   Towns and peoples [...]
         Famous rivers [...]
         Famous mountains [...]
         Islands [...]
         Extinct cities and peoples [...]
         Facts and inquiries and observations [...]

[Roman] Sources

Turanius Gracilis. Cornelius Nepos. Titus Livius. Cato the Censor. Marcus Agrippa. Marcus Varro. Deified Augustus. Varro Atacinus. Antias. Hyginus. Lucius Vetus. Pomponius Mela. Curio the Father. Caelius. Arruntius. Sebosus. Licinius Mucianus. Fabricius Tuscus. Lucius Ateius. <Ateius> Capito. Verrius Flaccus. Lucius Piso. Gellianus. Valerianus.

Foreign Sources

Artemidorus. Alexander Polyhistor. Thucydides. Theophrastus. Isidorus. Theopompus. Metrodorus of Scepsis. Callicrates. Xenophon of Lampsacus. Diodorus of Syracusae. Nymphodorus. Calliphanes. Timagenes.

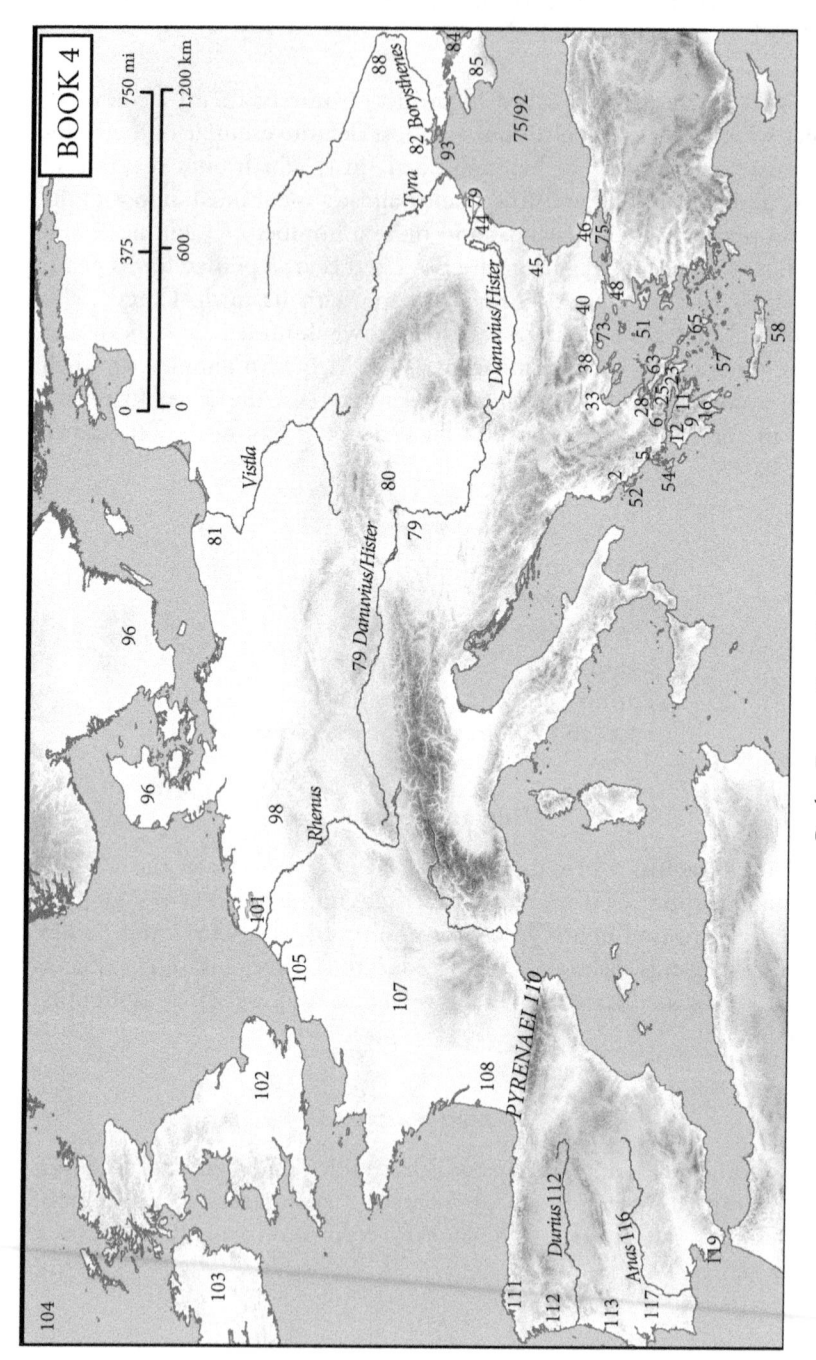

Book 4 Europe: East, North

BOOK 4

Europe: East, North

I–X Epirus, Achaia

1. Europa's third bay begins at the Acroceraunian mountains, ends at the Hellespont and is 1,925 miles round, excluding its smaller bays. It includes Epirus, Acarnania, Aetolia, Phocis, Locris, Achaia, Messenia, Laconica, Argolis, Megaris, Attica, Boeotia, and back on the other sea the same Phocis and Locris, Doris, Phthiotis, Thessalia, Magnesia, Macedonia, Thracia. All the myth-making of Graecia as well as its brilliant literature first shone out from this bay, so we shall linger over it briefly.

2. Epirus, as it is called overall, begins at the Ceraunian mountains. First in it the Chaones after whom Chaonia is named, then Thesproti, Antigonenses, place Aornos with its vapors that kill birds, Cestrini, Perrhaebi in possession of mount Pindus, Cassopaei, Dryopes, Selloe, Hellopes, Molossi with the temple of Zeus at Dodone and its famous oracle in their territory, mount Talarus acclaimed by Theopompus for 100 springs around its base.

3. Epirus itself extends to Magnesia and Macedonia; behind it are the Dassaretae mentioned above [3.145], a free people, then Dardani, a wild one. Triballi and Moesic peoples extend to the left side of the Dardani; their neighbors in front are Maedi and Denselatae, with Threces continuing beside them as far as the Pontus. Thus the ring that defends the heights of [mount] Rhodope and then of Haemus.

4. On the coast of Epirus the fortress Chimera in the Acroceraunian [mountains], and below it the spring Aquae Regiae, towns Maeandria and Cestria, Thesprotia's Thyamis river, colony Buthrotum, and Ambracius bay very well-known with its half-mile-wide inlet admitting a broad sheet of water 37 long, 15 wide. Into it discharges Acheron river which flows for

35 miles from lake Acherusia in Thesprotia and then, because of its mile-long bridge, is considered marvelous by people who marvel at every achievement of their own. On the bay the town Ambracia, Molossian rivers Aphas and Aratthus, Anactorican state, place Pandosia.

II

5. Towns of Acarnania (in the past called Curetis): Heraclia, Echinus and, right on the coast, Augustus' colony Actium famous for its temple of Apollo and for the free Nicopolitan state. Moving on from Ambracia's bay towards the Ionian [sea], the Leucadian shore comes next, cape Leucates, then the bay and actual Leucadia peninsula once called Neritis; it was separated from the mainland by its residents' efforts, but brought back by the force of winds building up masses of sand three stades long at a place called Dioryctos. Leucas town, once called Neritum, is here. Next the Acarnanian cities Alyzia, Stratos, Argos called Amphilochicum. Achelous river flowing from [mount] Pindus divides Acarnania from Aetolia, and with its constant deposits of soil attaches Artemita island to the mainland.

III

6. Aetolian communities: Athamanes, Tymphaei, Ephyri, Aenienses, Perrhaebi, Dolopes, Maraces, Atraces in whose land Atrax river flows into the Ionian sea. Calydon, an Aetolian town, is 7½ miles from the sea next to Evenus river. Then Macynia, Molycria, and behind them mount Chalcis and Taphiassus. Meanwhile, on the coast cape Antirrhium where Corinthian bay's mouth, under a mile wide, forms an inlet that divides the Aetoli from Peloponnesus. The cape projecting from the opposite side is called Rhium. But the Aetolian towns on Corinthian bay: Naupactos, Eupalimna and (inland) Pleuron, Halicarna. Notable mountains: Tomarus in Dodone, Crania in Ambracia, Aracynthus in Acarnania, and Achaton, Panaetolium, Macynium in Aetolia.

IV

7. Closest to the Aetoli are Locri also called Ozolae, a free people; town Oeanthe, harbor of Apollo Phaestius, Crisaeus bay. Inland, towns Argyna, Eupalia, Phaestum, Calamisus. Beyond, the Cirrhaean plains of Phocis, Cirrha town, Chalaeon harbor and, seven miles further inland at the foot of mount Parnasus, free town Delphi with its oracle of Apollo most famous worldwide.

8. Castalius spring, Cephisus river flowing by Delphi after rising in Lilaea city. In addition, there was once a town Crisa and, along with the Bulenses, Anticyra, Naulochum, Pyrrha, Amphisa (a free town), Tithrone, Tithorea, Ambrysus, Mirana and the region called Daulis. Then, at its furthest end, the bay's waters reach a corner of Boeotia, with its towns Siphae and Thebae also called Corsiae next to mount Helicon; Boeotia's third town Pagae (away from the sea here), from which the neck of Peloponnesus projects.

V

9. Peloponnesus (in the past called Apia and Pelasgia) is a peninsula more famous than any other land, situated between the two seas Aegean and Ionian, and shaped like a plane tree leaf because of its jagged indentations. According to Isidorus, its perimeter comes to 563 miles; at the same time he adds almost as much again to take bays into account. The narrow [neck] from which it extends is called Isthmos. Here the seas just mentioned press forward from opposite directions, north and east, and devour its entire width to the point where the rival onslaught of such mighty waters has eroded both sides down to a span of five miles, a narrow neck for Hellas to keep hold of Peloponnesus.

10. One side is called Corinthian bay, the other Saronic; on the former side Lecheae marks the end of the neck, on the latter Cenchreae. For ships of a size that prevents them being conveyed by cart it is a long and hazardous voyage round [from one side to the other]. This is why king Demetrius, dictator Caesar, *princeps* Gaius [Caligula] and Domitius Nero attempted to dig a channel for ships through the neck: a sacrilegious undertaking, as the death of each made plain.

11. In the middle of this span we have called Isthmos, the colony Corinthus (called Ephyra in the past) occupies a hillside site 60 stades from either coast. The citadel at the top, called Acrocorinthos, has a spring Pirene, and commands a view of the two opposite seas. The crossing from Leucas along Corinthian bay to Patrae is 88 miles. Patrae is a colony founded on the Peloponnesus' longest promontory opposite Aetolia and Evenus river where, as already mentioned [4.6], the span of the mouth is under a mile and the length of Corinthian bay then extends 85 to Isthmos.

VI

12. Achaia (from which the province takes its name) begins at Isthmos. In the past it was called Aegialos because of the cities laid out in a row on the shoreline. As we have said, first the Corinthians' harbor Lecheae, then Olyros, fortress of the Pellenaei, towns Helice and Bura where [earthquake] survivors took refuge after others had been swallowed up, Sicyon, Aegira, Aegium, Erineos. Inland, Cleonae, Hysiae.

13. Panhormus harbor and Rhium (already described [4.6]), a promontory five [miles] from Patrae as we mentioned above [4.6], place Pherae. Scioessa the best known of Achaia's nine mountains; spring Cymothoe. Beyond Patrae, town Olenum, colony Dyme, places Buprasium, Hyrmine, cape Araxus, Cyllenius bay, cape Chelonates 15 miles from Cyllene, Phlius fortress in a region called Araethyrea by Homer and later Asopis.

14. Then territory of the Elii, who used to be called Epioe. Elis itself in the interior, and 13 inland from Pylos the shrine of Olympian Zeus, which has controlled the calendar of Graecia because of its famous games. A town of the Pisaei was once here, with Alpheus river flowing by. Meanwhile on the coast, cape Ichthys, Alpheus river – navigable for six miles – towns Aulon, Leprium, cape Platanodes. All these face west,

VII

15. while to the south, Cyparissius bay, 75 round, with its city Cyparissus, towns Pylos, Mothone, place Helos, cape Acritas, Asinaeus bay named after town Asine, Coronaeus [bay] after Corone. These end with cape Taenarum. There, Messenia region and its 18 mountains, Pamisus river and, inland, Messene itself, Ithome, Oechalia, Arene, Pteleon, Thryon, Dorion, Zancle, each one famous at different times. This bay measures 80 round and 30 across.

VIII

16. After [cape] Taenarum the territory of the free Laconican people and a bay 106 round, 38 across. Towns: Taenarum, Amyclae, Pherae, Leuctra and (inland) Sparta, Therapne and where Cardamyle used to be, Pitane, Anthia, place Thyreae, Gerania. Mount Taygetus, Eurotas river, Aegilodes bay, town Psamathus. Gytheates bay named after its town [Gytheion], from where the passage to Creta island is the most reliable. All these end with cape Malea.

IX

17. The next bay extending to [cape] Scyllaeum is called Argolic; it is 50 across, 162 round. Towns: Boea, Epidaurus called Limera, Zarax, Cyphans harbor. Rivers Inachus and Erasinus, and between them Argos called Hippium, above the place Lerne, two [miles] from the sea; nine miles further on Mycenae and where Tiryntha is said to have been, and the place Mantinea. Mountains: Artemisius, Apesantus, Asterion, Parparus and 11 others. Springs: Niobe, Amymone, Psamathe.

18. Eighty miles from Scyllaeum to Isthmos. Towns: Hermione, Troezen, Coryphasium, Argos (called Inachium or Dipsium at different times), Schoenitas harbor. Saronicus bay was once surrounded by forests of oak, for which it is named because of the word for oak in ancient Graecia. Here, the town Epidaurum famous for its shrine of Aesculapius, cape Spiraeum, harbors Anthedus and Bucephalus, and (as we mentioned above [4.10]) Cenchreae on the other side of Isthmos with its shrine of Neptune renowned for quinquennial games.

19. What numerous bays there are ripping into Peloponnesus' coastline and seas raging at it, because indeed the Ionian crashes in from the north, Siculan is driven from the west, Cretic propelled from the south, Aegean from the north-east, and from the south-east Myrtoan which starts from Megaric bay and reaches the whole of Attica.

X

20. Most of [Peloponnesus'] interior is occupied by Arcadia; on all sides far from the sea, it was initially called Drymodes, later Pelasgis. Its towns: Psophis, Mantinea, Stymphalum, Tegea, Antigonea, Orchomenum, Pheneum, Pallantium after which the Palatium in Rome [is named], Megalepolis, Gortyna, Bucolium, Carnion, Parrhasie, Thelpusa, Melaenae, Heraea, Pylae, Pallene, Agrae, Epium, Cynaethae, Arcadian

Lepreon, Parthenium, Alea, Methydrium, Enispe, Macistum, Lampia, Clitorium, Cleonae. Between these [last] two towns is Nemea region, also called Bembinadia.

21. Mountains in Arcadia: Pholoe with its town, as well as Cyllene, Lycaeus where the shrine of Lycaean Zeus is, Maenalus, Artemisius, Parthenius, Lampeus, Nonacris and another eight insignificant ones. Rivers: Ladon flowing from Pheneum marshes, Erymanthus flowing down a mountain of the same name and into the Alpheus.

22. Remaining cities in Achaia to be mentioned: those of the Alipheraei, Abeatae, Pyrgenses, Paroreatae, Pharagenitae, Tortuni, Typanei, Thriusi, Tritienses. Domitius Nero gave freedom to all Achaia. The width of Peloponnesus from cape Malea to town Aegium on Corinthian bay extends for 190, but across from Elis to Epidaurum 125, and from Olympia to Argos through Arcadia 68; the distance from here [Olympia] again to Pylos has been stated above [4.14]. As if Nature were seeking to offset the sea's onslaught, the entire [Peloponnesus] is raised up with six and seventy mountains.

XI–XIII Graecia

23. Hellas, which we call Graecia, begins at the Isthmos narrows. In it first is Attica, called Acte in antiquity. It touches Isthmos with the part of it called Megaris from the colony Megara on the other side from Pagae. These two towns are situated where Peloponnesus stretches out, on either side of it, as if on the shoulders of Hellas; the Pagaei and even more the Aegosthenenses are associated with the Megarenses. On the coast: Schoenos harbor, towns Sidous, Cremmyon, the six[-mile]-long Scironian rocks, Gerania, Megara, Eleusis.

24. Both Oenoe and Probalinthos used to be there. Nowadays, 55 from Isthmos, are Piraeus and Phalera harbors, both joined by a wall to Athens five [miles] inland. This free state does not need its praises sung further; it has more than enough fame! In Attica: springs Cephisia, Larine, Callirroe Enneacrunos, mountains Brilessus, Aegialeus, Icarius, Hymettus, Lycabettus, place Ilisos. Cape Sunium 45 from Piraeus, cape Thoricos, former towns Potamos, Steria, Brauron, Rhamnus village, place Marathon, Thriasius plain, town Melita and Oropus on the Boeotian border.

XII

25. Here, Anthedon, Onchestos, free town Thespiae, Lebadea and – in its opinion no less famous than Athens – Boeotian Thebae, the home (they claim) of two deities, Liber and Hercules. They also assign the Muses a birthplace in the forest of Helicon. Associated with this Thebae too are the Cithaeron woodland, Ismenus river. In addition, Boeotia's springs: Oedipodia, Psamathe, Dirce, Epicrane, Arethusa, Hippocrene, Aganippe, Gargaphie. Its mountains, beside those already mentioned: Mycalesus, Hadylius, Acontius.

26. Remaining towns between the Megaric [region] and Thebae: Eleutherae, Haliartus, Plataeae, Pherae, Aspledon, Hyle, Thisbe, Erythrae, Glissa, Copae, Lamiae and Anichiae by Cephisus river, Medeon, Phlygone, Acraephia, Coronea, Chaeronea. On the coast, however, below Thebae: Ocalee, Eteonos, Scolos, Schoenos, Peteon, Hyrie, Mycalesos, Ireseum and Eleon, Ollarum, free community Tanagra and, right at the Euripus mouth (created by the island Euboea opposite), Aulis famous for the size of its harbor. Boeoti were called Hyantes in antiquity.

27. Next the Locri called Epicnemidii (and in the past Leleges), through whose territory Cephisus river reaches the sea. Towns: Opus, and thus Opuntian bay, Cynus. Phocis' one seaside [town] is Daphnus, but inland Larisa, Elatea and (on Cephisus' bank, as we have said [4.8]) Lilaea, as well as Cnemis and Hyampolis both facing Delphi. Back on the Locrian coast Larumna and Thronium near which Boagrius river flows to the sea. Towns Narycum, Alope, Scarphia. After them Maliacus bay (named after the local people), with its towns Halcyone, Aegonia, Phalara.

XIII

28. Then Doris with its [towns] Sparthos, Erineon, Boion, Pindus, Cytinum. Behind Doris is mount Oeta.

XIV–XV Thessalia

There follows Haemonia whose name has often changed – always after its king's name – at one time Pelasgis or Pelasgican Argos, [also] Hellas, and at another time Thessalia or Dryopis. One king born here was named Graecus, hence Graecia; another was Hellen, hence Hellenes. Homer had three names for the people here: Myrmidones and Hellenes and Achaeans. Those of them who live near Doris are called Phthiotae. Their towns are Echinus, then the narrow Thermopylae pass at the mouth of Sperchius river, which is why Heraclea four [miles] further is called Trechin. Mount Callidromus is there and the storied towns Hellas, Halos, Lamia, Phthia, Arne.

XV

29. In Thessalia, however, Orchomenus formerly called Minyius, and town Alimon alternatively called Holmon, Atrax, Palamna, Hyperia spring, towns Pherae – behind which Pieria extends towards Macedonia – Larisa, Gomphi, Thessalian Thebae, Pteleon forest, Pagasic bay, town Pagasa later called Demetrias, Tricca, plains of Pharsalus with a free state, Crannon, Iletia. Mountains of Phthiotis: Nymphaeus remarkable for Nature's landscaping efforts, Buzygaeus, Donacoessa, Bromiaeus, Daphusa, Chimarone, Athamas, Stephane.

30. Four and thirty [mountains] in Thessalia, the most famous being Cercetii, Pierian Olympus, Ossa with Pindus and Othrys opposite and home of the Lapithae; these incline west and Pelium east, all curving like a theater, with 75 cities in the bowl they overlook. Thessalia's rivers: Apidanus, Phoenix, Enipeus, Onochonus, Pamisus, Messeis spring, lake Boebeis and Penius best known of all, rising near Gomphi and flowing down a wooded gorge between Ossa and Olympus for a distance of 500 stades, half of it navigable.

31. Part of this course is called Tempe; five miles long and nearly one *iugerum* and a half wide, it has gently sloping hills rising out of sight to left and right, while straight ahead its own energy creates light. Here glides Penius bright, charming and melodious with gravel bed, grassy banks and bird-song. It meets Horcon river but does not merge; instead, [Penius] carries it a brief distance floating on top like oil (to cite Homer [*Il.* 2.751–755]), and then offloads it, refusing to mix its own silver stream with penal water to be used in uttering curses.

XVI Magnesia

32. Next to Thessalia is Magnesia with its Libethra spring, towns Iolcus, Ormenium, Pyrra, Methone, Olizon, cape Sepias, towns Castana,

Spalathra, cape Aeantium, towns Meliboea, Rhizus, Erymnae, Penius mouth, towns Homolium, Orthe, Iresiae, Pelinna, Thaumacie, Gyrton, Crannon, Acharne, Dotion, Melite, Phylace, Potniae.

The length straight across Epirus, Achaia, Attica and Thessalia is said to be 490, the width 297.

XVII Macedonia

33. Next Macedonia with its 150 communities, famous for two kings and its former world-empire; it was called Emathia in the past. Extending west as far as the peoples of Epirus behind the back of Magnesia and Thessalia, it is plagued by Dardani; Paeonia and Pelagonia protect its northern part from Triballi. Towns: Aegae where by tradition the kings are buried, Beroea, and Aeginium in the region called Pieria after the forest [of that name].

34. On the coast, Heraclea, Apilas river, towns Pydna, Aloros, Haliacmon river. Inland, Aloritae, Vallaei, Phylacaei, Cyrrestae, Tyrissaei, colony Pella, Stobi a town of Roman citizens, then Antigonea, Europus on Axius river and another town of that name through which Rhoedias flows, Scydra, Eordaea, Mieza, Gordyniae. Then on the coast, Ichnae, Axius river.

35. Along this border Dardani, Treres, Pieres live close to Macedonia. Beyond this [Axius] river peoples of Paeonia, the Paraxiaei, Eordenses, Almopi, Pelagones, Mygdones. Mountains: Rhodope, Scopius, Orbelus. Then in the fold of territory situated in front of them: Arethusii, Antiochienses, Idomenenses, Doberi, Aestrienses, Allantenses, Audaristenses, Morylli, Garresci, Lyncestae, Othryonei, and free Amantini and Orestae, colonies Bullidenses and Dienses, Xylopolitae, free Scotusaei, Heraclea Sintica, Tymphaei, Toronaei.

36. On the gulf's Macedonian coast: towns Chalastra and (inland) Piloros, Lete and, at the mid-point of the curving shore, Thessalonice with its free status (245 to here from Dyrrachium), Therme on Thermaic bay, towns Dicaea, Palinandrea, Scione, cape Canastraeum, towns Pallene, Phlegra. Mountains in the region: Hypsizonus, Epytus, Alcyon, Elaeuomne. Towns: Nissos, Phryxelon, Mendae and, on Pallenensian isthmus, what was once Potidaea nowadays colony Cassandria, Anthemus, Olophyxus,

37. Mecyberna bay, towns Myscella, Ampelos, Torone, Siggos, Stolos, and the 1½-mile-long channel with which the Persian king Xerxes severed mount Athos from the mainland. From flat ground the mountain itself projects 75 miles into the sea, [while] the perimeter at its base amounts to 150. There used to be a town on its peak called Acrothoon; nowadays there are Uranopolis, Palaehorium, Thyssus, Cleonae, Apollonia whose inhabitants are called Macrobii.

38. Town Assera and the other mouth of the isthmus, Acanthus, Stagira, Sithone, Heraclea and situated below it Mygdonia region, where Apollonia and Arethusa are farther back from the sea. Back on the coast Posidium and a bay with town Cermorus, free Amphipolis, Bisaltae people; then Strymon river (the Macedonian border) which rises on mount Haemus. That it streams into seven lakes before proceeding on its course is notable.

39. This is the Macedonia that once gained world power traversing Asia, Armenia, Hiberia, Albania, Cappadocia, Syria, Egypt, Taurus, Caucasus, then after securing the entire east controlled the Bactrians, Medes, Persians, and even conquered India, roaming in the footsteps of Father Liber and Hercules. This is the same Macedonia where our general Paulus Aemilius plundered and sold off 72 cities in a single day. [Just] two men were responsible for such different fates!

XVIII Thracia

40. There follows Thracia, among Europa's most powerful peoples, divided into 50 *strategiae*. Among its communities, those there should be no objection to mentioning are Denseletae and Maedi living on the right bank of Strymon river as far up as the Bisaltae mentioned above [4.38], and on the left bank Digerri and many divisions of Bessi as far as Mestus river which winds around the base of mount Pangaeus among Haleti, Diobessi, Carbilesi, and then Brigae, Sapaei, Odomanti. Odrysian peoples are at the source of the Hebrus, along which live Carbileti, Pyrogeri, Drugeri, Caenici, Hypsalti, Beni, Corpili, Bottiaei, Edoni.

41. In the same region are Sialetae, Priantae, Dolongae, Thyni, Coelaletae – Greater ones below [mount] Haemus, Lesser below [mount] Rhodope. Between them Hebrus river, and below Rhodope a town called Poneropolis in the past, then Philippopolis after its founder, and nowadays Trimontium after its location. It is a six-mile climb to the summit of Haemus. On its opposite side, sloping down towards the Hister [Danuvius], live Moesi, Getae, Aedi, Scaugdae and Clariae, and below them Arraei Sarmatians called Areatae, and Scythians and – around the coast of Pontus [Black Sea] – Moriseni and Sithoni, forebears of the prophet Orpheus.

42. Thus Hister borders [Thracia] to the north, Pontus and Propontis to the east, and Aegean sea to the south. On its coast, starting from the Strymon: Apollonia, Oesyma, Neapolis, Datos. Inland, colony Philippi (325 from Dyrrachium), Scotusa, state Topiros. Mestus river mouth, mount Pangaeus. Heraclea, Olynthos, free state Abdera, Bistones lake and people. There used to be Tirida, a sinister town because Diomedes stabled his horses there; nowadays there are Dicaea, Ismaron, place Parthenion, Phalesina, Maronea previously called Orthagurea.

43. Mount Serrium, Zone, then place Doriscum with space for 10,000 men – this is where Xerxes counted his army – Hebrus mouth, harbor of Stentor, free town Aenos with Polydorus' tomb, and what was once the Cicones' region. From Doriscum the coast curves for 112 miles to Macron Tichos, a place encircled by Melas river after which the bay is named. Towns: Cypsela, Bisanthe, Macron Tichos so named because a wall stretched between two seas from Propontis to Melas bay bars the advancing Cherronesus.

44. On the other side Thracia begins at the Pontic coast where Hister river joins the sea, an area where it has the loveliest cities: Histropolis [founded by] Milesians, Tomi, Callatis which was previously called Cerbatis, Heraclea. It had Bizone too until an earthquake struck; nowadays it has Dionysopolis (previously called Crunos), past which Zyras river flows. Scythians called Aroteres held this whole region. Their towns: Aphrodisias, Libistos, Zygere, Rhocobae, Eumenia, Parthenopolis, Gerania where there are said to have been Pygmaean people. Barbarians used to call them Catizi and believe that they were driven away by cranes.

45. On the coast after Dionysopolis is Odessus [founded by] Milesians, Pannysis river, town Ereta, Naulochus. On the top of mount Haemus, with its immense ridge sloping down to the Pontus, was town Aristaeum; nowadays on the coast, Mesembria, Anchialum where Messa had been. Astice region had the town Anthium; nowadays it is Apollonia.

Rivers: Panisos, Iuras, Tearus, Orosines. Towns: Thynias, Halmydesos, Deuelton with its marshes and now called Deultum Veteranorum, Phinopolis on the Bosporus. By some calculations it is 500 miles from the Hister mouth to the Pontus mouth, but Agrippa made it 60 more; then to the wall mentioned above [4.43] 150, and from it to Cherronesus 126.

46. But after the Bosporus, Lasthenes bay, Senum harbor and another called Mulierum. Cape Chryseon Ceras, on which the town Byzantium (previously called Lygos) has free status and is 711 miles from Dyrrachium. Such an expanse of land extends between the Hadriatic sea and Propontis.

47. Rivers: Bathynias and Pidaras or Athyras. Towns: Selymbria, Perinthus connected to the mainland by a 200-foot-wide [isthmus]. Inland, Bizye a citadel of Thracian kings and loathed by swallows because of Tereus' depravity, Caenica region, colony Flaviopolis where previously there was a town called Caela and, 50 miles from Bizye, colony Apros 189 from Philippi. Back on the coast, Erginus river; there used to be town Ganos. Lysimachea on the Cherronesus is now deserted too.

48. But at this point another Isthmos marks similar narrows with the same name and equal width. Likewise two cities have been established on the coast either side, Pactye on Propontis, Cardia on Melas bay. The latter gains its name from the lie of the land there, and both were later absorbed by Lysimachea five miles from Longi Muri [= Macron Tichos, 4.43]. On the Propontis side, Cherronesus had Tiristasis, Crithote, Cissa situated on Aegos river; nowadays it has Resisthos 22 miles from colony Apros and opposite the Parian colony.

49. As we said [cf. 2.205], the seven-stade-wide Hellespont dividing Europa from Asia has four cities lying opposite one another: Callipolis and Sestos in Europa, Lampsacus and Abydos in Asia. Then on Cherronesus cape Mastusia opposite Sigeum, and diagonally in front of it Cynossema (as Hecuba's tomb is named), the Achaeans' anchorage and tower, shrine of Protesilaus and, at the furthest end of Cherronesus called Aeolium, town Elaeus. Then those making for Melas bay [reach] the harbors Coelos and Panhormus, and Cardia mentioned above.

50. In this way Europa's third bay comes to an end. Thracia's mountains, besides those already mentioned: Edonus, Gygemeros, Meritus,

Melamphyllos. Rivers: Bargus which discharges into Hebrus, Syrmus. The length of Macedonia, Thracia and Hellespont was mentioned above [4.46]; some make it 720. Its width is 384.

51. The Aegean sea's name comes from an island, or really more a rock, which suddenly springs from the sea midway between Tenus and Chius, and is called Aex by Greeks because it looks like a she-goat. When sailing to Antandrus from Achaia you see it on the right, menacing and dangerous. Part of the Aegean is termed Myrtoan. This is named after a small island seen not far from Carystus on Euboea during a voyage to Macedonia from Geraestus. Romans have two names for all these seas: Macedonian wherever it touches Macedonia or Thracia, and Graeciensian where it laps Graecia. Greeks, however, also divide the Ionian sea into Siculan and Cretic (after the islands), and likewise [they term the stretch] between Samus and Myconus the Icarian. The remaining names come from bays we have mentioned.

XIX Islands off these lands, including:

52. So that is how the seas and peoples in Europa's third gulf are situated. Islands: Corcyra opposite Thesprotia, 12 miles from Buthrotum and 50 from the Acroceraunian [mountains], with a city also called Corcyra, a free state, and town Cassiope and temple of Jupiter Cassius. Ninety-seven long, it was called Scheria and Phaeacia by Homer, and also Drepane by Callimachus. Several [islands] around it, Othronos looking towards Italia and two Paxoe towards Leucadia, five [miles] away from Corcyra.

53. Not far from these and off Corcyra: Ericusa, Marathe, Elaphusa, Malthace, Trachie, Pythionia, Ptychia, Tarachie and, off Phalacrum a cape on Corcyra, the rock into which Ulysses' ship was transformed (or so the story goes) because they look the same. Numerous islands off Leucadia and Acarnania; among those called Teleboides (also Taphiae by local people): Taphias, Carnos, Oxia, Prinoessa. Off Aetolia the Echinades [islands]: Aegialia, Cotonis, Thia, Thera, Geoaris, Dionysia, Cyrnus, Chalcis, Pinara, Nystrus.

54. Off these in the open sea Cephallania and Zacynthus, both free, Ithaca, Dulichium, Same, Crocyle. Cephallania, once called Melaena, is ten miles from Paxos and 93 round. Despite its destruction by Romans, Same still has three towns. Between it and Achaia is Zacynthus splendid with its town and outstandingly fertile; once called Hyrie, it is 25 from the southern part of Cephallania. The remarkable mount Elatus is there; it measures 36 round.

55. Ithaca, with mount Neritus on it, is 15 away and measures 25 all round. From it to Araxus, a cape in Peloponnesus, 15. Off it in the open sea: Asteris, Prote and (35 to the south-east in front of Zacynthus) two Strophades alternatively called Plotae. Off Cephallania Letoia, off Pylos three Sphagiae, and an equal number of Oenussae in front of Messene.

56. In Asinaeus bay three Thyrides, in Laconican [bay] Theganusa, Cothon, Cythera with its town, called Porphyris in the past. This [island] is situated five miles from cape Malea, a risk to sail round because of the narrow strait. In Argolic [bay] Pityusa, Arine, Ephyre; opposite Hermionian territory Tricarenus, Aperopia, Colonis, Aristera; opposite Troezenian [territory] Calauria 50 away, Platea, Belbina, Lasia, Baucidias;

57. opposite Epidaurum Cecryphalos and Pityonesos six [miles] from the mainland. Fifteen from it Aegina (with free status), which is 19 to sail past and 20 from the Athenians' harbor Piraeus, called Oenone in the past. Lying off cape Spiraeum are Eleusa, Adendros, two Craugiae, two Caeciae, Selacosa; and Aspis seven [miles] from Cenchreae, and the four Methurides in Megaric bay, Aegila 15 from Cythera and 25 from Phalasarna town on Creta.

XX Creta

58. Creta itself, with one side facing south and the other north, extends east and west and is famous for its 100 cities. In Dosiades' opinion it was named after the nymph Crete, daughter of Hesperis, and in Anaximander's after a king of the Curetes; according to Philistides of Mallos and Crates it was first called Aeria, then later Curetis. Some have thought that Macaron[1] comes from its mild climate. Nowhere is it more than 50 wide, with its widest part around its middle; its length extends as much as 270, its perimeter 589. The longest side curves towards the Cretic sea (named after it), while its eastern cape Samonium projects opposite Rhodos, and its western Criu Metopon towards Cyrene.

---

[1] Another name for Creta.

59. Its distinguished towns: Phalasarna, Elaea, Cisamon, Pergamum, Cydonea, Minoium, Apteron, Pantomatrium, Amphimala, Rhithymna, Panhormum, Cytaeum, Apollonia, Matium, Heraclea, Miletos, Ampelos, Hierapytna, Lebena, Hierapolis and (inland) Gortyna, Phaestum, Gnosus, Polyrrhenum, Myrina, Lycastos, Rhamnus, Lyctus, Dium, Asium, Pyloros, Rhytion, Elatos, Pherae, Olopyxos, Lasos, Eleuthernae, Therapnae, Marathusa, Tylisos, as well as the still current memory of about 60 other towns. Mountains: Cadistus, Idaeus, Dictynnaeus, Corycus.

60. According to Agrippa, its cape called Criu Metopon is 125 from Cyrene's cape Phycus, while Cadistus is 80 from [cape] Malea in Peloponnesus. The island Carpathos is 60 west of cape Samonium and lies between it and Rhodos.

61. Remaining [islands] around it [Creta] and off Peloponnesus: two Corycoe, the same number of Mylae and (on the north side, with Creta to the right opposite Cydonea) Leuce and two Budroe, Dia opposite Matium, Onysia opposite cape Itanum, Leuce, Chrysea opposite Hierapytna, Gaudos. In the same area Ophiussa, Butoa, Ramnus and, after rounding Criu Metopon, three [islands] called Acusagorus. Off cape Samonium: Phocoe, Platiae, Stirnides, Naulochos, Harmedon, Zephyre.

62. Within Hellas, but still in Aegean sea, the Lichades, Scarphia, Corese, Phocasia and several others facing Attica without towns and therefore insignificant. But opposite Eleusis the famous Salamis; off it Psyttalia and Helene, five [miles] from Sunium. Then the same distance away, Ceos which some of us used to call Cea and Greeks also call Hydrusa, because it was wrenched from Euboea. Once it was 500 stades long, but later nearly four-fifths – the part facing Boeotia – was devoured by the sea too. Its remaining towns are Iulis, Carthaea. Coresus, Poeeessa have perished. According to Varro, very fine ladies' clothing used to be one of the island's exports.

XXI Euboea

63. Euboea itself has also been wrenched from Boeotia by a slight enough channel, Euripus, for the two to be linked by a bridge. It is notable for two southern capes, Geraestum facing Attica and Caphereus the Hellespont; there is one to the north, Cenaeum. Nowhere does it extend more than 40 wide, nor anywhere does it narrow to less than two [miles], but it extends the entire length of Boeotia from Attica all the way to Thessalia for 150, with a perimeter of 365.

64. On the Caphereus side it is 225 from Hellespont. Its cities in the past: Pyrrha, Porthmos, Nesos, Cerinthos, Oreos, Dium, Aedepsos, Oechalia. Nowadays: Chalcis which is opposite Aulis on the mainland, Geraestus, Eretria, Carystus, Oritanan Artemisium, Arethusa spring, Lelantus river and hot springs called Ellopiae. [Euboea] is nevertheless noted more for the marble of Carystus. In the past it was called Chalcodontis or Macris (so say Dionysius and Ephorus), or Macra (says Aristides), or Chalcis (says Callidemus) because copper was first found there, or Abantias (says Menaechmus), while poets typically call it Asopis.

XXII Cyclades

65. Beyond [Euboea] in the Myrtoan [sea] are many [islands], but the most renowned are Glauconnesos and Aegilia, and off cape Geraestum ones situated around Delos in a circle, hence their name Cyclades. The first of these, Andrus with its town, is ten [miles] from Geraestum and 38 from Ceos. Myrsilus says that it was called Cauros, then Antandrus, Callimachus [says] Lasia, others Nonagria, Hydrusa, Epagris; 93 round. After this, Tenus with its town one mile from Andrus and 15 from Delos; 15 across. Aristotle says that it was called Hydrusa because it has so much water, [but] some [call it] Ophiusa.

66. Other [islands]: Myconus with mount Dimastus 15 from Delos, Siphnus previously called Meropia and Acis 28 round, Seriphus 15 [round], Prepesinthus, Cythnos, and Delos by far the most famous, at the center of the Cyclades, well known for its temple of Apollo and its commerce. According to tradition, for a long time it floated around and – up to Marcus Varro's time – was the only island not to experience an earthquake. Mucianus said that it was struck twice by them. Aristotle says that it is called [Delos] because it surfaced and suddenly appeared. Aglaosthenes [calls it] Cynthia, others Ortygia, Asteria, Lagia, Chlamydia, Cynethus, Pyrpyle because fire was first found there. It is five miles round, and its highest point is mount Cynthius.

67. Next to it Rhene which Anticlides calls Celadusa, and also Artemite, Celadine. Syros (20 round according to ancient sources, 160 to Mucianus), Olearos, Paros (with its town 38 from Delos) famous for its marble, first called Platea, later Minoida. Seven and a half from it and 18 from Delos Naxus with its town once called Strongyle, then Dia, and later Dionysias because of its productive vineyards; some have named it Sicilia Minor or Callipolis. It is 75 miles round and half as large again as Paros.

XXIII Sporades

68. Those so far are still within the Cyclades, but the remainder which follow are among the Sporades. They are Helene, Phacusa, Nicasia, Schinusa, Pholegandros and, 38 miles from Naxus, Icaros which gives its name to the sea. It extends the same [distance (38)] in length and has two towns, with a third lost; in the past it was called Doliche and Macris and Ichthyusa. It is located 50 north-east of Delos and 35 from Samus; the strait between Euboea and Andrus is ten miles wide, and from it [Icaros] to Geraestum 112½ miles.

69. From here onwards it is impossible to maintain any order, so the remaining [islands] will be presented in a mass: Scyrus, Ios (18 from Naxus, revered for Homer's tomb, 22 long, previously called Phoenice), Odia, Oletandros. Gyara with its town (15 miles round, 62 from Andrus), Syrnos 80 from [Gyara], Cynethus, Telos famous for its perfume (called Agathusa by Callimachus), Donusa, Patmos 30 round,

70. Corassiae, Lebinthus, Gyrus, Cinara, Sicinus previously called Oenoe, Heraclia also called Onus, Casos also called Astrabe, Cimolos also called Echinusa, Melos with its town called Mimblis by Aristides, Zephyria by Aristotle, Mimallis by Callimachus, Siphis and Acytas by Heraclides, the most circular of the islands. Buporthmos, Machia, Hypere (once called Patage or Platage by others and nowadays Amorgos), Polyaegas, Sapyle, Thera (called Calliste when it first emerged). Therasia was later wrenched from it, and Automate (also called Hiera) then surfaced between the two, as did Thia in our time near them. Ios is 25 miles from Thera.

71. There follow Lea, Ascania, Anaphe, Hippuris. Astypalaea, a free state 88 round (125 from Cadistus on Creta and 60 from Platea, then 38 to Caminia), Azibintha, Lamse, Atragia, Pharmacusa, Thetaedia, Chalcia, Calymna with a town, Cous, Eulimna 25 from Carpathos, which has given its name to the Carpathian sea. South-west from there to Rhodos

50, from Carpathos to Casos seven, from Casos to cape Samonium on Creta 30. In the Euboean Euripus moreover, close to its first entrance four Petaliae islands, and at its exit Atalante. The Cyclades and Sporades lie bounded to the east by Asia's Icarian shores, to the west by Attica's Myrtoan ones, to the north by the Aegean sea, and to the south by the Cretic and Carpathian ones, in an area 700 long by 200 wide.

72. In front of Pagasic bay are Euthia, Cicynethus, Scyrus mentioned above [4.69] but most remote of the Cyclades and Sporades, Gerontia, Scandira; [in front of] Thermaeus [bay] Iresia, Solymnia, Eudemia, Nea sacred to Minerva. Athos has four [islands] in front of it: Peparethus with its town once called Euoenos, nine miles away, Sciathus 15 away, Imbrus with its town 88 away. This is 22 miles from Mastusia on Cherronesus, 62½ round, and with Ilissos river for water.

73. Lemnus from there 22, then 87 to Athos, 115½ miles round, with towns Hephaestia and Myrina, where on the solstice Athos casts its shadow over the forum. Six miles from it [Lemnus] free Thasus, once called Aeria or Aethria. From there, Abdera on the mainland is 22 and Athos 62½, the same distance away as the free island Samothrace opposite the Hebrus, 32 from Imbrus, 22½ miles from Lemnus, 38 from the coast of Thracia; 35 round, it rises one mile high with mount Saoces, and of all the islands is the most harborless. Callimachus calls it by its ancient name Dardania.

74. Between Cherronesus and Samothrace, about 15 miles from both, Halonesos; beyond it Gethone, Lamponia, Alopeconnesus not far from Coelos the harbor of Cherronesus, and some other unimportant [islands]. Names of uninhabited islands in this bay may also be given, so far as they can be found: Auesticos, Sarnos, Cissyros, Charbrusa, Calathusa, Scyllia, Dialeon, Dictaea, Melanthia, Dracanon, Arconesos, Diethusa, Ascapos, Capheris, Mesate, Aeantion, Pateronnesos, Pateria, Calathe, Neriphus, Pelendos.

XXIV Hellespont, Pontus, Maeotis

75. The fourth of Europa's great bays begins from Hellespont and ends at the mouth of Maeotis. The shape of the entire Pontus should be briefly outlined so that its parts are more easily grasped. A vast sea situated in front of Asia and shut out from Europa by the projecting Cherronesus

coastline, it hacks into these lands with a narrow channel and, as has been said [4.49], separates Europa from Asia with a width of [only] seven stades. The first part of the narrows is called Hellespont; here the Persian king Xerxes led his army across a bridge constructed from ships. Then the slender Euripus extends a distance of 86 [miles] to Priapus, a city in Asia, where Alexander the Great crossed.

76. From there the sea widens and then contracts again into a strait. The expansive part is called Propontis, the narrow one Thracian Bosporus; it is 500 paces wide where Xerxes' father Darius led his forces across on a bridge. Its entire length from Hellespont is 239 [miles]. Then the vast sea Pontus Euxinus, once called Axenus, seizes lands that recoil far and by greatly bending its shores curves back in the shape of horns; projecting from them on either side, it stretches out to make the exact shape of a Scythian bow. Mid-curve it is joined to the mouth of lake Maeotis. This mouth is called Cimmerian Bosporus, 2½ miles wide.

77. According to Polybius, the distance in a straight line between the two Bospori, Thracian and Cimmerian, is 500. The perimeter of the Pontus is actually 2,150 according to Varro and ancient sources generally. Nepos Cornelius adds 350 miles, Artemidorus makes it 2,919, Agrippa 2,560, Mucianus 2,425. There is a similar disagreement over the length of Europa's coast, measured as 1,479 by some, 1,100 by others.

78. Marcus Varro calculates as follows: 187½ miles from Pontus mouth to Apollonia, the same from there to Callatis, 125 to Hister mouth, 250 to Borysthenes, 375 miles to Cherronesus (town founded by Heracleotae), 212½ to Panticapaeum alternatively called Bosporus and the furthest point on the coast of Europa – giving a total of 1,337½. Agrippa [reckons] 560 from Byzantium to Hister river, then 638 to Panticapaeum. Lake Maeotis, which takes in Tanais river – flowing from the Ripaean mountains and forming the remotest border between Europa and Asia – is said to be 1,406 miles around, or alternatively 1,125. There is agreement that from the [lake's] mouth to Tanais mouth is 275 in a straight line. Those who live along its bay as far as Histropolis have been mentioned above in the treatment of Thracia [4.44–45]. Hister's mouths begin there.

79. It rises in Germania in the range of mount Abnova opposite Rauricum, a Gallic town, and with the name Danuvius flows for many miles beyond the Alpes and past countless peoples; then, with a hugely increased volume of water, it is called Hister as soon as it reaches Illyricum. After taking in 60 rivers, almost half of them navigable, it discharges into the Pontus in six vast streams. Its first mouth is called Peuces (close by is an island called Peuce), after which the nearest channel is named; this is sucked into a 19-mile-wide marsh. From this same channel and above Histropolis a lake is created, 63 miles round, called Halmyris. The second mouth is called Naracustoma, the third Calon Stoma (by the island Sarmatica), the fourth Pseudostoma (including the island Conopon Diabasis), and thereafter are Borion Stoma and Psilon Stoma. Each of these mouths is said to be so powerful that the sea is overwhelmed for a distance of 40 miles and water drawn from it there is fresh.

XXV–XXVI Dacia, Sarmatia, Scythia

80. From this point all the peoples are generally Scythian, although different groups have occupied coastal areas, in one instance Getae (called Daci by Romans), in another Sarmatians (called Sauromatae by Greeks) and those of them called Hamaxobii or Aorsi, in another Lower Scythians (including ones of slave origin) or Trogodytae, and then Alani and Rhoxolani. In the uplands between the Danuvius and Hercynian forest as far as the Pannonian winter-quarters at Carnuntum and the German borderland there, Sarmatian Iazyges occupy the plains and lowland, while Daci – expelled from there by them – occupy the mountains and forests up to Pathissus river.

81. From the Marus (or if it is the Duria separating [the Daci] from the Suebi and Vannius' kingdom) Basternae hold the opposite side and thereafter other Germans. Agrippa reported that this entire region from Hister to the ocean is 1,200 miles long, and four short of 400 wide from the Sarmatian wilderness to Vistla river. The Scythian name spreads in every direction towards the Sarmatians and Germans but this ancient name has lasted only for the most distant of those peoples, ones who live almost unknown to the rest of mankind.

XXVI

82. After Hister: towns Cremniscoe, Aepolium, Macrocremni mountains, famous Tyra river which bestows its name on the town previously called Ophiusa. In the same [river], Tyragetae inhabit an extensive island; it is 130 from Hister's Pseudostoma mouth. Then the Axiacae named after their river [Axiaces], and beyond them Crobyzi, Rhode river, Saggarius bay, Ordesos harbor and, 120 from Tyra, Borysthenes river and lake and people with the same name and a town 15 miles inland from the sea, its ancient names being Olbiopolis and Miletopolis.

83. Back on the shore the Achaeans' harbor, Achilles' island famous for the hero's tomb, and 125 miles from it a peninsula in the shape of a sword projecting diagonally called Dromos [racecourse] Achilleos because of his training. Agrippa said it is 80 long. Scythian Sardi and Siraci occupy this entire area. Next a wooded region that gives its name to the Hylaeum sea, which reaches it; its inhabitants are called Enoecadioe. Beyond, Panticapes river which marks a divide between nomads and farmers, then Acesinus. There are those who say that Panticapes merges with Borysthenes below Olbia, but according to more meticulous sources it is Hypanis which does so. Those who place the latter somewhere in Asia have committed a really grave error.

84. The sea advances in a great gulf until it and Maeotis are only five miles apart, surrounding a vast area and many peoples. It is called Carcinites bay. Pacyris river and towns Navarum, Carcine. Behind, lake Buces discharges into the sea through a canal. Buces itself is shut off from Coretus, a bay in lake Maeotis, by a rocky outcrop. It takes in the Buces, Gerrhus, Hypanis rivers, which enter from different areas. For Gerrhus separates Basilidae and nomads, while Hypanis flows by nomads and Hylaei into Buces through a manmade channel, and into Coretus through a natural one. This region of Scythia is named Sindica.

85. But Taurica [Crimea] begins from Carcinites [bay] where it was once surrounded by sea and is nowadays level ground; then it rises in steep ridges. It has 30 communities, 23 of them living inland; those in its six towns: Orgocini, Characeni, Assyrani, Stactari, Acisalitae, Caliordi. Scythotauri occupy the ridge itself. They are confined to the west by Cherronesus Nea, and to the east by Scythian Satarci. On the coast after Carcine the towns Taphrae right at the narrows of the peninsula, then Heraclea Cherronesus granted free status by the Romans; called Megarice in the past, it has special distinction within this whole region because it maintains Greek traditions; there is a five-mile-long wall round it.

86. Next cape Parthenium, Placia a state of the Tauri, Symbolum harbor, cape Criu Metopon which projects out to the middle of Euxinus [Black

Sea] opposite cape Carambis in Asia (the two being 170 apart); the shape of a Scythian bow is formed for this reason in particular. After it, numerous harbors and lakes of the Tauri. The town Theodosia 125 miles from Criu Metopon but 165 from Cherronesus. Beyond it were the towns Cytae, Zephyrium, Acrae, Nymphaeum, Dia.

87. Panticapaeum, founded by Milesians, still stands as by far the strongest [town] right at the Bosporus entrance, 87½ miles from Theodosia, 2½ miles (as we said [4.76]) from the town Cimmerium situated across the strait. This width here separates Asia from Europa, one that quite normally can be crossed on foot when the strait is frozen over. The width of Cimmerian Bosporus 12½ miles. Its towns are Hermisium, Myrmecium and the island Alopece within it. The distance along Maeotis from the furthest isthmus at the place called Taphrae to Bosporus mouth amounts to 260.

88. After Taphrae the interior of the mainland is occupied by Auchetae in whose territory Hypanis rises, Neuroe in whose territory Borysthenes rises, Geloni, Thyssagetae, Budini, Basilidae and Agathyrsi with their blue hair. Above them nomads, then Anthropophagi and – after [lake] Buces above Maeotis – Sauromatae and Essedones. Along the coast as far as the Tanais, Maeotae (the lake is named after them) and finally, behind them, Arimaspi. Next Ripaean mountains and the region called Pterophoros because of the constant fall of snow resembling feathers; this part of the world is doomed by Nature's course, plunged in dense fog and producing only freezing [temperatures] and reserves of icy north wind.

89. Behind these mountains and beyond the north wind there are, if we believe it, the happy people called Hyperboreans; they live to extreme old age and are famous for legendary wonders. Here are believed to be the world's poles and the outer limits of the planets' orbits, where there is six months of daylight and one day of darkness – though not, as has been ignorantly claimed, from the spring equinox until autumn. Here the sun rises [only] once a year, at the spring solstice, and sets once, at the winter solstice. A sunny region with a productive climate free of any damaging wind. Woods and groves are their homes; both individually and collectively they worship the gods, with strife and grief completely unknown. Death only comes after a full life, when they hold a banquet – their old age thoroughly soaked in luxury – and then leap into the sea off some crag; to them, this is the most desirable form of burial.

90. Some [sources] have placed them not in Europa but on the nearest stretch of Asian shore, because [people] of similar character and environment are there, called Attaci. Others have placed them midway between both suns – one setting in the antipodes, and our sunrise – which is quite impossible with such a vast expanse of sea in between. According to sources who have placed them in some region other than one with six months of daylight, they sow in the morning, reap at midday, pick fruit from the trees at sunset, and retire to caves for the night.

91. We should not doubt the existence of these people when so many authors report that they regularly send their first-fruits to Delos for Apollo, who they worship in particular. Virgins used to bring these, and for a number of years they were respected according to the universal code of hospitality; but once that trust was breached, it became their custom to deposit their offerings at the nearest border with their neighbors, who would convey them to those next to them, and so on all the way to Delos. Later even this custom lapsed. According to Marcus Agrippa, the territories of Sarmatia, Scythia, Taurica and the whole region from Borysthenes river is 980 long, 716 wide. I do not regard measurements for this part of the world as exact.

XXVII Islands of Pontus

Maintaining the sequence already organized, the remainder of this bay should be noted; in fact we have [already] specified its seas.

92. Hellespont has no islands in Europa worth mentioning. In the Pontus, 1½ from Europa and 14 from the mouth, the two Cyaneae alternatively called Symplegades; in legend these are said to have collided with one another because, with such minute distance between them, anyone entering [Pontus] directly sees them as a pair, but with a slight shift of sightline gains an impression of them merging. On this side of Hister 80 miles from Thracian Bosporus one of the Apolloniatae, from which Marcus Lucullus brought a [statue of] Capitoline Apollo. We have [already] mentioned the islands in Hister's mouths [4.79].

93. In front of Borysthenes is Achilles' island mentioned above [4.83], also called Leuce and Macaron. A contemporary calculation places it 140 from Borysthenes, 120 from Tyra, and 50 from Peuce island; it is about ten miles round. Remaining [islands] in Carcinites bay: Cephalonesos,

Spodusa, Macra. Before leaving Pontus, I should not fail to mention a widely held opinion that all the interior seas spring from this source and not from the Gaditan strait; the claim is plausible insofar as the current always flows out of the Pontus and never back into it.

94. Next we must move on to describe Europa's fringes; after crossing the Ripaean mountains we must keep to the left along the shore of the northern ocean until arriving at Gades. In this zone there are said to be several islands without names; Timaeus said there is one off Scythia called Baunonia a day's journey away, where in spring-time amber is cast up by the waves. No certainty about the rest of these shores. Reports have made the northern ocean known. After Parapanisus river where [the ocean] reaches Scythia, Hecataeus calls it Amalcius, a name that in those people's language means 'frozen'.

95. Philemon says that the Cimbri named it Morimarusa (thus dead sea) from there [Parapanisus river] all the way to cape Rusbeae and then on to Cronian [sea]. Xenophon of Lampsacus says that three days' sail from the Scythian coast is an island of immense size called Balcia, which Pytheas calls Basilia. Also recorded are Oeonae whose inhabitants live on birds' eggs and oats; other [islands] called Hippopodes, where men are born with horse feet; and others, Phanesii, whose inhabitants, otherwise nude, have huge ears that cover their entire bodies.

96. Clearer reports then begin with the Inguaeones people, the first in Germania. Here mount Saevo, immense and as high as the Ripaean range, forms an enormous bay as far as the Cimbrians' cape, which is called Codanus and teems with islands, the best-known of them being Scatinavia. Its size has yet to be discovered, and only part of it (so far as known) is inhabited by the Hilleviones people in 500 villages; they call it another world. Aeningia is reckoned to be no smaller.

97. Some sources say that these [regions] are inhabited as far as Vistla river by Sarmatians, Venedi, Sciri, Hirri, [and that there is] a bay called Cylipenus with an island Latris at its mouth, and then another bay, Lagnus, bordering the Cimbri. Cimbrians' cape projects far into the sea to form a peninsula called Tastris. Then 23 islands known to the Roman army, Burcana the most notable of them, called Fabaria by us because a vegetable [like the *faba*] grows wild there; Glaesaria too as [our] soldiers call it from its amber, called Austeravia by barbarians; and also Actania.

XXVIII–XXIX Germania

98. The full length of the sea-coast as far as Scaldis river in Germania is inhabited, but to state the size of the area is impossible, given the serious differences among reports. Greek authors and some of ours have stated 2,500 for the coast of Germania. Agrippa says that with Raetia and Noricum it is 636 long and 248 wide, although [really] the width of Raetia alone is almost greater, but of course it was [only] conquered around the time of his death [12 BCE]. There was only detailed information about Germania many years after, and even then not about all of it.

99. If a guess is allowed, [I think] the coast will be not much shorter than the Greeks thought and the figure Agrippa gave for its length. Five kinds of Germans: Vandili, who include Burgodiones, Varinnae, Charini, Gutones. The second kind: Inguaeones who include the Cimbri, Teutoni and Chaucian peoples.

100. Nearest to the Rhenus the Istuaeones who include Sicambri. Inland, Hermiones who include Suebi, Hermunduri, Chatti, Cherusci. The fifth group Peucini, Basternae who border Daci mentioned above [4.80]. Notable rivers flowing into the ocean: Guthalus, Visculus or Vistla, Albis, Visurgis, Amisis, Rhenus, Mosa. The Hercynian range, second to none in grandeur, extends across the interior.

XXIX

101. Within Rhenus itself, the most notable island is that of the Batavi and Cannenefates, nearly 100 long; also those of the Frisii, Chauci, Frisiavones, Sturii, Marsaci, spread out between Helinium and Flevum. These are the names of the mouths into which Rhenus gushes, dispersing into lakes northward, and westward into Mosa river; as the center mouth in between these, it retains an insignificant channel to bear its own name.

XXX Ninety-six islands in the Gallic ocean, including Britannia

102. Situated opposite this location the island Britannia, well-known from accounts by Greeks and ourselves, lies to the north-west, a substantial distance across from Germania, Gallia, Hispania, by far the largest parts of Europa. It was itself named Albion, while all [the islands] which we are just about to mention were called Britanniae. The shortest crossing is 50 from Gesoriacum on the coast of the Morini people. Pytheas and Isidorus report that its perimeter is 4,875; almost 30 years ago now, the Roman military undertook a reconnaissance there, although only as far as the locality of Calidonia's forest. Agrippa believes it to be 800 long, 300 wide, and Hibernia the same width but 200 less in length.

103. It is situated above [Britannia], 30 away by the shortest crossing from the Silures people. None of the remaining [islands] is said to have a perimeter greater than 125. There are 40 Orcades with narrow channels separating them, seven Haemodae, 30 Hebudes and – between Hibernia and Britannia – Mona, Monapia, Riginia, Vectis, Silumnus, Andros; below, Samnis and Axanthos, and opposite them scattered across the Germanic sea are the Glaesiae, which very recently Greeks have called Electrides because amber comes from there.

104. Remotest of all those to be mentioned is Thule where, as we have pointed out [2.187], there are no nights at the summer solstice when the sun is passing through the constellation Cancer, and on the other hand no days during the winter solstice. Indeed there are those who believe this to be the case for six straight months. The historian Timaeus says that it is a six-day sail inward from Britannia to the island Ictis where tin is found, and that Britanni sail there in wicker boats covered with hides. There are reports of other [islands] too, Scandiae, Dumna, Bergi and Berrice largest of all, where voyages to Thule start. From Thule it is a single day's sail to the frozen sea that some call Cronium.

XXXI Belgica Gallia

105. The whole of Gallia known by the single name Comata is divided into three different communities, who are separated by rivers for the most part. From Scaldis to Sequana is Belgica, from there to Garunna is Celtica also called Lugdunensis, and from there to the slopes of the Pyrenaei mountains Aquitanica previously called Aremorica. Agrippa calculated the whole coast to be 1,750, and all of Gallia between the Rhenus and Pyrenaei and the ocean and the Cebenna and Jures mountains (separating off Gallia Narbonensis) to be 420 long, 318 wide.

106. Starting from Scaldis, Texuandri with their several names live in the outer areas, then Menapi, on the coast Morini associated with Marsaci in the district called Gesoriacus, Britanni, Ambiani, Bellovaci, Bassi. Inland,

Catoslugi, Atrebates, free Nervi, Veromandui, Suaeuconi, free Suessiones, free Ulmanectes, Tungri, Sunuci, Frisiavones, Baetasi, free Leuci, Treveri (who used to be free) and allied Lingones, allied Remi, Mediomatrici, Sequani, Raurici, Helveti, and the colonies Equestris and Raurica. The peoples of Germania living along Rhenus in the same province are Nemetes, Triboci, Vangiones, colony of Agrippinensis (among Ubii), Guberni, Batavi, and those on islands of Rhenus we have mentioned [4.101].

XXXII Lugdunensis Gallia

107. In Gallia Lugdunensis are Lexovii, Veliocasses, Caleti, Veneti, Abrincatui, Ossismi, the famous Liger river, and also an even more remarkable peninsula that projects into the ocean from the edge of Ossismian territory, 625 round, 125 wide at its neck. Beyond it, Namnetes and, inland, allied Aedui, allied Carnuteni, Boi, Senones, Aulerci also called both Eburovices and Cenomani, free Meldi, Parisi, Tricasses, Andecavi, Viducasses, Bodiocasses, Venelli, Coriosuelites, Diablinti, Riedones, Turones, Atesui, free Segusiavi in whose territory is the colony Lugdunum.

XXXIII Aquitanica Gallia

108. To Aquitanica belong Ambilatri, Anagnutes, Pictones, free Santoni, free Bituriges also called Vivisci, Aquitani (hence the province name), Sediboviates. Then Convenae settled together in a town, Begerri, Tarbelli Quattuorsignani, Cocosates Sexsignani, Venami, Onobrisates, Belendi, Pyrenean woodland and below it Monesi, Oscidates Montani, Sybillastes, Camponi, Bercorcates, Pinpedunni, Lassuni, Vellates, Torvates, Consoranni, Ausci, Elusates, Sottiates, Oscidates Campestres, Successes, Latusates, Basaboiates, Vassei, Sennates, Cambolectri Agessinates.

109. Moreover, joined with Pictones are free Bituriges called Cubi, then Lemovices, free Arverni, free Vellavi, Gabales. In the other direction, bordering Narbonensis province, Ruteni, Cadurci, Nitiobroges and Petrocori separated from Tolosani by Tarnis river.

The seas around the coast are the northern ocean up to Rhenus, Britannic between Rhenus and Sequana, Gallic between it and Pyrenaei. Several islands of the Venesti, both those called Veneticae and Uliaros in the Aquitanic bay.

XXXIV Nearer Hispania along the ocean

110. Hispania begins at Pyrenaei cape and is not only narrower than Gallia but also than itself, as we have said [3.29], because of the huge pressure applied to it from the ocean on one side and from the Hiberic sea on the other. The Pyrenaei range itself spreads out from the equinox sunrise [due east] to the winter sunset [south-west] and makes the Hispaniae shorter on the northern side than the southern. The nearest coast is the location of Nearer, also known as Tarraconensis. After Pyrenaei along the ocean: forests of the Vascones, Olarso, towns of the Varduli, Morogi, Menosca, Vesperies, Amanum harbor where colony Flaviobrica is nowadays.

111. The region of nine states of the Cantabri, Sauga river, Victoria Juliobrigensium harbor (Hiberus rises 40 miles from here), Blendium harbor, Orgenomesci (a Cantabrian people) and their harbor Veseiasueca, Astures region, town Noega. Paesici on a peninsula and then in the Lucensian *conventus*, starting from Navia river, Albiones, Cibarci, Egi, Varri also called Namarini, Adovi, Arroni, Arrotrebae. Celticum cape, Florius and Nelo rivers. Celtici also called Neri and Supertamarci on whose peninsula are three altars of Sestius dedicated to Augustus, Copori, town Noeta, Celtici also called Praestamarci, Cileni. Among the islands, Corticata and Aunios should be named.

112. After the Cileni, in the Bracares' *conventus*, Helleni, Grovi, fortress Tyde, all of Greek origin. Siccae islands, town Abobrica, Minius river four [miles] wide at its mouth, Leuni, Seurbi, the Bracares' town Augusta, Gallaecia above them. Limia river, Durius river, one of the largest in Hispania, rising among the Pelendones and flowing by Numantia, then through Arevaci and Vaccaei, separating Vettones from Asturia and Gallaeci from Lusitania, and here also keeping apart Turduli and Bracares. The entire region from Pyrenaei just described full of gold, silver, iron, lead and tin mines.

XXXV Lusitania

113. Lusitania begins after the Durius. The ancient Turduli, Paesuri, Vagia river, town Talabrica, town and river Aeminium, towns Conimbriga, Collippo, Eburobrittium. Then there projects out to sea a cape like an enormous horn which some have called Artabrum, others Magnum, and many Olisiponensian after its town; it splits land, sea and sky. With it ends the side of Hispania, and after making a turn its front begins.

114. North of here the Gallic ocean, and west in the other direction the Atlantic ocean. Some accounts say the cape extends 60, others 90; plenty say that from here to Pyrenaei is 1,250, and they locate here the Artabres people – a blatant error, because these never existed. What they have done is to locate here (with a change of spelling) the Arrotrebae, whom we mentioned [4.111] just before Celticum cape.

115. There have also been errors about well-known rivers. According to Varro, it is 200 from Minius (which we mentioned above [4.112]) to Aeminius, which some consider to be elsewhere and name Limaea – called in antiquity [river of] Oblivion and the subject of many tales. Tagus is 200 from the Durius, with Munda between them. Tagus is famous for its gold-bearing sands. One hundred and sixty further on, cape Sacrum leaps out from about the half-way point of Hispania's front. According to Varro, from it to half-way along Pyrenaei amounts to 1,400,

116. but only 126 to Anas (where we have set the boundary between Lusitania and Baetica [3.6]), with an additional 102 from Gades. Peoples: Celtic Turduli and Vettones around Tagus, Lusitani between Anas and [cape] Sacrum. Notable towns on the coast after Tagus: Olisipo famous for its mares impregnated by the west wind, Salacia also called Urbs Imperatoria, Merobrica. Cape Sacrum and another [cape] Cuneus, towns Ossonoba, Balsa, Myrtilis.

117. The whole province is divided into three *conventus*, the Emeritensian, Pacensian, Scalabitan; 45 communities in total, including 5 colonies, 1 municipality of Roman citizens, 3 with ancient Latin rights, and 36 tribute-payers. Colonies: Augusta Emerita located on Anas river, the Metellinensian, Pacensian, Norbensian also called Caesarina; Castra Servilia and Castra Caecilia are associated with it. The fifth [colony] is Scalabis also called Praesidium Julium. The municipality of Roman citizens is Olisipo also called Felicitas Julia. Towns with ancient Latin rights: Ebora also called Liberalitas Julia, and Myrtilis and Salacia which we have mentioned [4.116].

118. Among the tribute-payers, there should be no objection to naming (besides those already mentioned in the names for Baetica [3.14–15]): the

Augustobrigenses, Aeminienses, Aranditani, Arabricenses, Balsenses, Caesarobrigenses, Caperenses, Caurienses, Colarni, Cibilitani, Concordienses, Elbocori, Interannienses, Lancienses, Mirobrigenses also called Celtici, Medubrigenses also called Plumbari, Ocelenses, Turduli also called Bardili, and Tapori.

Agrippa said that Lusitania including Asturia and Gallaecia stretches 540 long, 536 wide. All of Hispania from one cape of Pyrenaei around to the other along the entire sea-coast is thought to measure 2,924, or 2,600 according to others.

XXXVI Islands in the Atlantic sea

119. Opposite Celtiberia are several islands: those called Cassiterides by Greeks because of their abundance of lead, and the Six Gods (which some have called Fortunatae) in the area of the cape of the Arrotrebae. But at the very start of Baetica, 25 miles from the mouth of the strait, Gades, 12 long, three wide according to Polybius. Its distance from the mainland at the nearest point is less than 700 feet, while elsewhere over seven [miles]. Its expanse is 15. It has a town of Roman citizens who are called Augustani of Urbs Julia Gaditana.

120. About 100 paces away on the side facing Hispania is another island, one mile long and one wide, where the town of Gades used to be. Ephorus and Philistus call it Erythea, Timaeus and Silenus call it Aphrodisias, but its own people call it Juno's island. Timaeus says that the larger island was called Potinusa after its wells; we call it Tartesos, and Punic people call it Gadir after the Punic word for fence. It was called Erythea because their ancestors, Tyri, were said to have originated from the Erythrean sea. There are those who think that Geryones – whose cattle Hercules stole – lived on this island. In the opinion of others it was a different island across from Lusitania, and there is one there which they call by the same name.

XXXVII Dimensions of Europa in its entirety

121. Having finished the circuit of Europa we must offer a total figure, so that those who wish to know one will not be left without it. Artemidorus and Isidorus reported its length from Tanais to Gades as 8,714. At a time when the size of Europa had still to be ascertained, Polybius wrote that its width from Italia to the ocean was 1,250.

122. As we have said [3.43], [the length] of Italia itself to the Alpes is 1,020, and from there to Lugdunum and on to the Britannic harbor of the Morini – which appears to be Polybius' line of measurement – is 1,169. But a more reliable and longer dimension in a straight line from the Alpes again to the north-west and to Rhenus mouth through the legionary camps in Germania is 1,243.

From here then Africa and Asia will be the subject.

****

Totals   Towns and peoples [...]
         Famous rivers [...]
         Famous mountains [...]
         Islands [...]
         Extinct cities and peoples [...]
         Facts and inquiries and observations [...]

[Roman] Sources

Cato the Censor. Marcus Varro. Marcus Agrippa. Deified Augustus. Varro Atacinus. Cornelius Nepos. Hyginus. Lucius Vetus. Mela Pomponius. Licinius Mucianus. Fabricius Tuscus. Ateius Capito. Ateius the philologist.

Foreign Sources

Polybius. Hecateaus. Hellanicus. Damastes. Eudoxus. Dicaearchus. Timosthenes. Eratosthenes. Ephorus. Crates the grammarian. Serapion of Antioch. Callimachus. Artemidorus. Apollodorus. Agathocles. Timaeus of Sicilia. Myrsilus. Alexander Polyhistor. Thucydides. Dosiades. Anaximander. Philistides of Mallos. Dionysius. Aristides. Callidemus. Menaechmus. Aglaosthenes. Anticlides. Heraclides. Philemon. Xenophon. Pytheas. Isidorus. Philonides. Xenagoras. Astynomus. Staphylus. Aristocritus. Metrodorus. Cleobulus. Posidonius.

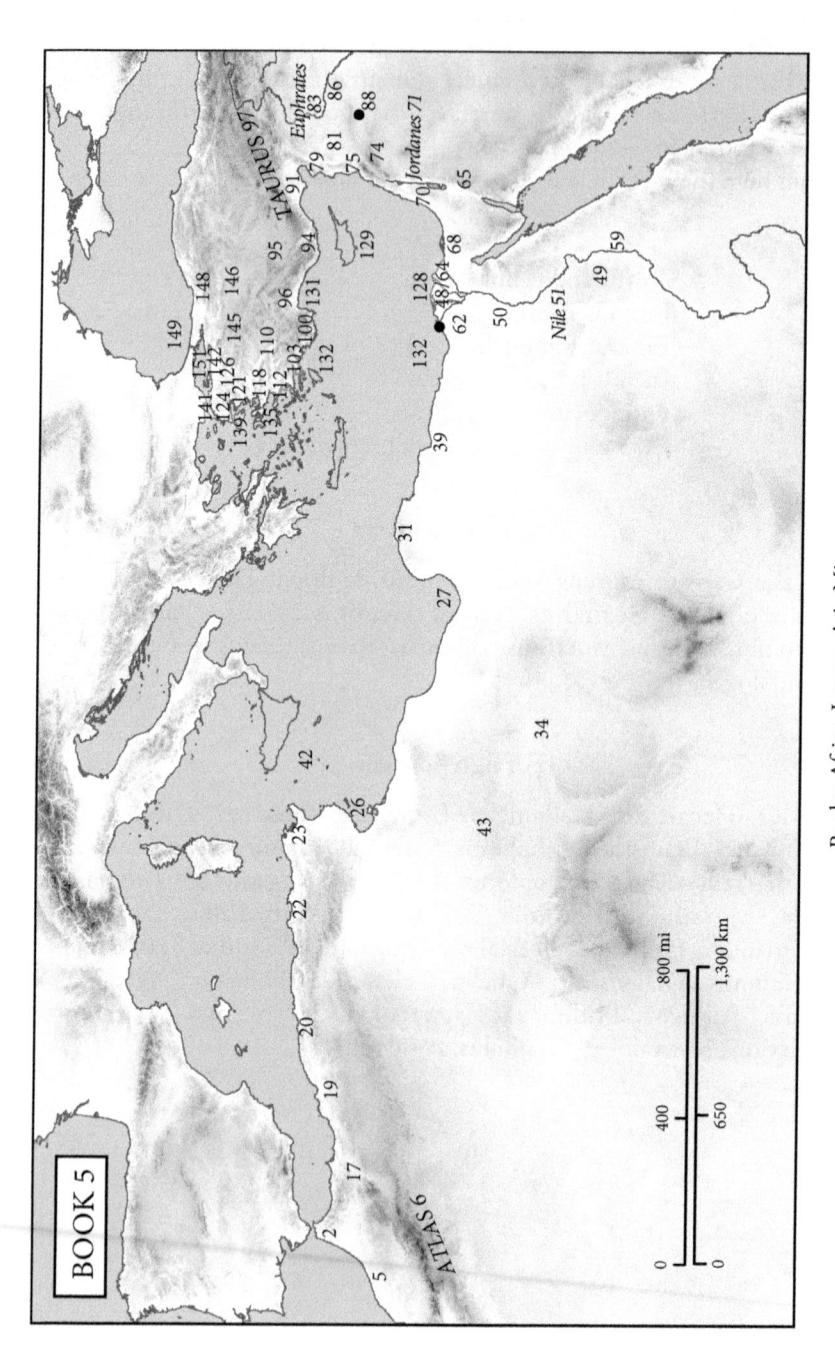

Book 5 Africa, Levant, Asia Minor

BOOK 5

Africa, Levant, Asia Minor

I Mauretaniae

1. Greeks called Africa 'Libya' and the sea in front of it 'Libycan'. It is bordered by Egypt. With a long shoreline extending diagonally from the west, there is no other part of earth broken by fewer bays. The names of its communities and towns are for the most part unpronounceable except in their own languages; other than that, they typically live in fortresses.

2. The first lands are called Mauretaniae; they remained kingdoms until Germanicus' son, Gaius Caesar [Caligula], brutally divided them into two provinces. The furthest cape in the ocean is named Ampelusia by Greeks. Beyond Hercules' pillars were the towns Lissa and Cottae; nowadays there is Tingi, founded originally by Antaeus and thereafter called Traducta Julia by Claudius Caesar when he made it a colony. By the shortest crossing it is 30 from Baelo, a town in Baetica. Along the ocean coast, 25 from [Tingi], Julia Constantia Zilil, a colony of Augustus which was removed from [Mauretanian] royal control and instructed to look to Baetica in matters of justice. From there, 32, Lixus made a colony by Claudius Caesar and the subject of the most remarkable ancient tales.

3. Here are Antaeus' palace, the site of his contest with Hercules, and the gardens of the Hesperides. In addition, there pours in from the sea a winding channel which nowadays is regarded as having been like a dragon on guard. [The channel] completely surrounds an island, the only ground in the area not flooded by sea-surges because it is rather more elevated. On the [island], too, stands an altar of Hercules, but – except for some olive trees – there is nothing of the gold-bearing grove that features in the story.

4. Surely the fantastic lies circulated by Greeks about these matters and Lixus river are not so surprising to anyone who considers that lately we [Romans] have told some hardly less extraordinary tales about the same things. That this city was very powerful, greater than Carthago Magna, and moreover that it lies opposite [Carthago],[1] and that the distance between it and Tingi is almost impossible to measure: all this and more Cornelius Nepos believed very enthusiastically.

5. From Lixus, 40 inland, is Babba, another colony of Augustus called Julia Campestris, and a third one, 75 miles away, called Banasa, with the further name Valentia. Thirty-five from there, Volubile town situated at the same distance from both seas.[2] But on the coast, 50 from Lixus, the magnificent and navigable Sububus river flows by Banasa colony. The same number of miles from [Banasa] Sala town, located on a river of the same name, close to wilderness and infested by herds of elephants, but even more so by the Autoteles people through whose land lies the route to Atlas, Africa's most remarkable mountain.

6. It is reported to rise into the sky from the midst of sands. The side which faces the [Atlantic] ocean shore (hence its name) is rugged and parched. The mountain is also shady, wooded and well-watered by gushing springs on the side facing Africa, where fruits of all kinds sprout so naturally that desires are never left unsatisfied.

7. None of its inhabitants is seen during the day; everything is silent except for dread of the desert; an unspoken reverence enters the hearts of those who approach, together with a fear of the mountain's height stretching above the clouds and close to the moon's orbit. At night it sparkles with numerous fires, is filled with the lasciviousness of Aegipanes [Goat-Pans] and Satyrs, and resounds with the music of flutes and pipes and with the sound of drums and cymbals. Famous authors have given these accounts, along with the labors accomplished there by Hercules and Perseus. The distance to it is immense and uncertain.

8. There were also records made by a Carthaginian commander Hanno, who was ordered to investigate the circumference of Africa when the Punic state was at its height. Most authors (Greeks and ours) have

---

[1] i.e. on the same latitude.
[2] Atlantic and Mediterranean.

followed his account, spreading all kinds of myths, including the vast number of cities he founded there, although no memory or trace of any survives.

9. When Scipio Aemilianus held a command in Africa, the annalist Polybius was equipped with a fleet by him to circumnavigate and explore that part of the world. [Polybius] reported that to the west of the mountain Africa produces tracts full of wild animals as far as Anatis river <…> 496 <…>. Agrippa says that Lixus is 205 from it, and that Lixus is 112 from the Gaditan strait. Then the bay called Sagigi, a town on cape Mulelacha, Sububa and Salat rivers, Rutubis harbor 224 from Lixus, then Sol's cape, Rhysaddir harbor, Gaetulan Autoteles, Quosenum river, Selatiti and Masathi peoples, Masath river, Darat river in which crocodiles are born.

10. Then a 616[-mile] bay enclosed by a cape of mount Braca which extends to the west and is called Surrentium. After this, Salsum river beyond which are Aethiopian Perorsi and behind them Pharusii. Inland next to the latter, Gaetulan Darae; but on the coast Aethiopian Darathitae and Banbotum river filled with crocodiles and hippopotamuses. From there an uninterrupted range of mountains runs all the way to the one we shall call Theon Ochema. From there to cape Hesperu a sailing-time of ten days and nights. [Polybius] locates Atlas in the center of this space, although according to everyone else it lies in the most distant part of Mauretania.

11. The first Roman campaigns in Mauretania occurred during Claudius' principate, after king Ptolemy had been executed by Gaius Caesar [Caligula] and his freedman Aedemon sought vengeance; by all accounts, [the Romans] reached mount Atlas in the course of pursuing the barbarians as they fled. The glory of having reached as far as Atlas belonged not just to senators who have held the consulship and were in command at the time, but also to Roman *equites* who served as governors there subsequently.

12. As we said, there are five Roman colonies in this province, and it can be considered accessible; but experience outweighs this impression in a generally most misleading way, because dignitaries who have no interest in establishing the truth but are ashamed of their ignorance do not mind lying, and there is no type of error more likely to gain acceptance than one backed by some lofty figure. I am truly less surprised by a certain lack of awareness on the part of men of equestrian rank, even though subsequently they enter the senate, than that some luxuries should escape notice: after all, they are felt to make the strongest, most effective impact, with forests scoured for ivory and citrus-wood, and every Gaetulan rock for sources of purple dye.

13. The natives, however, say that Asana river is on the coast 150 from the Salat; it has brackish water but is notable for its harbor. Next, a river called Fut, from which it is 200 to Addiris – this is generally understood to be the name for Atlas in their language – with a river called Ivor along the way. Here a sanctuary stands out near traces of a land that was once inhabited, and the remains of vineyards and palm groves.

14. Suetonius Paulinus, whom we saw as consul [66 CE], was the first Roman leader to make a crossing that went as far as some miles beyond mount Atlas. He made the same report as others do about its great height. He reported further that its lowest slopes are filled with dense, lofty forests of trees of an unknown kind remarkable for their height, free of knots, and glossy; their foliage resembles that of cypresses, except for the strong odor; they are covered with a thin wool from which it is possible – with skill – to make clothing like that of silk. The peak is covered with deep snow even in summer.

15. After a ten-day march[, Suetonius says], he reached this point and then continued onward to a river called Ger, then on through deserts of black sand, broken here and there by towering rocks that appeared to be scorched – a region rendered uninhabitable by the heat, even though his experience of it occurred in winter. The people who inhabit the nearest tracts – teeming with all kinds of elephants and wild beasts and snakes – are called Canarii [Dog People], since their diet does not differ from that of those animals and since they share with them the entrails of wild beasts.

16. It is well known that the next people are Aethiopians called Perorsi. Ptolemy's father Juba, who was the first to rule both Mauretaniae and is remembered more for his outstanding scholarship than for his rule, gave a similar report about Atlas, and went on to say that a plant grows there called 'euphorbea' after the name of his doctor who discovered it. In a volume devoted exclusively to it, he bestows exceptional praise on the value of its milky juice for improving eyesight and providing protection against snakes and every poison. But this is enough – more than enough – about Atlas.

17. The province Tingitana is 370 long. Peoples in it: at one time the foremost were Mauri, hence the name [Mauretania]. Most writers have called them Maurusii; thinned out by wars, they have been reduced to a few families. Next used to be Masaesyli, but they have been wiped out in much the same way. Nowadays Gaetulan peoples dominate, Baniurae and Autoteles (by far the most powerful) and Nesimi, who were once a sub-group of theirs [Autoteles] but split from them to become a people in their own right located in the Aethiopians' direction.

18. Elephants come from the mountainous east of the province, as well as from mount Abila and from those called Seven Brothers because they are more or less the same height. The latter are joined to Abila and overhang the strait. From here on the coast of the internal sea [Mediterranean] the navigable Tamuda river where there was once a town, Laud river also navigable, Rhysaddir town and harbor, navigable Malvane river.

19. Siga town, Syphax's royal capital, lies opposite Malaca in Hispania, and is already in the second Mauretania; for a long time the two of them were named after kings, with the further being called Bogutiana and the other, nowadays Caesariensis, after Bocchus. After [Siga], Portus Magnus a town of Roman citizens so-called from its size, Mulucha river (the border between Bocchus and the Masaesyli), Quiza Cenitana a town of foreigners, Arsennaria (which has Latin rights) three [miles] from the sea,

20. Cartenna a colony of Augustus settled by the Second Legion, Gunugu another of his colonies established by a praetorian cohort, Apollo's cape and there its best-known town, Caesarea once called Iol, Juba's royal capital, granted the status of colony by Deified Claudius; on his orders, too, veterans established Oppidum Novum, and Tipasa was given Latin rights. Likewise, emperor Vespasian granted the same rights to Icosium. Rusguniae a colony of Augustus, Rusucurum honored with citizenship by Claudius, Rusazus a colony of Augustus, Saldae another of his colonies. Likewise Igilgili,

21. Tucca a town located by [both] the sea and Ampsaga river. Inland, Aquae a colony of Augustus, likewise Succhabar, likewise Tubusuptu, states Timici and Tigavae, rivers Sardaval, Aves, Nabar, Macurebi people, Usar river, Nababes people. Ampsaga river is 322 from Caesarea. The two Mauretaniae are 1,038 long, 467 wide.

II Numidia

22. After the Ampsaga is Numidia, famous for the name of Masinissa. Greeks called this country Metagonitis, and Numidians 'Nomads' because of how frequently they change pasture while carrying their *mapalia*, their homes in other words, all over the place on wagons. Towns: Chullu, Rusiccade and – 48 [miles] inland from there – a colony called Cirta Sittianorum, and another inland, Sicca, and Bulla Regia a free town. Meanwhile on the coast, Tacatua, Hippo Regius, Armua river, Thabraca a town of Roman citizens, Tusca river the border of Numidia. Nothing notable originates here except Numidic marble and wild beasts.

III Africa

23. After Tusca, Zeugitan region and what is properly called Africa. Three capes – Candidum, then Apollo's facing Sardinia, Mercury's facing Sicilia – plunging into the sea form two bays: [first] the Hipponiensian very close to the town called Hippo Dirutus or by Greeks Diarrhytum because of its irrigation channels. Very close by, but further from the seashore, Theudalis a tax-exempt town.

24. Then Apollo's cape, and in the second bay Utica a town of Roman citizens famous for Cato's death, Bagrada river, place Castra Cornelia, colony Carthago on the remains of Carthago Magna, and colony Maxula, towns Carpi, Misua and free Clypea on Mercury's cape, as well as free Curubis, Neapolis. Next is another distinct part of Africa where those called Libyphoenices live: Byzacium. This is the name for the region measuring 250 miles round, an exceptionally fertile one where the soil gives farmers a hundredfold return.

25. Here the free towns Lepcis, Hadrumetum, Ruspina, Thapsus. Then Thenae, Aves, Macomades, Tacape, Sabrata on the edge of Lesser Syrtis. From the Ampsaga to this point, the length of Numidia and Africa is 580, width (so far as known) 200. The part we have called Africa is divided into two provinces, Vetus and Nova; they are separated by a trench dug by the second Africanus and kings all the way to Thenae, a town 216 from Carthago.

IV Syrtes

26. A third bay is divided into a pair made treacherous by the shallows and tides of the two Syrtes. Polybius says that it is 300 from Carthago to the nearer of the two, Lesser, which has a 100-mile-deep bight and a 300[-mile] circumference. Using the stars as a guide, however, there is also a land route to it through deserts of sand and serpents. Next come tracts filled with quantities of wild beasts, and in the interior a desolate elephant-country, then vast deserts, and beyond these the Garamantes, a 12-day journey from the Augilae.

27. Above them were the Psylli people, above them lake Lycomedes surrounded by deserts. The Augilae themselves are located more or less in the center between Aethiopia which faces west and the region lying between the two Syrtes, at an equal distance between both. But along the shore it is 250 between the two Syrtes. Here the Oeensian state, Cinyps river and region, towns Neapolis, Taphra, Habrotonum, a second Lepcis called Magna. From there Greater Syrtis, 625 miles round with a 312-deep bight. The Cisippades people live nearby.

28. The inmost part of the bay used to be the coast of Lotophages (called alternatively Machroae); it extended to the Altars of the Philaeni which are made of sand. After these, and not far upcountry, a vast swamp that gets its name from Triton river which flows into it; Callimachus called it Pallantias and said that it was on the nearer side of Lesser Syrtis, but many authors locate it between the two Syrtes. The cape which encloses Greater [Syrtis] is called Borion; beyond it Cyrenaica province.

29. From Ampsaga river and as far as this border, Africa has 516 communities under Roman rule. That includes six colonies – those already mentioned,[3] along with Uthina and Thuburbi. There are 15 towns of Roman citizens, among which and inland should be noted the Absuritan, Abutucensian, Aboriensian, Canophic, Chiniavensian, Simittuensian, Thunusidiensian, Thuburnicensian, Thibidrumensian, Tibigensian,

---

[3] Cirta and Sicca (5.22), Carthago and Maxula (5.24).

Uchitan (two, Greater and Lesser), Vagensian; one town with Latin rights, the Uzalitan; one tribute-paying town at Castra Cornelia;

30. Thirty free towns: among those inland should be noted the Achollitan, Aggaritan, Avittensian, Abziritan, Canopitan, Melizitan, Materensian, Salaphitan, Thusdritan, Thisicensian, Thunisensian, Theodensian, Tagesensian, Sigensian, Ulusubburitan, another Vagensian, <...>, Zamensian. Of the remaining number, most could fairly be termed not so much states as peoples, such as Nattabudes, Capsitani, Musulami, Sabarbares, Massili, Nicives, Vamacures, Cinithi, Musuni, Marchubi, and all Gaetulia as far as Nigris river which separates Africa from Aethiopia.

V Cyrenaica

31. Cyrenaica, also called Pentapolitan region, is famous for the oracle of Hammon (situated 400 miles from Cyrene), the Sun's spring and especially for its five cities: Berenice, Arsinoe, Ptolemais, Apollonia and Cyrene itself. Berenice, which lies on the furthest tip of Syrtis, was once named after the Hesperides mentioned above [5.3], an indication of how Greek tales can shift. Not far in front of the town, Lethon river and the sacred grove where the Gardens are reputed to be. It is 375 from Lepcis.

32. From [Berenice] to Arsinoe called Teuchira 43, and then 22 to Ptolemais whose ancient name was Barce. A further 40 on, cape Phycus projects out into the Cretic sea 350 miles away from cape Taenarum in Laconica, but only 125 from Creta. After that Cyrene 11 miles from the sea; 24 from Phycus to Apollonia; 90 to Cherronesus; 216 from there to Catabathmus.

33. Marmaridae live spread out almost from the region of Paraetonium across to Greater Syrtis. After them Acrauceles, and already on the Syrtis coast Nasamones who in the past Greeks called Mesammones because of their location situated amid sands. Cyrenaic territory extending 15 miles from the seashore is considered good for the growth of trees, and for the same distance further inland good for growing cereals, but beyond that over an area 30 wide and 250 long good only for the silphium plant.

34. After Nasamones live Asbytae and Macae. Beyond them, Garamantes an 11-day journey west from Greater Syrtis. These people in turn are surrounded by sands, yet have no difficulty finding water from wells dug to a depth of about two cubits, because this is where there are pools of the overflow from Mauretania. They construct their houses with salt mined from their mountains as if with stone. From [the Garamantes] it is a seven-day journey south-west to the Trogodytae; the only trade with them is for a gemstone imported from Aethiopia, which we call carbuncle.

35. Phazania interrupts the African desert country already mentioned beyond Lesser Syrtis, and faces it. Here we have subjugated the Phazanian people and the cities Alele and Cilliba, as well as Cidamus from the region of Sabrata. From these a mountain stretches a long way from east to west; it is called Ater [Black] by our writers because of its natural condition which appears scorched or burned by the sun's reflection.

36. Beyond it, desert, then Thelge a town of Garamantes, as well as Dedris where there is a spring that flows with boiling water from midday until midnight, and with freezing water for the same length of time until midday [the next day]. Then Garama, the very famous capital of the Garamantes, who have been completely subdued by the Roman military, an achievement that gained a triumph for Cornelius Balbus [19 BCE], the one and only foreigner to be awarded one as well as Roman status; in fact, after being born in Gades, he and his great-uncle Balbus were granted Roman citizenship. And the remarkable thing is that our writers recorded the towns mentioned above that he captured. Moreover, names and images of all kinds of other peoples and towns except Cidamus and Garama featured in his triumphal procession, going past in this order:

37. town Tabudium, Niteris people, town Miglis Gemella, Bubeian people or town, Enipi people, town Tuben, mountain called Niger, Nitibrian <people>, town Rapsa, Visceran people, town Decri, Nathabur river, town Thapsagum, Tamiagi people, town Boin, town Pege, Dasibari river, then the following towns: Baracum, Bulba, Halasit, Galsa, Balla, Maxalla, Cizania, mount Giri preceded by a placard stating that gems come from there.

38. Until nowadays the route to the Garamantes was impassable because the wells – which those with local knowledge are aware do not need to be dug deep – were filled with sand by this people's brigands. During the recent war fought with the Oeenses early in emperor Vespasian's reign, a shortcut taking four days was discovered; this route is called *Praeter Caput Saxi* [By Rock's Head].

The last place in Cyrenaica is called Catabathmus, a town and valley with a sudden drop. From Lesser Syrtis to this boundary the length of Cyrenaic Africa is 1,060 and its width, so far as known, 910.

VI

39. The region that follows it and borders Egypt is called Mareotis Libya. Marmarides, Adirmachidae and then Mareotae occupy it. The distance from Catabathmus to Paraetonium 86. Inland in this region is Apis, a place renowned in Egyptian religion. From there to Paraetonium 62½, then 200 to Alexandria. The width is 169. Eratosthenes reports that it is a journey of 525 by land from Cyrene to Alexandria.

40. Agrippa made the whole of Africa from the Atlantic sea, including Lower Egypt, 3,080 long. Polybius and Eratosthenes, who are considered very painstaking, made it 1,100 from the ocean to Carthago Magna, and from there to Canopus, Nile's nearest mouth, 1,688; Isidorus made it 3,697 from Tingi to Canopus, Artemidorus 40 less than Isidorus.

VII Islands off Africa

41. The seas here encompass very few islands. Best known is Meninx, 25 long, 22 wide, called Lotophagitis by Eratosthenes. It has two towns, Meninx on the African side and Phoar on the other; it is located 1½ miles from the cape on the righthand side of Lesser Syrtis. One hundred miles from here opposite the left-hand [cape] Cercina with its free city of the same name; 25 long and half as wide at most, but no more than five at its narrowest. The very small [island of] Cercinitis faces Carthago and is linked to it by a bridge.

42. Roughly 50 miles from these [islands] Lepadusa, six long; then Gaulos, and Galata where the soil destroys scorpions, creatures that are the scourge of Africa. It is said that they die on Clupea too, opposite which is Cossora with a town. Two Aegimoeroe opposite Carthago's bay. The Arae ['altars'], really rocks more than islands, mainly between Sicilia and Sardinia. There are authors who claim that even they were once inhabited, but they have since sunk.

VIII Africa's hinterland

43. The expanse of Africa's interior facing south and above the Gaetuli, after an intervening desert, is inhabited first of all by Libyan Egyptians, then by Leucoe-Aethiopians. Above them Aethiopian peoples, Nigritae named after the [Nigris] river already mentioned [5.30], Gymnetes Pharusii and then, at the ocean's edge, Perorsi mentioned above [5.10 and 16] at the border of Mauretania. To the east of all these, vast deserts all the way to the Garamantes and Augilae and Trogodytae. The most reliable reckoning is that of those who place two Aethiopiae above the deserts of Africa; Homer [*Odyssey* 1.23–24] above all says that the Aethiopians are divided in two, facing east and west.

44. Nigris river is identical in character to the Nile. It produces reed and papyrus and the same animals, and it rises at the same time. Its source is between the Aethiopian Tarraelii and Oechalicae whose town is Magium. Some have situated the Atlantes in the middle of the desert, and next to them half-wild Aegipanes and Blemmyae as well as Gamphasantes and Satyrs and Himantopodes.

45. If we are to believe it, the standards of the Atlantes as humans have slipped. They have no names for one another; as they contemplate the sun's rising and setting, they utter dreadful curses in view of how fatal it is to themselves and their fields; and unlike the rest of mankind they have no visions in their sleep. Trogodytae excavate caves for their homes, eat snake-flesh, and screech rather than speak, so they lack the ability to engage in conversation. Garamantes do not practice marriage and so live promiscuously with women. Augilae only worship the dead. Gamphasantes, who go naked and have no experience of war, do not associate with any other people.

46. Blemmyae are said to have no heads; rather, their mouths and eyes are attached to their chests. Except for their appearance, Satyrs have no human characteristics. The appearance of Aegipanes is just as it is commonly painted. Himantopodes have clubfeet and naturally creep along with them. Pharusii, once Persians, were Hercules' companions when he travelled to the Hesperides.

There is nothing more worth mentioning about Africa.

IX Egypt: Chora [district of Lower Egypt], Thebaid, Nile

47. Asia is joined to [Africa]. Timosthenes says it extends 2,638 miles from the [Nile's] Canopic mouth to the mouth of Pontus [Black Sea]. According to Eratosthenes, it is 1,545 from the mouth of Pontus to the mouth of lake Maeotis, while Artemidorus and Isidorus say that the whole distance to Tanais river, including Egypt, is 5,013¾. It has many seas named after those who live around them, so they will be described together.

48. The inhabited region next to Africa is Egypt: its interior extends southward to where Aethiopians range behind it. The course of the Nile, divided into left and right branches, defines its lower part, with the Canopic mouth separating it from Africa and the Pelusiac from Asia in a 170-mile span. This is why some have classified Egypt as an island, because the Nile divides itself in such a way as to form a triangle-shaped landmass; as a result, many have called Egypt by the name of the Greek letter Delta. The distance from where the single river course first splits into branches is 146 to the Canopic mouth, and 156 to the Pelusiac.

49. The uppermost part which borders Aethiopia is called Thebais. It is divided into town prefectures called [in Greek] *nomes*: the Ombite, Apollonopolite, Hermonthite, Thinite, Phaturite, Coptite, Tentyrite, Diospolite, Antaeopolite, Aphroditopolite, Lycopolite. The region near Pelusium has the Pharbaethite, Bubastite, Sethroite, Tanite nomes. Remaining [nomes]: the Arabic, Hammoniac (extending to the oracle of Jupiter Hammon), Oxyrynchite, Leontopolite, Athribite, Cynopolite, Hermopolite, Xoite, Mendesian, Sebennyte, Cabasite, Latopolite, Heliopolite, Prosopite, Panopolite, Busirite, Onuphite, Saite, Ptenethus, Ptemphus, Naucratite, Metelite, Gynaecopolite, Menelaite, the region of Alexandria and Libyan Mareotis.

50. The Heracleopolite [nome] is a 50-mile-long island of Nile, on which there is also a town named after Hercules. There are two Arsinoite; they and the Memphite reach all the way to the furthest point of the Delta. Bordering it on the African side are two Oasite [nomes]. Certain [authors] change some of these names and substitute other nomes, such as the Heroopolite and Crocodilopolite. Between the Arsinoite and Memphite there used to be a lake, 250 [miles] round or, according to Mucianus, 450 round and 50 feet deep; it was manmade and called

Moeris after the king who had created it. From there, it is 62 miles to Memphis, once capital of the kings of Egypt; from there, a 12-day journey to the oracle of Hammon, and 15 to the Nile-divide (what we have called the Delta [5.48]).

X

51. After rising from obscure sources, Nile flows through torrid desert for an immense distance and has only been explored in peacetime circumstances rather than during the wars which have led to the discovery of all other lands. It originates, so far as king Juba was able to determine, in a mountain of lower Mauretania not far from ocean where it immediately forms a stagnant lake called Nilidis. The following types of fish are found there: alabeta, coracinus, silurus. Also as proof, you can see today the crocodile which [Juba] dedicated in the temple of Isis at Caesarea. In addition, it has been observed that Nile rises whenever snow and rain have saturated Mauretania.

52. After pouring out of this lake, the river flows impatiently through sandy, barren terrain and then conceals itself for a journey of several days, before eventually bursting out into another, larger lake in the territory of the Masaesyli, a people of Mauretania Caesariensis; as if it had taken a survey of society here, it makes sure to maintain the same species. After being swallowed up in the desert sands for a second time, it is buried for another 20 days as far as the nearest Aethiopians, where again it senses mankind and bursts out, most probably at the spring called Nigris.

53. From there it separates Africa from Aethiopia, and even though it does not immediately [have any] communities, it nonetheless teems with wild creatures and beasts as well as productive forests. Where it divides the Aethiopians in two it is called Astapus, which in those peoples' language means 'water flowing from darkness'. It sheds countless islands, some of them of such vast dimensions that, despite a swift flow, it takes at least five days to pass them. The channel circling to the left of Meroe, most famous of the islands, is called Astabores, that is 'branch of water coming out of darkness', while the one to the right is called Astosapes which means 'flank'. It is not called Nile before all the river's channels reunite in a single course. Even then for several miles it is called by its former name Giris, and Homer [*Odyssey* 4.477] calls the entire course Aegyptus, while others call it Triton.

54. Now and then the river is thrust against islands and finally, roused forward by these same obstructions, it is surrounded by mountains; nowhere is it more rapid. Rushing waters propel it to a place in Aethiopia called Catadupi; here, at the last cataract, because of the immense noise it seems not to flow with a great crash between the oncoming rocks, but to tumble. Thereafter [the river] is gentle, its waters broken and its force tamed; it is also rather exhausted by the distance it has traveled, even though it does disgorge from multiple mouths into the Egyptian sea. Yet still, for a definite period, with its high-water level extended the length of Egypt, it floods the land productively.

55. Various causes for this rising of the river have been proposed, but the most plausible ones are either counter-pressure from the Etesiae [winds] which blow from the opposite direction at that time of year, causing the sea to be beaten into the river-mouths, or summer rainfall in Aethiopia because the same Etesiae drive clouds there from the rest of the world. The mathematician Timaeus advanced an arcane explanation: its source is called Phiala, and [the river] itself is sunk in subterranean channels and gives off vapor from the steaming rocks that conceal it. The sun, however, coming close at this time of year, draws it out by force of heat, causing it to rise and overflow, and then hide itself to avoid being swallowed up.

56. He says that this happens from the rising of Canis when the sun enters into Leo,[4] at which point the star [Canis] stands directly above the source with the result that shadows disappear altogether in this region. The general view, by contrast, is that the river's flow is greater when the sun withdraws to the north, which happens when it is in Cancer and Leo,[5] and so the river evaporates less then. After the sun returns back to Capricorn[6] and the south pole, there is more evaporation and in consequence the river flows more moderately. But anyone convinced by Timaeus' idea that it is possible [for the river to be drawn out by the sun] should consider the point that the absence of shadows in that region at that time of year is in fact a year-round phenomenon.

57. Nile begins to rise at the new moon after the solstice;[7] it has a cautious, moderate flow while the sun passes through Cancer, and is at

---

[4] Late July to late August.
[5] Late June to late August.
[6] Late December to late January.
[7] Late June.

full surge when the sun passes through Leo; when the sun is in Virgo,[8] the river-flow diminishes just as it had risen. According to Herodotus [2.19.2], it has fully withdrawn within its banks by the hundredth day when the sun is in Libra.[9] The weight of opinion is that kings or prefects have no right to sail on the river when it is rising. Wells marked with scales record the extent of its rises.

58. The ideal rise is 16 cubits. When less than that, the water does not irrigate everything; when more, it takes too long for the water to drain away. In that event, with the soil waterlogged, the sowing season is cut short; the opposite case, with the soil still parched, offers no opportunity to sow. The province reflects upon both outcomes. It senses famine at a rise of 12 cubits, and even at 13 cubits there is a feeling of hunger, but 14 cubits prompt smiles, 15 confidence, 16 ecstasy. To date, the highest rise has been 18 cubits during the principate of Claudius [41–54 CE], while the lowest was five during the Pharsalus campaign, with the river, like a prodigy, recoiling at Magnus' death.[10] Once the water has leveled off, the sluices are opened, and it can pour in. Sowing occurs as each piece of land is freed. This river is unique in that it does not give off any breeze.

59. The river enters Egyptian territory on the border with Aethiopia at Syene: this is the name of the peninsula one mile in circumference where there is a fort on the Arabian side, and opposite it four islands and Philae, 600 miles from where Nile divides (hence, as we said [5.48], it is called Delta). Artemidorus stated this distance and in the same work said that there were 250 towns, whereas Juba put the distance at 400 miles, while Aristocreon said that it is 750 from Elephantine to the sea. Elephantine is an inhabited island, located four miles up from the last cataract and 16 downstream from Syene, terminus of Egyptian navigation, 585 miles from Alexandria; the authors just cited have been this much mistaken. This is where Aethiopic ships meet; their vessels are collapsible and can be conveyed by portage whenever they reach cataracts.

---

[8] Late August to late September.
[9] Late September to late October.
[10] After his defeat by Julius Caesar at Pharsalus (Greece) in 48 BCE, Pompey (Magnus) fled to Egypt but was killed on arrival there.

XI

60. The glory of its antiquity aside, Egypt boasted 20,000 inhabited cities during the reign of Amasis [569–525 BCE]. Nowadays, too, there is a great density of them even if they are undistinguished. Apollo's one, however, is much visited; then Leucothea, Diospolis Magna also called Thebes and remarkable for its famous 100 gates, Coptus a trade-center for Indian and Arabian goods very close to Nile, then Venus' town and in turn those of Jupiter and Tentyris and, below it, Abydus famous for Memnon's palace and Osiris' temple, 7½ miles from the river on the Libyan side.

61. Then Ptolemais and Panopolis and another [town] of Venus, and Lycon on the Libyan side where mountains border Thebais. After these, towns of Mercury, of the Alabastri, of the Canes, and of Hercules mentioned above [5.50]. Then that of Arsinoe and the already mentioned Memphis [5.50]. Between it and the Arsinoite nome on the Libyan side, towers called pyramids, and the labyrinth constructed on lake Moeris without the use of wood, and Crocodiles' town. In addition the one inland place of real note: Sol's town bordering on Arabia.

62. However, Alexandria, on the shore of the Egyptian sea, merits praise: it was founded by Alexander the Great on the African side [of the river] 12 miles from the Canopic mouth, near lake Mareotis. Previously the place was named Rhacotes. The architect Dinochares, a man with a remarkable range of talents, laid it out. The occupied area is 15 miles in extent and the shape of a Macedonian cloak, with indentations along its circuit and projecting corners to the left and right sides; nonetheless, even then a fifth of the site was given over to the royal palace.

63. Lake Mareotis off the southern part of the city conveys commercial traffic from the interior via a canal from the Canopic mouth; within it there are many islands too. According to Claudius Caesar, it is 30 across, 250 in circumference. Others say that it extends to 40 *schoeni* in length – where one *schoenus* equals 30 stades, this makes a length of 150 miles – and has an identical width.

64. On the lower Nile there are many towns of note, especially those which have bestowed their names on mouths – though not on all of them, because 12 are identified and there are four more called 'false' mouths. But the best-known seven are the Canopic near Alexandria, then the Bolbitine, Sebennytic, Phatmitic, Mendesic, Tanitic and finally Pelusiac. In addition, Butos, Pharbaethos, Leontopolis, Athribis, Isis' town, Busiris, Cynopolis, Aphrodites, Sais, Naucratis – hence some use the name Naucratitic for the mouth that others call Heracleotic, preferring that mouth over the Canopic, which is very close by.

XII Arabia near the Egyptian sea

65. Beyond the Pelusiac [mouth] is Arabia, which extends to Red sea and to the land called Beata, famous for its fragrance and wealth. Arabians here are called Catabanes and Esbonitae and Scenitae. Except where it reaches the Syrian borderlands, it is a barren land with mount Casius its only notable feature. The neighboring Arabians are Canchlei to the east and Cedrei to the south, both of whom are then next to Nabataeans. Red sea has two bays bending towards Egypt, one called Heroopoliticus, the other Aelaniticus. The distance between the two towns Aelana and Gaza (on our sea [Mediterranean]) is 150. Agrippa says that Arsinoe, a town on Red sea, is 125 miles across the desert from Pelusium. Such a short distance there separates such very different environments.

XIII Idumaea, Syria, Palaestine, Samaria

66. Next along the shore lies Syria, once the greatest of lands, and distinguished by very many names. For the part bordering the Arabians was called Palaestine and Judaea and Coele, then Phoenice and inland Damascena. Even further south is Babylonia, also called Mesopotamia between the Euphrates and Tigris. On the far side of Taurus mountains is Sophene, with Commagene close to it and Armenia Adiabene (once called Assyria) further away; and Antiochia where it borders Cilicia.

67. [Syria's] length between Cilicia and Arabia is 470 miles; its width from Seleucia Pieria to town Zeugma on Euphrates 175. Those who make a more precise division maintain that Phoenice is surrounded by Syria, and that [the sequence is] the Syrian seashore (part of which is Idumaea and Judaea), then Phoenice, then Syria. The entire sea off this coast is called Phoenician sea. To their great glory, the Phoenician people invented the alphabet and the arts of astronomy, navigation and warfare.

XIV

68. After Pelusium, Chabrias' camp, mount Casius, shrine of Jupiter Casius, tomb of Pompey the Great, Ostracine. The boundary of Arabia is 65 miles from Pelusium. Next Idumaea begins, and Palaestine at the outlet of lake Sirbonis, which some have said is 150 round. Herodotus [2.6] says that it extended to mount Casius, but nowadays it is merely a swamp. Towns: Rhinocolura and inland Rhaphea, Gaza and inland Anthedon, mount Argaris; the coastal region is Samaria; free town Ascalon, Azotos, two Iamneae, one of them inland.

69. Phoenician Iope, which is said to predate the flood, sits upon a hill in front of which lies a rock where marks left by Andromeda's chains can be seen. The mythical Ceto is worshiped there. Then Apollonia, [and] Strato's Tower – also called Caesarea as founded by King Herod, but nowadays the colony Prima Flavia established by emperor Vespasian – on the frontier of Palaestine, 189 miles from the border of Arabia. Then Phoenice. Inland towns of Samaria: Neapolis previously called Mamortha, Sebaste on a mountain, and Gamala higher up.

XV Judaea

70. Judaea extends far and wide above Idumaea and Samaria. Its part next to Syria is called Galilaea, while nearest to Arabia and Egypt is Peraea covered with rough mountains and separated from the other parts of Judaea by Jordanes river. The rest of Judaea is divided into ten *toparchies*, which we will give in order: those of Hiericous [Jericho] filled with palms and well-watered by springs, Emmaus, Lydda, the Iopic, Acrabaten, Gophanitic, Thamnitic, Betholeptephenian, Orine where Hierosolyma [Jerusalem] was by far the most famous city of the east and not just of Judaea, and Herodium with its famous town of the same name.

71. Jordanes river rises from Paneas spring which, as will be explained [5.74], provides Caesarea with a further name. A pleasant river winding in whichever way the terrain allows, and offering itself to those living there while seeming reluctant to flow towards lake Asphaltites [Dead Sea] with its dreadful character. Eventually [the river] is absorbed, and its acclaimed waters are lost as they merge with the [lake's] toxic ones. So, at the first opportunity the valleys offer, it pours itself out into a lake which is usually called Genesara. Sixteen miles long, six wide, it has pleasant towns around it, Julias and Hippo to the east, Tarichea to the south (which is also the name that some call the lake), and Tiberias with its bracing hot springs to the west.

72. [Lake] Asphaltites produces nothing except bitumen, from which it gets its name. It rejects the body of any living creature; even bulls and camels float, and by all accounts there is nothing that sinks in it. It is more than 100 miles long, and in width between as much as 75, as little as six. Nomadic Arabia faces it from the east, Machaerus (once second only to Hierosolyma as a citadel of Judaea) from the south. On the same side is Callirrhoe ['beautifully flowing'], a hot spring with healing properties; its very name declares the glory of its waters.

73. To the west, Esseni live far enough away from the shore to avoid harm. They are a solitary people, and by comparison with all others in the world they are amazing because they lack women (having completely renounced sex) and money, and they socialize with palm trees. Every day without fail this collection of vagabonds gains a fresh infusion from great throngs who are weary of living and attracted to their way of life by the fickleness of fortune. So, incredibly, from time immemorial a people into which no-one is born keeps on living. This is how productive others' dissatisfaction with life proves for them! Below them used to be a town Engada, second only to Hierosolyma in its fertility and in its groves of palm trees, but nowadays it too is a heap of ashes. Then comes Masada, a fortress on a cliff, not far from lake Asphaltites. Judaea extends no further.

XVI [Decapolitan region]

74. Next to it on the Syrian side is the Decapolitan region, which takes its name from the number of its towns, although not every authority lists the same [ten]. The first is nevertheless Damascus, with its lush water-meadows that drain Chrysorrhoas river, [then] Philadelphia, Rhaphana (all these extend into Arabia), Scythopolis previously called Nysa (after Father Liber's nurse, whom he buried there after settling Scythians), Gadara past which the Hieromix flows, and Hippo (already mentioned [5.71]), Dion, water-rich Pella, Garasa, Canatha. Between and around these cities are tetrarchies, each one like its own kingdom.

They are also joined into kingdoms: Trachonitis, Paneas (the location of Caesarea with its spring already mentioned [5.69, 71]), Abila, Arca, Ampeloessa, Gabe.

XVII Phoenicians

75. From here we must return to the coast and Phoenice. There used to be Crocodiles' town, [now] there is a river. The cities Dorum and Sycaminum are a memory. Cape Carmelum and on a mountain a town with the same name, once called Acbatana. Next Getta, Geba, Pacida or Belus river which along its narrow banks produces the kind of rich sand used for glass-making. This river flows out of Cendebia marsh at the base of Carmelum. Next, Ptolemais a colony of Claudius Caesar once called Acce, town Ecdippa, cape Album.

76. Tyrus was once an island separated [from the mainland] by a very deep strait 700 paces wide, but nowadays, thanks to Alexander's siegeworks, it is attached to it. It was once famous for giving birth to the cities Lepcis, Utica, and in particular Carthago – Rome's rival for empire and world-domination – and even Gades which was founded beyond the [known] world. Nowadays it is well-known only for its shellfish and purple dye. Its perimeter, including Palaetyrus, is 19 [miles]; the town itself occupies 22 stades. Then towns Sarepta and Ornithon, and Sidon producer of glass and mother-city of Thebae in Boeotia.

77. Behind [Sidon] rises mount Libanus, extending 1,500 stades to Zimyra, which is also called Coele Syria. Facing it and of equal length, with a valley in between, stretches mount Antilibanus; [the two ranges] were once linked by a wall. In the interior beyond the latter, Decapolitan region and its associated tetrarchies already mentioned [5.74], as well as the entire expanse of Palaestine.

78. Along the coast and below Libanus: Magoras river, colony Berytus called Felix Julia, Lion's town, Lycus river, Palaebyblos, Adonis river, towns Byblos, Botrys, Gigarta, Trieris, Calamos, Tripolis (occupied by Tyrians and Sidonians and Aradians), Orthosia, Eleutheros river, towns Zimyra, Marathos and opposite them Arados (the town and island occupying seven stades, situated 200 paces off the mainland), the region in which the mountains already mentioned end, and after some intervening plains mount Bargylus.

XVIII Syria Coele, Syria Antiochia

79. Here Phoenice ends and Syria begins again. Towns: Carne, Balanea, Paltos, Gabala, the cape with free Laodicea on it, Diospolis, Heraclea, Charadrus, Posidium. Then the cape of Syria's Antiochia. Inland, free Antiochia itself, also called 'near Daphne' and split by Orontes river. Free Seleucia, called Pieria, on a cape.

80. Above it a mountain with the same name as another, Casius. Its great height makes it possible to observe the sun rising through the darkness during the fourth watch [at dawn], so that by simply turning around both day and night are visible at the same time. The winding route to the summit is 19 miles; measured vertically, it stands four high. On the coast, Orontes river which rises between Libanus and Antilibanus near Heliopolis, town Rhosos, and behind it the so-called 'Syrian Gates' in a space between the Rhosian and Taurus mountains. On the coast, town Myriandrus [and] mount Amanus with the town Bomitae on it. This mountain separates Cilicia from Syria.

XIX

81. Now the interior should be described. Coele has Apamea separated from the tetrarchy of the Nazerini by Marsyas river; Bambyce alternatively called Hierapolis, but by Syrians Mabog (the monster Atargatis, called Derceto by Greeks, is worshiped here); Chalcis also called 'on Belus', hence the name Chalcidene for the region, the most fertile in Syria; and after that Cyrrestic Cyrrus, the Azetae, Gindareni, Gabuleni, two tetrarchies called Tigranucomatae, the Hemeseni, Hylatae an Ituraean people consequently called Baethaemi,

82. Mariamnitani, tetrarchy called Mammisea, Paradisus, Pagrae, the Penelenitae, two Seleuciae (besides the one already mentioned [5.67, 79]) one called 'on Euphrates', the other 'on Belus', Cardytenses. Leaving aside what will be described along with Euphrates, Syria has the Arethusii, Beroeenses, Epiphanenses on Orontes, Laodiceans called 'on Libanus', Lysiadi, Larisaei, as well as 17 tetrarchies divided into kingdoms with barbarous names.

XX Euphrates

83. This is the most appropriate point to mention the Euphrates. According to recent eyewitnesses, it rises in Caranitis, a prefecture of Greater Armenia; Domitius Corbulo says on mount Aga, while Licinius Mucianus says at the foot of a mountain called Capotes, 12 miles above Zimara. Initially, its name is Pyxurates. It flows first through Derzene, then Anaetica, and divides the regions of Armenia from Cappadocia.

84. Dascusa is 75 miles from Zimara. From there it is navigable for 50 to Sartona, then 24 to Melitene in Cappadocia, and ten to Elegea in Armenia; by this point it has gained the tributaries Lycus, Arsanias, Arsanus. At Elegea the Taurus range, even though 12 miles wide, stands in its way but does not stop it. [The river] is called Omma where it plunges into the range, and then Euphrates where it has burst through and becomes rocky and turbid.

85. From there, three *schoeni* wide, it divides the Arabian region called Orroeon on the left from Commagene on the right, although even where it breaks through the Taurus range it tolerates a bridge. At Claudiopolis in Cappadocia it shifts its course westward. Here for the first time in the struggle Taurus diverts [the river], and despite being overcome and cut through, it overpowers the river by other means, blocks it, and forces it to turn south. So this primordial struggle ends in a draw, with [the river] going where it wants to go, but [the range] blocking its chosen route. After cataracts it is navigable again; 40 miles from there lies Samosata, capital of Commagene.

XXI

86. Arabia, mentioned above, has the towns Edessa once called Antiochia, Callirrhoe named after its spring, Carrhae famous for Crassus' disaster [53 BCE]. Adjoining it is the prefecture of Mesopotamia, which originated with the Assyrians. Here towns Anthemusia and Nicephorium. Next Arabians called Praetavi with their capital at Singara. Below Samosata, on the Syrian side, Marsyas river flows. Commagene ends at Cingilla, and the state of the Imenei begins. Towns on the Euphrates river-bank are Epiphania and Antiochia called 'on Euphrates', as well as Zeugma 72 miles from Samosata and famous as a Euphrates crossing-point. Apamea, on the opposite bank, is linked to [Zeugma] by a bridge built by Seleucus, who founded both towns.

87. The people bordering Mesopotamia are called Rhoali. However, the towns Europus and Thapsacus are in Syria <...> ; once called <...>, but nowadays Amphipolis, Arabian Scenitae. So [Euphrates] flows all the way to the place Sura where, veering east, it leaves behind Syria's Palmyrene deserts which extend as far as the city Petra and region called Arabia Felix.

88. The city Palmyra is famous for its location, rich soil and delightful springs. Its fields are surrounded on every side by a vast ring of sand as if Nature were isolating it from the world. Having its own destiny, situated as it is between the two foremost powers, Rome and Parthia, it is always the first concern of both whenever a rupture develops. It is 337 miles from Parthian Seleucia which is called 'on Tigris', 203 from the nearest spot on the Syrian seashore, and 27 less from Damascus.

89. Below the Palmyrene desert is the region Thelendena, and the already mentioned Hierapolis and Beroea and Chalcis [5.81–82]. Beyond Palmyra, Hemesa also claims a part of this desert, as does Elatium which is half the distance Petra is from Damascus. Nearest to Sura is Philiscum, a Parthian town on Euphrates. From there a ten-day sail to Seleucia, and roughly the same to Babylon.

90. Five hundred and ninety-four miles from Zeugma, Euphrates divides around the village Masice; the left branch advances towards Mesopotamia through Seleucia itself, and after flowing round it pours into Tigris. The channel more to the right heads towards Babylon, once capital of Chaldaea; after passing through its center and through [a place] called Mothris, it breaks up into marshes. But, like the Nile, [Euphrates] also raises its level at fixed days with little variation, and floods Mesopotamia when the sun reaches the 20th degree of Cancer.[11] The level begins to reduce when the sun crosses into Virgo from Leo, and is completely back to normal once the sun is in the 29th degree of Virgo.[12]

XXII Cilicia and neighboring peoples

91. But we should return to the Syrian coast nearest Cilicia: Diophanes river, mount Crocodilus, Gates of mount Amanus, the Androcus, Pinarus, Lycus rivers, Issic bay, town Issus also called Alexandria, Chlorus

---

[11] The sun is in Cancer from late June through late July.
[12] The sun exits Leo in late August and Virgo in late September.

river, free town Aegaeae, Pyramus river, Gates of Cilicia, towns Mallos, Magirsos. And inland: Tarsus, Aleian plains, towns Casyponis, free Mopsos on the Pyramus, Tyros, Zephyrium, Anchiale,

92. Saros and Cydnus rivers, the latter cutting through Tarsus a free city far from the sea. Celenderitis region with a town, place Nymphaeum, Cilician Soloe nowadays Pompeiopolis, Adana, Cibyra, Pinare, Pedalie, Alae, Selinus, Arsinoe, Iotape, Dorion, and near the sea Corycos with a town and harbor and cave of the same name. Then Calycadnus river, cape Sarpedon, towns Holmoe, Myle, the cape and town of Venus closest to the island Cyprus.

93. Still on the mainland: towns Mysanda, Anemurium, Coracesium and Melas river, the old boundary of Cilicia. Inland should be mentioned Anazarbeni nowadays Caesarea Augusta, Castabala, Epiphania previously Oenoandos, Eleusa, Iconium and, beyond Calycadnus river, Seleucia also called 'Tracheotis' relocated away from the coast, where it was called Hermia. In addition inland, the Liparis, Bombos and Paradisus rivers, mount Imbarus.

XXIII Isaurica, Omanades

94. Everyone joins Pamphylia to Cilicia, having taken no account of the Isauric people. Their inland towns: Isaura, Clibanus, Lalasis; [their territory] runs down to the sea at Anemurium mentioned above [5.93]. In the same way everyone overlooked their neighbors the Omanades whose inland town is Omana. There are 44 other fortresses concealed among rugged valleys.

XXIV Pisidia

The Pisidians, once called Solymi, live on the crest of a mountain; their colony Caesarea also called Antiochia. Towns Oroanda, Sagalessos.

XXV Lycaonia

95. Lycaonia surrounds them. For jurisdiction it turns to the province of Asia where Philomelienses, Thymbriani, Leucolithi, Pelteni, Tyrienses assemble. Also assigned [to that jurisdiction] is the Lycaonian tetrarchy which borders part of Galatia and has 14 states; Iconium its best-known city. The well-known ones in Lycaonia itself are Thebasa in the Taurus [range], Hyde on the border between Galatia and Cappadocia. From [the Lycaonia] side, above Pamphylia, come Milyae, a people of Thracian origin whose town is Arycanda.

XXVI Pamphylia

96. Pamphylia was previously called Mopsopia. The Pamphylian sea adjoins the Cilician. Towns: Side, and Aspendus on a mountain, Plantanistum, Perga; cape Leucolla, mount Sardemisus; Eurymedon river runs near Aspendus, Catarractes near Lyrnessus and Olbia. The last place on its coast is Phaselis.

XXVII Mount Taurus

97. Next to [Pamphylia] are the Lycian sea and Lycian people, along with an immense bay. Mount Taurus, advancing from the eastern shore, marks it off with cape Chelidonium. It is enormous and oversees countless peoples. Where it first rises from the Indian sea, its right side faces north, while its left faces south; it also extends west, and would split Asia in half if seas did not stop it from overwhelming the land. So it withdraws northward and winds as it undertakes a tremendous journey, as if the natural order were purposely and repeatedly blocking it with one sea after another, in one direction the Phoenician and Pontic, in another the Caspian and Hyrcanian, and opposite them lake Maeotis.

98. And so under such pressure it twists between these obstacles, and triumphantly cuts a winding escape route all the way to the shared ridges of the Ripaean mountains, ones with many names and well-known for new ones wherever it advances: its first section is called Imaus, then Emodus, Paropanisus, Circius, Cambades, Pariades, Choatras, Oreges, Oroandes, Niphates, Taurus and (where it exceeds even itself) Caucasus and (where it repeatedly flings out branches as provocation for the seas) Sarpedon, Coracesius, Cragus and Taurus again.

99. Even when the range splits and makes way for people, it still asserts its unity through the term Gates, which in one place are called Armenian, in another Caspian, in another Cilician. Moreover, despite being subdued and avoiding seas as well, it fills itself in all directions with the names of a tremendous number of peoples: on the right it is called

Hyrcanius and Caspius, on the left Parihedrus, Moschic, Amazonic, Coraxic, Scythic, while in Greek the whole range is called Ceraunius.

XXVIII Lycia

100. So in Lycia the town Simena on its cape, mount Chimaera which spews fire at night, the state Hephaestium which itself has a mountain range frequently ablaze. A town Olympus used to be there, nowadays there are the mountain ones Gagae, Corydalla, Rhodiapolis. Near the sea Limyra with its river into which the Arycandus flows, mount Masicytus, Andria state, Myra, towns Aperlae and Antiphellos (once called Habesos), and Phellos in a glen; then Pyrrha, as well as Xanthus, 15 from the sea with a river of the same name; then Patara previously Pataros, and Sidyma on a mountain, cape Cragus.

101. Beyond, a bay as large as the one before; Pinara is there, as well as Telmesus which marks the boundary of Lycia. Lycia once had 70 towns, but nowadays has 36. The best-known of these (besides those already mentioned): Canas, Candyba where Eunias grove is praised, Podalia, Choma past which flows Aedesa river, Cyaneae, Cadyanda, Lisa, Melanoscopium, Tlos, Telandrus. Inland [Lycia] also includes Cabalia and its three cities: Oenoanda, Balbura, Bubon.

102. After Telmesus the Asiatic or Carpathian sea and what is properly called Asia. Agrippa divided it into two parts. One part he surrounded with Phrygia and Lycaonia to the east, Aegean sea to the west, Egyptian sea to the south, Paphlagonia to the north; this part he made 475 long, 320 wide. The borders of the other part – 575 long, 325 wide – he made Lesser Armenia to the east, Phrygia, Lycaonia, Pamphylia to the west, the Pontic province to the north, the Pamphylian sea to the south.

XXIX Caria

103. Caria is next on the coast, then Ionia and beyond it Aeolis. Caria has Doris at its center and surrounds it as far as the sea on both sides. In it: cape Pedalium, Glaucus river, its tributary Telmedius, towns Daedala, Crya (founded by refugees), Axon river, town Calynda. Indus river, rising in the Cibyratic range, gains 60 perennial rivers and well over 100 streams.

104. Free town Caunus, then Pyrnos, Cressa harbor situated 20 miles from Rhodos island, place Loryma, towns Tisanusa, Paridon, Larymna, Thymnias bay, cape Aphrodisias, town Hydas, Schoenus bay, Bubassus region. There used to be a town Acanthus alternatively named Dulopolis. On a cape is free Cnidus, Triopia, then Pegusa and one called Stadia. Thereafter Doris begins.

105. First, however, [Caria's] back side and its inland jurisdictions should be noted. One of these is called the Cibyratic with its town belonging to Phrygia. Here 25 states assemble – Laodicea being the best-known city. Established along Lycus river, Asopus and Caprus flow to its sides; it was first called Diospolis, then Rhoas. Of the rest in this *conventus*, the ones there should be no objection to naming are the Hydrelitae, Themisones, Hierapolitae. Another *conventus* is named after Synnas; here assemble the Lycaones, Appiani, Eucarpeni, Dorylaei, Midaei, Julienses and a further 15 insignificant communities.

106. A third group goes to Apamea previously called Celaenae, then Cibotos. Lying at the foot of mount Signia, it is encircled by the Marysas, Obrima, Orba rivers, tributaries of Maeander. Here the Marsyas emerges, first rising only to be submerged again soon afterward. Where [Marsyas] challenged Apollo in playing the flute is called Aulocrene ['pipe fountain']; this is the name of the valley ten miles from Apamea for those making for Phrygia. From this *conventus* it should be appropriate to name the Metropolitae, Dionysopolitae, Euphorbeni, Acmonienses, Pelteni, Sibliani, and a further nine insignificant ones.

107. On the bay of Doris: Leucopolis, Hamaxitos, Eleus, Etene; then the Carian towns Pitaium, Eutane, Halicarnasus to which Alexander the Great subjected six towns: Theangela, Side, Medmassa, Uranium, Pedasum, Telmisum. There is an inhabited area between two bays, Ceramic and Iasian. Then Myndus and where Palaemyndus was, Nariandos, Neapolis, Caryanda, Termera, free Bargylia, and town Iasus from which Iasian bay gains its name.

108. Caria shines brighter because of the famous places in its interior. The towns here, after all, include free Mylasa, Antiochia situated where the towns Symmaethos and Cranaos used to be; nowadays the Maeander and Morsynus flow round it. Meandropolis, too, used to be in this region. There is Eumenia situated on Cludrus river, Glaucus river, town

Lysias, and Otrus, Berecyntian territory, Nysa, Trallis (also called Euanthia and Seleucia and Antiochia). By it flows Eudon river, through it Thebaites;

109. some say that Pygmies once lived there. In addition, there are Thydonos, Pyrrha, Eurome, Heraclea, Amyzon, free Alabanda from which a *conventus* takes its name, Stratonicea, Labrayndus, Ceramus, Troezene Phorontis. More distant but included in the same district are the Orthosienses, Alindenses, Euhippini, Xystiani, Hydissenses, Apolloniatae, Trapezopolitae, free Aphrodisienses; besides these are Coscinus, Harpasa situated on Harpasus river which flowed by Trallicon too when it existed.

XXX <Lydia>

110. Watered by the winding bends of Maeander river, Lydia stretches above Ionia, borders Phrygia to the east, Mysia to the north, and encircles Caria with its southern part. In the past it was called Maeonia. It is most famous for Sardis, founded on the side of mount Tmolus previously called Timolus. The Pactolus and also the Chrysorrhoas flow from it, and it is the Tarnus' source. The state itself was called Hyde by the Maeonii and is famous for lake Gygaeum.

111. Nowadays the jurisdiction here is called Sardiana and, in addition to those already mentioned, there assemble here Macedonian Cadieni, Philadelphini, and those very Maeonii settled on Cogamus river at the foot of Tmolus, Tripolitani (also called Antoniopolitae) who are watered by the Maeander, Apollonihieritae, Mysotimolitae and other insignificant peoples.

XXXI Ionia

112. Ionia, beginning at Iasian bay, has a more indented coastline. In it first Basilicus bay, cape and town Posideum, oracle of the Branchidae (called nowadays of Apollo at Didyma) 20 stades from the shore and a further 180 to Miletus, Ionia's capital, called Lelegeis and Pityusa and Anactoria in the past, founder of over 90 cities on every sea, indisputably the city of Cadmus who first instituted the practice of writing prose.

113. Maeander river rises from a lake on mount Aulocrene and flows by very many towns as well as being swelled by numerous tributaries;

because of its curves and bends it could often be thought to be flowing backward. It first wanders through Apamea region, then through that of Eumenia and then the Hyrgaletic plains and finally Caria. Its calm flow spreads the richest sediment over the soil everywhere here. It gently slips into the sea ten stades from Miletus. Then mount Latmus, towns Heraclea with a mountain that has the same name in Carian, Myuus which legend relates was originally founded by Ionians who came from Athens, Naulochum, Priene on a stretch of coast called Trogilia, Gaesus river, a region sacred to all Ionians, hence its name Panionia.

114. Next Phygela founded by refugees, as its name indicates; there also used to be a town Marathesium. Above these Magnesia, distinguished by its further name 'on Maeander' and founded by Magnesia in Thessalia. It is 15 miles from Ephesus and three further from Trallis. In the past it was called Thessaloche and Mandrolytia, and its seaside location allowed it to appropriate the Derasidae islands from the sea.

115. Inland also Thyatira – near which Lycus flows – has had in the past the further names Pelopia and Euhippia. On the coast the oracle-site Ephesus, built by Amazons and given multiple names over time: Alope during the Trojan War, then Ortygia, Amorge; it was also called Zmyrna with the further name Trachia, as well as Haemonion and Ptelea. It is established on mount Pion and provided with water by the Caystrus which rises in the Cilbiani range and is swollen by many streams and by lake Pegaseum, a creation of Phyrites river. The resulting accumulation of silt builds-up land and has left the island Syrie situated in the middle of a plain. In the city, the spring Callippia and two both called Selinus flowing from different directions to surround the temple of Diana.

116. After Ephesus another oracle-site, this one belonging to the Colophonians; and inland Colophon itself, past which the Halesus flows. Then the shrine of Apollo at Claros, Lebedos and what used to be Notium town, cape Corynaeum, mount Mimas extending out 150 miles and then sloping down into the adjoining plains. This is where Alexander the Great ordered the flat land to be bisected for a distance of 7½ miles in order to connect two bays and to make an island of Erythrae together with Mimas.

117. Nearby where there were the towns Pteleon, Helos, Dorion are nowadays Aleon river, Corynaeum, Mimas' cape, Clazomenae, Parthenie and the Hippi which, when islands, were called Chytrophoria. Alexander had them attached to the mainland two stades away. Inland, Daphnus and Hermesia no longer exist, nor does Sipylum once called Tantalis, capital of Maeonia, where lake Sale is nowadays. Also extinct is Archaeopolis which replaced Sipylum, as are Colpe which later replaced [Archaeopolis], and Libade which replaced [Colpe].

118. To come back then from the interior, 12 miles further along the coast Zmyrna, founded by an Amazon and restored by Alexander; it delights in Meles river, whose source is not far away. Asia's finest mountains more or less sprawl across this region: Mastusia is behind Zmyrna and Termetis connected to the base of Olympus ends at Draco, Draco at Tmolus, Tmolus at Cadmus, and that one at Taurus.

119. From Zmyrna, Hermus river cuts through plains and gives them its name. Its source is near Dorylaeum, a state of Phrygia; it gains many tributaries, including Phryx which separates [Phrygia] (to whose people it gives a name) from Caria; Hyllus and Cryon which themselves are swelled by rivers of Phrygia, Mysia, Lydia. At the [Hermus] estuary there used to be a town Temnos, but nowadays at the end of the bay are Myrmeces rocks, town Leuca (on a cape that used to be an island), and Phocaea at the border of Ionia.

120. Much of Aeolis too (described below [5.121–123]) is part of the Zmyrnaean *conventus*, as are Macedonians called Hyrcani and Magnetes from Sipylus. But assembling at Ephesus, Asia's other shining light, are the more remote Caesarienses, Metropolitae, Lower and Upper Cilbiani, Mysomacedones, Mastaurenses, Briullitae, Hypaepeni, Dioshieritae.

XXXII Aeolis

121. Next is Aeolis (once called Mysia), and Troas which borders Hellespont. Here, after Phocaea, Ascanius harbor. Then there used to be Larisa and now Cyme, Myrina which calls itself Sebastopolis, and inland, Aegaeae, Attaleia, Posidea, Neon Tichos, Temnos. But on the coast, Titanus river and the state named after it. There also used to be Grynia which is nowadays only a harbor but was once connected to an island; also town Elaea and Caicus river flowing from Mysia, town Pitane, Canaitis river.

122. Extinct are Canae, Lysimachea, Atarnea, Carene, Cisthene, Cilla, Cocylium, Thebe, Astyre, Chrysa, Palaescepsis, Gergitha, Neandros.

Nowadays there is the state Perperene, Heracleotes region, town Coryphas, Grylios and Ollius rivers, Aphrodisias region (formerly Politice Orgas), Scepsis region, Euenus river on the banks of which Lyrnesos and Miletos are extinct. In this region mount Ida, and on the coast Adramytteos (once called Pedasus and from which the bay and *conventus* are named), the Astron, Cormalos, Crianos, Alabastros, Hieros rivers (the last flowing from Ida); inland, mount Gargara and a town with the same name.

123. Back on the shore, Antandros (previously called Edonis), then Cimmeris, Assos also called Apollonia; there also used to be a town Palamedium; cape Lectum separates Aeolis and Troas. There also used to be a state Polymedia, Chrysa and another Larisa. The temple of Zminthium still stands, but inland Colone is extinct. Adramytteos is the business center for the Apolloniatae from Rhyndacus river, Eresi, Miletopolitae, Poemaneni, Macedonian Asculacae, Polichnaei, Pionitae, Cilician Mandacandeni, Mysians called both Abretteni and Hellespontii, and for other insignificant peoples.

XXXIII Troas and neighboring peoples

124. The first place in Troas is Hamaxitos, then Cebrenia, and Troas itself once called Antigonia but nowadays Alexandria a Roman colony, town Nee, navigable Scamander river, and once on a cape Sigeum town; then Achaeans' harbor into which Xanthus flows after merging with Simoeis, and Palaescamander which first forms a marsh. There is no trace of the other rivers made famous by Homer – Rhesus, Heptaporus, Caresus, Rhodius. Granicus flows into Propontis from a different region. However, as in the past, the small state Scamandria still exists, as does tax-exempt Ilium [Troy] 2½ miles from its harbor, the focus of all those storied events.

125. Beyond the bay are the Rhoetean shores occupied by towns Rhoeteum and Dardanium and Arisbe. There also used to be a town Achilleon established near Achilles' tomb first by Mytilenaeans and later by Athenians, where [Achilles'] fleet had been stationed at Sigeum. There also used to be Aeantion founded by Rhodians at the opposite end of the bay, where Ajax was buried 30 stades away from Sigeum, and just where his fleet had been stationed. Above Aeolis and part of Troas is the inland [place] called Teuthrania, a Mysian possession in antiquity. Caicus river (already mentioned [5.121]) rises here. A populous community in its own right was here, especially when the whole region was called Mysia.

126. In it, Pioniae, Andera, Idale, Stabulum, Conisium, Teium, Balce, Tiare, Teuthrania, Sarnaca, Haliserne, Lycide, Parthenium, Gambre, Oxyopum, Lygdamum, Apollonia and Pergamum, Asia's most famous city by far. The Selinus runs through it and the Cetius around it, flowing down from mount Pindasus. Not far away on the coast, as already noted [5.121], is Elaea. Pergamene is the name for this region's jurisdiction. Assembling there are Thyatireni, Mossyni, Mygdones, Germeni, Hierocometae, Perpereni, Tiareni, Hierolophenses, Hermocapelitae, Attalenses, Pandenses, Apollonidenses, and other unremarkable states.

127. From Rhoeteum it is 70 stades to the small town Dardanium, and a further 18 to cape Trapeza where Hellespont's swift current begins.

Eratosthenes says that the following peoples of Asia are extinct: Solymi, Leleges, Bebryces, Colycantii, Tripsedi. Isidorus mentions Arimi and Capreatae at the place where Apamea was founded by king Seleucus between Cilicia, Cappadocia, Cataonia, Armenia; it was initially called Damea ['subdued'] because he had overcome the most ferocious peoples.

XXXIV 212 islands off Asia, including:

128. Of the islands off the coast of Asia, the first is at the Nile's Canopic mouth, and is said to be named after Canopus, Menelaus' helmsman. Second, Pharos, a colony of the dictator Caesar, joined to Alexandria by a bridge. At one time it was a day's journey from Egypt, but nowadays it guides ships' courses at night with a lighthouse. Treacherous shoals mean that Alexandria is approachable from the sea by only three channels: Steganus, Posideum, Taurus. Then in Phoenician sea lies Paria – the whole of it a town where according to the story Andromeda was thrown to a monster; also Arados already mentioned [5.78]. Even though the sea between it and the mainland is 50 cubits deep, Mucianus informs us that fresh water is drawn from a spring right at the bottom through a leather pipe.

XXXV Cyprus

129. The islands in the Pamphylian sea are insignificant, but the Cilician sea has one of the five biggest, Cyprus lying opposite Cilicia to the west and Syria to the east; nine kingdoms were once established there.

Timosthenes reported that its perimeter measures 427½ miles, but Isidorus' figure is 375. Artemidorus gives the distance between the two capes Clidae and Acamas (lying to the west) as 162½; Timosthenes says 200. Philonides says it was called Acamantis in the past, Xenagoras says Cerastis and Aspelia and Amathusia and Macaria, Astynomus says Cryptos and Colinias.

130. Fifteen towns there: Nea Paphos, Palaepaphos, Curias, Citium, Corinaeum, Salamis, Amathus, Lapethos, Soloe, Tamasos, Epidaurum, Chytri, Arsinoe, Carpasium, Golgoe. There also used to be Cinyria, Marium, Idalium. It is 50 from Anemurium in Cilicia. The expanse of sea here is called Aulon Cilicius. In it also lie the island Elaeusa and four Clides off the cape opposite Syria. Back off the one in the other direction, Stiria [island], as well as Hiera Cepia opposite Nea Paphos, Salaminiae opposite Salamis.

131. In the Lycian sea three barren ones off Cyprus: Idyris, Telendos, Attelebussa; also Dionysia previously called Charaeta. Then opposite cape Taurus the same number of Chelidoniae, a deadly threat to mariners. After these Leucolla with a town, the Pactyae: Lasia, Nymphais, Macris, Megista whose state is extinct. Then many insignificant islands, but – opposite Chimaera – Dolichiste, Choerogylion, Crambusa, Rhoge, eight Xenagora, two Daedaleon, three Cryea, Strongyle, and opposite Sidyma the Antiochi and Lagusa facing Glaucus river, Macris, Didymae, Melanoscope, Aspis, as well as Telandria whose town is extinct, and Rhodusa very close to Caunus.

XXXVI Rhodos, Cous

132. But the loveliest is free Rhodos, 125 round (or 103 if we believe Isidorus instead), populated by the cities Lindus, Camirus, Ialysus, and now Rhodos. According to Isidorus, it is 583 from Egypt's Alexandria, or 469 according to Eratosthenes, or 500 according to Mucianus; 176 from Cyprus. In the past it has been called Ophiusa, Asteria, Aethria, Trinacria and Ombria, Petreessa, Atabyria after a king, then Macaria, Aithaloessa.

133. Rhodians' islands: Carpathus which gives its name to the sea, Casos, Hagne, Eulimna, Nisyros 15½ from Cnidus and called Porphyris in the past; also Syme similarly situated half-way between Rhodos and Cnidus.

Measuring 37½ round, it offers eight welcoming harbors. In addition to these, other islands around Rhodos: Cyclopian Steganon, Cordylusa, four Diabatae, Hymos, Chalce with a town, Teutlusa, Narthecusa, Dimastos, Progne and (off Cnidus) Cisserusa, Therionarcia, Calydne with three towns, Notion, Istros, Mendeteros, and town Ceramus on Arconnesus. Along the Carian coast: those called Argiae, 20 of them, as well as Hyetusa, Lepsia, Leros.

134. Best-known in this bay, however, is Cous, 15 from Halicarnasus, 100 round, usually believed to have been called Merope, but Staphylus calls it Cea, Dionysius first Meropis, then Nymphaea. Mount Prion is here, and Nisyros [island], called Porphyris in the past, is thought to be a broken-off piece of it. From here, Caryanda with its town, and Pidossus close to Halicarnasus. In Ceramic bay: Priaponesos, Hipponesos, Pserima, Lepsimandus, Passala, Crusa, Pyrrhaethusa, Sepiusa, Melano, and just off the mainland, [the island] called Cinaedopolis, where king Alexander abandoned perverts.

XXXVII Samus

135. The Ionian coast has the Trageae and Corseae [islands], Icaros (already mentioned [4.68]), Lade previously called Late, and some insignificant ones including two Camelitae near Miletus, three Trogiliae near Mycala, Psilion, Argennon, Sandalium, free Samus 87½ round or, according to Isidorus, 100. Aristotle says that it was first called Parthenia, later Dryusa, then Anthemusa; Aristocritus mentions the further names Melamphyllus, then Cyparissia; others call it Parthenope (after the nymph) and Stephane. On it the Imbrasus, Chesius, Hibiethes rivers, Gigartho and Leucothea springs, mount Cercetius. Nearby islands include Rhypara, Nymphaea, Achillea.

XXXVIII Chius

136. Equally famous, 94 from [Samus], is free Chius with its town. Ephorus calls it by its ancient name Aethalia; Metrodorus and Cleobulus call it Chia after the nymph Chione; others from [the Greek word for] snow, and also Macris and Pityusa. Mount Pelinnaeus is there, [the source of] Chian marble. According to ancient accounts, its perimeter comes to 125, but Isidorus adds nine. It is located between Samus and Lesbos, directly opposite Erythrae.

137. Closest to it are Thallusa sometimes written Daphnusa, Oenusa, Elaphitis, Euryanassa, Arginusa with its town. These are really around Ephesus, as also the Anthinae (called Pisistratus' [islands]), Myonessos, Diarrheusa (the towns on both are extinct), Pordoselene with its town, Cerciae, Halone, Commone, Illetia, Lepria, Aethre, Sphaeria, Procusae, Bolbulae, Pheate, Priapus, Syce, Melane, Aenare, Sidusa, Pele, Drymusa, Anhydros, Scopelos, Sycusa, Marathusa, Psile, Perirrheusa and many of no significance.

138. Famous Teos with its town in the open sea, 72½ from Chius and the same distance from Erythrae. Near Zmyrna are the Peristerides, Carteria, Alopece, Elaeusa, Bacchium, Aspis, Psyra, Crommyonesos Megale; off Troas the Ascaniae, three Plateae, then Lamiae, two Plitaniae, Plate, Scopelos, Gethone, Arthedon, Coele, Lagusae, Didymae.

XXXIX Lesbos

139. Very famous Lesbos, 65 from Chius, has been called Himerte and Lasia, Pelasgia, Aegira, Aethiope, Macaria. It used to be renowned for having nine towns. Of these, Pyrrha has been swept away by the sea, Arisbe was wrecked by an earthquake, Antissa was absorbed by Methymna, which itself within 37 miles has nine cities of Asia as neighbors. Both Agamede and Hiera are extinct; but Eresos and Pyrrha remain, as does free Mytilene, which has been powerful for 1,500 years.

140. According to Isidorus, the entire island is 168 round, but ancient sources say 195. Its mountains are Lepetymnus, Ordymnus, Macistus, Creone, Olympus. The shortest distance to the mainland is 7½ miles. The nearby islands are Sandalium and the five Leucae, including Cydonea with its hot spring. Four miles from Aege [promontory on the mainland] are the Arginusae then Phellusa, Pedna. Beyond Hellespont and facing Sigeum on the seashore lies Tenedus, which has been called Leucophrys as well as Phoenice and Lyrnesos; it is 56 away from Lesbos, 12½ from Sigeum.

XL Hellespont, Mysia

141. From there Hellespont becomes violent, assaults the sea, and bores a breach with its crests until it separates Asia from Europa. We have already mentioned cape Trapeza [5.127]. Ten miles from it the town Abydos where the straits are seven stades wide. Then town Percote, as well as Lampsacus called Pityussa in the past, colony Parium (called Adrastia by Homer), town Priapus, Aesepus river, Zelia, Propontis (this is the name for where the sea widens out), Granicus river, Artace harbor where a town used to be.

142. Further on, the island that Alexander joined to the mainland, on which Milesians founded a town [now] called Cyzicum, but once called Arctonnesos, as well as Dolionis and Didymis; mount Didymus towers above it. Next, towns Placia, Artace, Scylace with the so-called Mysian mount Olympus behind, state Olympena, rivers Horisius and Rhyndacus once called Lycus. Proceeding from its source in lake Artynia near Miletopolis, it draws in Macestos river and plenty of others, and divides Asia from Bithynia.

143. This [region] used to be called Cronia, then Thessalis, then Malianda and Strymonis. Homer called [the people] Halizonae since they were surrounded by sea. There used to be a huge city named Atussa, nowadays there are 12 states, among them Gordiu Come called Juliopolis. On the coast: Dascylos, then Gelbes river, and the inland town Helgas also known as Germanicopolis or alternatively as Boos Coete (like Apamea nowadays known as Myrlea of the Colophonians), Echeleus river – the ancient limit of Troas and start of Mysia.

144. Thereafter a bay with Ascanium river, town Bryalion, Hylas and Cios rivers, and a town with the same name [Cios]. It used to be a center for trade with the neighboring district of Phrygia and was founded by Milesians, but on a site called Ascania of Phrygia. So now is the ideal point to deal with that region.

XLI Phrygia

145. Phrygia is located above Troas and the communities already mentioned from cape Lectum [5.123] to Echeleus river. In the north it borders on Galatia, in the south on Lycaonia, Pisidia, Mygdonia, and to the east it touches Cappadocia. Its best-known towns besides those already mentioned: Ancyra, Andria, Celaenae, Colossae, Carina, Cotiaion, Ceraine, Conium, Midaium. According to some writers, Mysians, Phrygians, Bithynians are named after Moesi and Brygi and Thyni who migrated from Europa.

XLII Galatia and neighboring peoples

146. Galatia should be mentioned now too. It is situated above Phrygia and controls territory that was mostly once part of Phrygia, as well as the old Phrygian capital Gordium. That place has been settled by Galli called Tolostobogii and Voturi and Ambitouti, while Trogmi occupy the region of Maeonia and Paphlagonia. To the north and east [of Galatia] extends Cappadocia; Tectosages and Toutobodiaci occupy its most fertile part. While these are the peoples here, there are 195 communities and tetrarchies in all. Towns: Ancyra (belonging to Tectosages), Tavium (to Trogmi), Pisinuus (to Tolostobogii).

147. Well-known [peoples] besides these: Attalenses, Arassenses, Comamenses, Hydenses, Hieronenses, Lystreni, Neapolitani, Oeandenses, Seleucenses, Sebasteni, Timonianenses, Thebaseni. Galatia also borders Cabalia in Pamphylia and the Milyae around Baris, as well as Cyllanicum and Pisidia's Oroandic district, and Lycaonia's Obizene area. Besides those already mentioned,[13] its rivers are Saggarium and Gallus, after which priests of the Mother of the Gods [Cybele] are named.

XLIII Bithynia

148. Now the rest of the coast: inland from Cios in Bithynia is Prusa, founded by Hannibal at the foot of [mount] Olympus. From here, 25 miles to Nicaea, with lake Ascanius in between. Then Nicaea (previously called Olbia) situated on the furthermost bay of Ascanius, and Prusias, as well as another [Prusias] at the foot of mount Hypius. There used to be Pythopolis, Parthenopolis, Coryphanta. On the coast are the Aesius, Bryazon, Plataneus, Areus, Aesyros, Geudos (also called Chryssorrhoas) rivers, the cape where a town Megarice used to be. Hence the bay was called Craspedites because this town was like the fringe of a garment. There used to be Astacum too, which accounts for Astacenus bay. There also used to be a town Libyssa where nowadays only Hannibal's tomb remains.

149. Nicomedia, a famous city of Bithynia, is on the far end of the bay. Cape Leucatas, which encloses the other end of Astacenus bay, is 37½ miles from Nicomedia. Back where the landmasses come closer together

---

[13] Caystrus (5.115), Rhyndacus (5.123, 142), Cios (5.144).

there are more narrows which extend to the Thracian Bosporus. On these free Calchadon 62½ miles from Nicomedia and previously called Procerastis, next Colpusa, and thereafter the town of Caeci ['blind people']; its name signifies an inability to choose a location, given that Byzantium, an altogether far more advantageous site, is [only] seven stades away. Otherwise, inland in Bithynia: the Apamean colony, the Agrillenses, Juliopolitae, Bithynion; rivers: Syrium, Laphias, Pharnucias, Alces, Serinis, Lilaeus, Scopas, Hieros which divides Bithynia and Galatia.

150. Beyond Calchadon used to be Chrysopolis, then Amycopolis and the bay which still bears its name, where Amycus harbor is situated. Then cape Naulochum, Estiae, Neptune's temple. The Bosporus, half a mile wide, again separates Asia from Europa, 12½ miles from Calchadon. From there 8¾ miles to where the narrows begin, where a town Spiropolis used to be. Thyni inhabit the entire coast, Bithynians the interior.

This is the end of Asia and the 282 communities counted from the border of Lycia to this point. We already noted [4.76] that the distance from Hellespont and Propontis to Thracian Bosporus is 239 miles. Isidorus says that it is 322½ miles from Calchadon to Sigeum.

XLIV [Islands in Propontis]

151. Islands in Propontis off Cyzicum: Elaphonnesus also once called Neuris and Proconnesus, the source of Cyzicene marble; then follow Ophiusa, Acanthus, Phoebe, Scopelos, Porphyrione, Halone with its town, Delphacie, Polydora, Artacaeon with its town. There is also Demonnesos opposite Nicomedia, as well as Thynias (called Bithynia by barbarians) beyond Heraclea and facing Bithynia. There is also Antiochia and Besbicos 18 miles round and opposite the Rhyndacus mouth. There is also Elaea and two Rhodusae, Erebinthote, Megale, Chalcitis, Pityodes.

****

Totals	Towns and peoples [...]
	Famous rivers [...]
	Famous mountains [...]
	Islands, 118
	Extinct cities and peoples [...]
	Facts and inquiries and observations [...]

[Roman] Sources

Agrippa. Suetonius Paulinus. Marcus Varro. Varro Atacinus. Cornelius Nepos. Hyginus. Lucius Vetus. Mela. Domitius Corbulo. Licinius Mucianus. Claudius Caesar. Arruntius. Livius the son. Sebosus. Records of Triumphs.

Foreign Sources

King Juba. Hecataeus. Hellanicus. Damastes. Dicaearchus. Baeton. Timosthenes. Philonides. Xenagoras. Astynomus. Staphylus. Dionysius. Aristotle. Aristocritus. Ephorus. Eratosthenes. Hipparchus. Panaetius. Serapion of Antioch. Callimachus. Agathocles. Polybius. Timaeus the mathematician. Herodotus. Myrsilus. Alexander Polyhistor. Metrodorus. Posidonius who [wrote] *Periplous* or *Periegesis* [in Greek]. Sotades. Pyrrandrus. Aristarchus of Sicyon. Eudoxus. Antigenes. Callicrates. Xenophon of Lampsacus. Diodorus of Syracusae. Hanno. Himilco. Nymphodorus. Calliphanes. Artemidorus. Megasthenes. Isidorus. Cleobulus. Aristocreon.

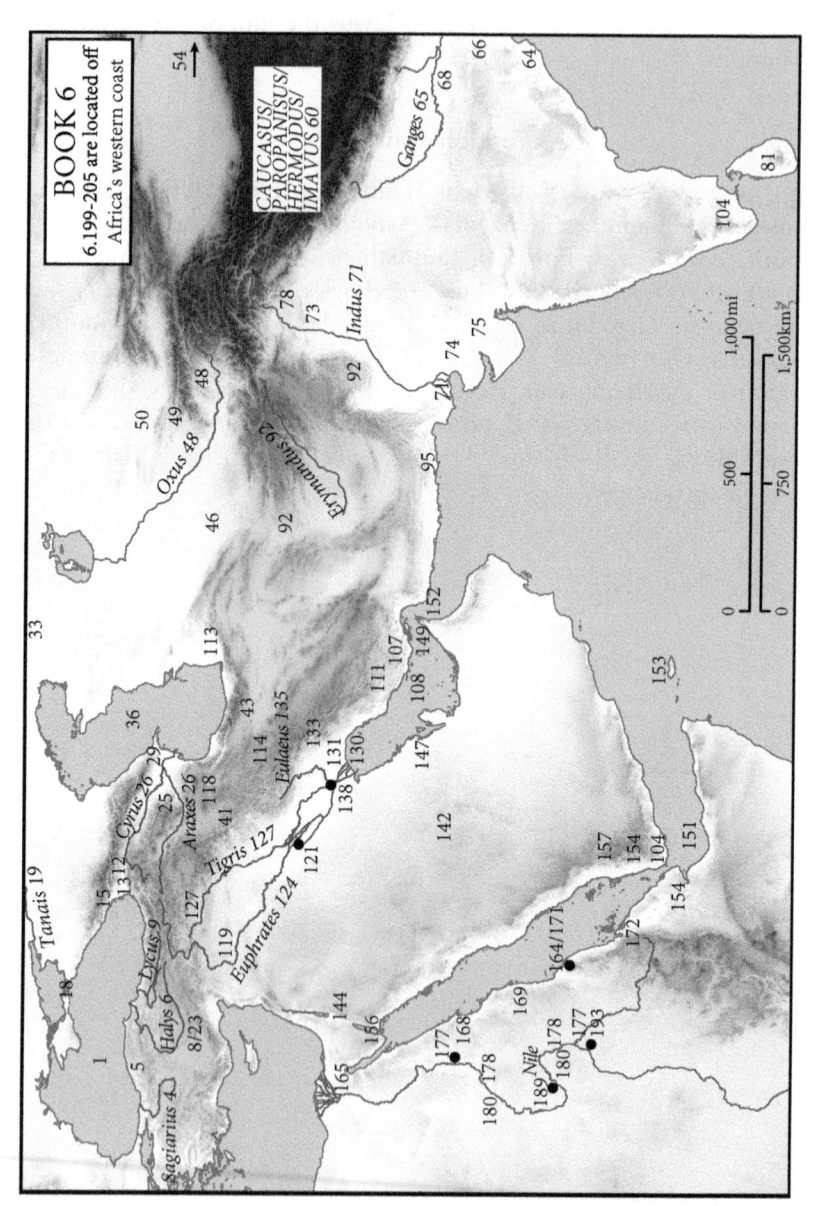

Book 6 The East, India

BOOK 6

The East, India

I Pontus, Mariandyni

1. Previously Pontus Euxinus [Black Sea] was called Axinus because of its inhospitable roughness. Thanks to Nature's peculiar jealousy infinitely indulging the sea's greed, it is spread between Europa and Asia. The ocean had not been satisfied by having surrounded the earth and with even less justification sweeping away part of it, nor by bursting into severed mountains and tearing Calpe from Africa to have submerged so much greater an area than it left remaining, nor by engulfing yet more land to have filled Propontis through Hellespont. Even beyond Bosporus, still unsatisfied, it widens out into another vast [sea] to be joined in its devastation by the Maeotian lakes.

2. That this happened against earth's will is demonstrated by how many narrows there are and by the tightness of the passages [cut] despite Nature's resistance – 875 paces [wide] at Hellespont, at the two Bospori a crossing even fordable by cattle (hence the name of both). These sibling breaches even have some identical features in common: birds singing and dogs barking on one side are audible on the other, and humans can even talk to one another, with the conversation continuing between two worlds except when swept away by wind.

3. Some make Pontus 1,438½ [miles] from Bosporus to lake Maeotis, but Eratosthenes 100 less. Agrippa makes Calchadon to the Phasis 1,000, and from there to Cimmerian Bosporus 360. In general, I give distances as measured in my own time (even when there has been disagreement over the very mouth of Cimmerian [Bosporus]).

4. So after Bosporus' narrows is Rebas river which some have called Rhesum. Then Syris, Calpas harbor, Sangaris a famous river. It rises in

Phrygia, receives mighty tributaries including Tembrogius and Gallus, and is mostly called Sagiarius; the Coralius where the Mariandyni begin, bay and town Heraclea situated on Lycus river (200 from Pontus' mouth), Acone harbor scary because of aconite poison, Acherusia cavern, the Paedopides, Callichorum, Sonautes rivers, Tium town 38 miles from Heraclea, Billis river.

II Paphlagones

5. Beyond it the people of Paphlagonia, which some have called Pylaemenia, are surrounded to the rear by Galatia, Mastya town [founded by] Milesians; then Cromna (to which place Nepos Cornelius links the Eneti, from whom he thinks it credible that the similarly named Veneti in Italia are descended), Sesamon town nowadays Amastris, mount Cytorus 63 miles from Tium, towns Cimolis, Stephane, Parthenius river.

6. Extending far out, cape Carambis is 325 or 350 (in the opinion of some) from Pontus' mouth, and the same distance from the Cimmerian [mouth] or (as some prefer) 312½. There also used to be a town with the same name, and then another called Armine; nowadays there is colony Sinope 164 from Cytorus, Varecum river, Cappadocian people, town Caturia Gazelum, Halys river flowing down from the foot of [mount] Taurus through Cataonia and Cappadocia,

7. towns Gangre, Carusa, free Amisum 130 from Sinope, and a bay of the same name indented so far as to make Asia a peninsula, it being a mere 200 overland to Issic bay in Cilicia. It is said that in this entire region there are only three peoples rightly called Greek: the Dorian, Ionian and Aeolian; the rest are barbarian. Eupatoria, a town founded by Mithridates, used to be linked to Amisum; after his defeat both were called Pompeiopolis.

III Cappadoces

8. Inland Cappadocia has Claudius Caesar's colony Archelais past which flows Halys river, towns Comana on the Salius, Neocaesarea on the Lycus, Amasia on the Iris in Gazacena region. Sebastia and Sebastopolis in Colopene; these are small yet equal to those mentioned above. Elsewhere Melita founded by Samiramis not far from the Euphrates,

Diocaesarea, Tyana, Castabala, Magnopolis, Zela and, at the foot of mount Argaeus, Mazacum nowadays called Caesarea.

9. The part of Cappadocia lying in front of Greater Armenia is called Melitene, in front of Commagene is Cataonia, in front of Phrygia is Garsauritis, Sargaurasana and the Cammaneni, and in front of Galatia is Morimene where they are divided by Cappadox, the river from which [Cappadocians] gained their name after being called Leucosyri previously. Lycus river divides Lesser Armenia from Neocaesarea mentioned above [6.8]. Inland there is also the famous Coeranus, but on the coast after Amisum the town and river Chadisia, Lycastum where Themiscyrene region begins, Iris river with its tributary Lycus. Inland the state Zela famous for Triarius' disaster [67 BCE] and Gaius [Julius] Caesar's victory [47 BCE].

IV Themiscyrene region and its peoples. Heniochi

10. On the coast Thermodon river, rising at the fortress called Phanoria, flows by the foot of mount Amazonius. There used to be a town with the same name and five others: Amazonium, Themiscyra, Sotira, Amasia, Comana nowadays Matium.

11. Genetae and Chalybes peoples, town of the Cotyi, Tibareni and Mossyni peoples who tattoo their bodies, Macrocephali people, Cerasus town, Cordule harbor, Bechires and Buxeri people, Melas river, Machorones people, Sideni, and Sidenum river lapping town Polemonium 120 from Amisum. Then Jasonium and Melanthium rivers, as well as Pharnacea town 80 from Amisum, fortress and river Tripolis, Philocalia, Liviopolis which has no river, and free Trapezus 100 from Pharnacea and enclosed by a huge mountain.

12. Beyond it, Armenochalybes people and Greater Armenia 30 miles away. On the coast before Trapezus is Pyxites river, and beyond it Sanni Heniochi people, Absarrum river with a fortress named after it at its narrows 140 from Trapezus. Behind the mountains here is Hiberia, on the coast Heniochi, Ampreutae, Lazi, rivers Acampseon, Isis, Nogrus, Bathys, Colchian peoples, Matium town, Heracleum river and a cape with the same name, and Phasis Pontus' best-known [river].

13. It rises among the Moschi, is navigable even by large vessels for 38½ miles, then a long way further by smaller ones, and crossed by 120 bridges. It used to have several towns along its banks, the most famous being Tyndaris, Circaeum, Cygnus, and Phasis at its mouth. Most famous, however, was Aea 15 miles from the sea, where the mighty Hippos and Cyaneos rivers run into it from opposite directions. Nowadays it has only Surium, named after a river flowing into [Phasis] at the point to which we [just] said large vessels could navigate. [Phasis] also receives other rivers remarkable for their size and number, among them Glaucus. At [Phasis'] mouth there is an unnamed island 70 from Absarrum.

14. Then another river, Charien, the Saltiae people called Phthirophagi in antiquity, and another, the Sanni, Chobum river which flows from the Caucasus through the Suani, then Rhoan [river], Egritice region, rivers Sigama, Thersos, Astelephus, Chrysorrhoas, Absilae people, Sebastopolis fortress 100 from Phasis, Sanigae people, Cygnus town, Penius river and town. Then Heniochan peoples with their many names.

V Colica region and peoples. Achaean peoples. Other peoples in the same region

15. Colica region lies below Pontus where the Caucasus range veers towards the Ripaean mountains (as already mentioned [5.98]), one side sloping towards Euxinus [sea] and [lake] Maeotis, the other to the Caspian and Hyrcanian seas. The rest of the shoreline is occupied by the savage Melanchlaeni and Coraxi tribes, with a now deserted Colchian city Dioscurias by Anthemus river. As Timosthenes related, it was once so famous that 300 tribes with their different languages came down here. In later times our [Roman] business was conducted there through 130 interpreters.

16. There are those who think that it was founded by charioteers of Castor and Pollux, Amphitus and Telchius, from whom it is largely agreed the Heniochan people are descended. The town Heracleum is 100 away from Dioscurias, 70 from Sebastopolis. Achaeans, Mardi, Cercetae and, after them, Seraci, Cephalotomi. Deep into this region is Pityus, an extremely wealthy town looted by Heniochi. Behind it, Epagerritae a Sarmatian community in the Caucasus range, after them Sauromatae.

17. Mithridates had fled to them during the principate of Claudius, and reported that next to them on the eastern side are Thali who spread to the mouth of Caspian sea which dries up when the tide ebbs. But on the coast next to the Cercetae Icarus river, and Achaeans with the town and river Hieron 136 from Heracleum. Then cape Crunoe after which Toretae hold a steep ridge, state Sindica 67½ from Hieron, Setheries river.

The entrance to Cimmerian Bosporus 88½ from there.

VI Cimmerian Bosporus

18. But the length of the peninsula itself extending out between Pontus and lake Maeotis is no more than 67½ miles, its width nowhere less than two *iugera*; it is called Eon. The actual coast of Bosporus on both Asian and European sides curves towards Maeotis. Towns at the entrance Hermonasa, then Cepoe [founded by] Milesians, next Stratoclia as well as Phanagoria and the almost deserted Apaturos; at the extreme end of the mouth, Cimmerium previously called Chimerion.

Then lake Maeotis said to be in Europa.

VII Maeotis. Peoples around Maeotis

19. Maeotici, Vali, Serbi, Serrei, Scizi, Gnissi inhabit [the area] after Cimmerium. Then Sarmatians live where Tanais river with its twin mouths flows. They are said to be descendants of Medes and are themselves divided into many branches. First, Gynaecocratumenian Sauromatae who marry Amazons; then Naevazae, Coitae, Cizici, Messeniani, Costoboci, Zecetae, Zigae, Tindari, Thussagetae, Tyrcae all the way up to rough wilderness with forested valleys, beyond which Arimphaei extend to the Ripaean mountains.

20. Scythians call the Tanais itself Silis, the Maeotis Temarunda, which means 'mother of the sea'. Also a town at Tanais mouth. Cares occupied the nearby territory first, then Clazomeni and Maeones, thereafter Panticapaeenses. These are the peoples said to be around Maeotis up to the Ceraunian mountains:

21. from the coast Napitae and, above them, Essedones linked to Colchians by mountain peaks, then Camacae, Orani, Autacae, Mazamacae, Cantiocaptae, Agamathae, Pici, Rymosoli, Acascomarci and, at the Caucasus range, Icatalae, Imadochi, Rami, Anclacae, Tydii, Carastasei, Authiandae; Lagous river coming down from Cathei mountains and, where Opharus flows into it, Cauthadae, Opharitae peoples;

rivers Menotharus and Imityes coming from Cissii mountains; below, Acdei, Carnae, Uscardei, Accisi, Gabri, Gegari and, around the Imityes' source, Imityi and Apartaei.

22. Others [say] that Scythian Auchetae, Athernei, Asampatae have spread here, and that they completely wiped out the Tanaitae and Napaei. According to some reports, Ocharius river runs through the Cantici and Sapaei, while Tanais went through the Satharchei Hertichei, Spondolici, Synhietae, Anasi, Issi, Cataeetae, Tagorae, Caroni, Neripi, Agandaei, Meandaraei, Satharchei Spalaei.

VIII Cappadoces

23. The inner coast, places, rivers and inhabitants have been covered; now a huge inland area is to be treated, where there is no denying that my account will often differ from old ones because I have been deeply concerned to learn about recent activity in this area from Domitius Corbulo and kings sent from there as suppliants or kings' children sent as hostages. We shall begin, however, with the Cappadocian people.

24. Of all Pontic [peoples] they extend furthest into the interior, passing on the left side Lesser and Greater Armenia as well as Commagene, on the right all the peoples in Asia already mentioned. They spread over a great number of communities, climb very energetically to the east and the Taurus range, cross Lycaonia, Pisidia, Cilicia, proceed above Antiochia region and stretch as far as its Cyrrestica district in the part called Cataonia. So from here the length of Asia amounts to 1,250, its width 640.

IX Lesser Armenia, Greater Armenia

25. Greater Armenia begins at the Parihedri mountains; as mentioned [5.83], it is separated from Cappadocia by Euphrates river and, where Euphrates moves away, from Mesopotamia by the no less famous Tigris river. Both rivers spread themselves out and form the start of Mesopotamia, situated between the two; Orroean Arabians occupy that intervening area. So [Mesopotamia] pushes its boundary direct into Adiabene where it is hemmed in by a mountain range crossing it; on the left it expands its width to Cyrus river, crossing the Araxes, and its length all the way to Lesser Armenia, from which it is separated by

Absarrum river flowing into Pontus and by Parihedri mountains which emit Absarrum.

X Cyrus river, Araxes river

26. Cyrus rises in Heniochi mountains, alternatively called Coraxici; Araxes in the same mountain as Euphrates (although six miles apart), and after a boost from Usis river it is conveyed into Caspian sea (according to general opinion) by Cyrus. The best-known towns in Lesser [Armenia] are Caesarea, Aza, Nicopolis; in Greater [Armenia] Arsamosata close to Euphrates, Carcathiocerta close to Tigris, Tigranocerta on a height, and Artaxata in flat country near the Araxes.

27. Aufidius said that the total size [of Armenia] is 5,000, and according to Claudius Caesar its length from Dascusa to the edge of Caspian sea is 1,300 miles, its width from Tigranocerta to Hiberia half that. Without question, it is divided into 120 prefectures with barbaric names called *strategiae*, some of which once were even separate kingdoms. Cerauni mountains enclose it to the east, but not immediately, and likewise Adiabene region.

28. Cepheni occupy the space lying between them; after them Adiabeni occupy the range beyond, and in the glens closest to Armenia are Menobardi and Moscheni. Tigris and impassable mountains surround Adiabene. The region to its left belongs to Medes up to where Caspian sea can be seen. As we shall explain in the appropriate place [6.36], this [sea] is inflow from ocean and wholly surrounded by Caucasus mountains. The inhabitants along the border with Armenia will now be mentioned.

XI Albania, Hiberia and nearby peoples

29. Albanian people occupy the entire plain from the Cyrus onwards, then Hiberians separated from them by Ocazanes river flowing down from Caucasus mountains and into Cyrus. Significant towns are Cabalaca in Albania and, in Hiberia, Harmastus next to a river, Neoris. Thasie region and Thriare all the way to the Parihedri mountains. Beyond are Colchian wilderness (where Armenochalybes live on the side facing the Ceraunian [mountains]), and the region of the Moschi over to Hiberus river, a tributary of Cyrus. Below them are Sacasani and then

Macerones as far as Absarrum river. This is how the plains and slopes are settled. Back beyond the border with Albania wild Silvi peoples along the entire mountain frontage, and below them the Lupenii, then Diduri and Sodi.

XII Caucasian Gates

30. After these are the Caucasian Gates, often very mistakenly called Caspian. A huge work of Nature, mountains were suddenly split apart and gates with iron beams installed here; a river that makes an abominable stench flows midway below them. On this [Roman] side of them on a crag a fortress called Cumania, fortified to block countless peoples passing through. [So] at this spot the world is bisected by gates, immediately opposite the Hiberian town Harmastus. After the Caucasian Gates and through Gurdinii mountains Valli, Suani, unconquerable peoples who nonetheless mine for gold. After these and all the way up to Pontus numerous Heniochan peoples, then Achaeans. This, then, is how it is in one of the most famous parts of the world.

31. Some have said that there are no more than 375 miles between Pontus and Caspian sea; Cornelius Nepos' figure is 250: here too [cf. 6.7] such narrows threaten Asia. Claudius Caesar reported the distance from Cimmerian Bosporus to Caspian sea as 150, and Nicator Seleucus was thinking about digging through here just when he was killed by Ptolomaeus Ceraunus [281 BCE]. It is generally agreed to be 200 from Caucasian Gates to Pontus.

XIII Islands in the Pontus

32. Islands in Pontus: Planctae (or Cyaneae or Symplegades), then the Apollonia called Thynias to distinguish it from the one in Europa – one mile from the mainland, three round – and, opposite Pharnacea, Chalceritis which Greeks called Aria and is sacred to Mars, a place where birds beating their wings assaulted strangers.

XIV Peoples around the Scythian ocean

33. With everything in the interior of Asia now covered, attention should shift over the Ripaean mountains and proceed to the right along the shore of ocean. Reaching Asia from three separate directions, it is called

Scythian in the north, Eoan in the east, Indian in the south, as well as being variously divided according to its bays and its inhabitants' multiple names. But a great part of Asia also faces north and has extensive wilderness because of its harmful freezing climate.

34. From the furthest north-north-eastern point to where the sun first rises in summer are Scythians. Apart from them and beyond the start of the north-north-eastern point some have placed Hyperboreans, although most say they are in Europa. Further on we first know about Lytharmis, a cape in Celtica, [and] Carambucis river where the climate loses its energy and the Ripaean mountain range peters out; here too, we are told, are some Arimphaei, a people little different from Hyperboreans.

35. They live in the woods and eat berries. Long hair is disapproved of for men as well as women, and they have pleasant manners. So, according to the accounts, they are considered holy and are unharmed even by wild communities nearby; this applies not only to them, but also to anyone who has taken refuge with them. Directly beyond them Scythians, Cimmerians, Cissi, Anthi, Georgi and Amazonian people all the way up to Caspian and Hyrcanian sea.

XV Caspian and Hyrcanian sea

36. For it even bursts into Asia's rear from Scythian ocean; it is named after its numerous inhabitants, but best-known by two, Caspian and Hyrcanian. Clitarchus thinks it is no smaller than Pontus Euxinus. Eratosthenes also gives a measurement from the south-east along the coast of Cadusia and Albania as 5,400 stades; from there through the Anariaci, Amardi and Hyrcani to the mouth of Zonus river as 4,800; and from there to Jaxartes' mouth as 2,400, all of which makes 1,575 miles. Artemidorus' figure is 25 miles less.

37. Agrippa said that Caspian sea and its surrounding peoples (Armenia included) are bounded to the east by Seric ocean, to the west by Caucasus range, to the south by Taurus [range], and to the north by Scythian ocean, stretching – so far as is known – 480 long, 290 wide. But there are some who give the perimeter of the entire sea from the strait as 2,500 miles.

38. According to Marcus Varro, [the sea] rushes in through a narrow and protracted long mouth, and where it begins to expand in width – just as it descends from the mouth towards lake Maeotis – it becomes slanted with crescent horns like a sickle. The first bay is called Scythian because Scythians live on both sides and maintain contact across the narrows, having on one side Nomads and Sauromatae (with many names) and on the other Abzoae (with just as many). From the entrance on the right Udini, a Scythian community, occupy the actual point of the mouth, then along the coast Albani descended (we are told) from Jason; hence the sea there is called Albanian.

39. They are a people spread over the Caucasus mountains and coming down to Cyrus river, the border between Armenia and Hiberia, as already mentioned [6.29]. Above its coastal locations and those of the Udini people extend Sarmatians, Uti, Aorsi, Aroteres, behind whom are Sauromatian Amazons already noted [6.19]. Running down to the sea through Albania are Casus and Albanus rivers, then Cambyses rising in Caucasus mountains, then Cyrus in the Coraxici, as we stated [6.26]. According to Agrippa, the whole coast from the Casus makes landing impossible for 425 miles because of very high cliffs. After the Cyrus it begins to be called Caspian sea; Caspi live there.

40. This is the place to correct an error that many have made, even after serving recently in Armenia with Corbulo. For they have given the name 'Caspian' to the Gates of Hiberia which we have said [6.30] are called 'Caucasian', and even drawings of the landscape sent from there have this name written on them. Also, the Caspian Gates were said to be the goal of *princeps* Nero's threat, although this was aimed at the pass [or Gates; see 6.43] which extends through Hiberia into Sarmatian territory, where intervening mountains make access to Caspian sea almost impossible. There are, however, other [Gates] associated with Caspian peoples which cannot be identified except through accounts by Alexander the Great's followers.

XVI Adiabene

41. For the Persians' kingdoms, which nowadays we regard as Parthian, are elevated by Caucasus range between two seas, Persian and Hyrcanian. As we have said [6.28], along the slopes of both sides Cephenia is joined to the front part of Greater Armenia (turning towards Commagene) as well as to Adiabene, the start of Assyrian [territory], part of it being Arbilitis where Alexander defeated Darius, very close to Syria.

42. Macedonians named this whole territory Mygdonia because of its similarity [to Mygdonia in Macedonia]. Towns, Alexandria as well as Antiochia which they call Nesebis; it is 750 miles from Artaxata. There also used to be the once very famous Ninus situated on the Tigris and facing the setting sun. On the rest of the front part [of Greater Armenia] extending to Caspian sea Atrapatene, separated from Armenia's Otene region by the Araxes. Its town Gaza 450 miles from Artaxata and the same distance from Ecbatana [founded by] Medes; Atrapateni are a part of them.

XVII Media, Caspian Gates

43. King Seleucus founded Ecbatana, capital of Media, 750 miles from Seleucia Magna, 20 from the Caspian Gates. Remaining Median towns, Phisganzaga and Apamea also called Rhagiane. The reason for the name 'Gates' is the same as above [6.40]: the range is interrupted by an eight-mile-long narrow pass, barely wide enough to move a single line of wagons, the whole creation manmade. To right and left are beetling crags that seem scorched; the region is waterless for 28 miles, and the narrow passage is blocked by a build-up of salt fluid from the rocks which runs off the same way. In addition, the quantity of snakes makes it impossible to travel through except in winter.

44. Connected to the Adiabeni are Carduchi as they were once called, nowadays Cordueni, with Tigris flowing past them. By them the so-called *'Roadside'* [in Greek] Pratitae who hold the Caspian Gates. Meeting them on the other side are Parthia's desert and Citheni range. Next a very charming location, also Parthian, called Choara. Two Parthian cities here – once bulwarks against Medes – Calliope and on another crag Issatis. But the actual Parthian capital Hecatompylos is 133 miles from the Gates; so the Parthian kingdoms too are shut off by barriers.

45. Immediately after exiting the Gates, all the way to the shore are Caspian people who have given their name to Gates and to the sea. The left [side] is mountainous. Back from this people to Cyrus river is said to be 225 miles, and 700 by going from that same river to the Gates. For this has been made the base-point of Alexander the Great's journeys, the distance from these Gates to the Indian border being given as 15,690 stades, and to the town Bactra also called Zariasta, 3,700, with 5,000 from there to Jaxartes river.

XVIII Peoples around the Hyrcanian sea

46. The region facing eastward after the Caspi is called Apavortene where there is a place famous for its fertility, Dareium; next peoples: Tapyri, Anariaci, Staures, Hyrcani from whose shores – after Sideris river – the same [Caspian] sea begins to be called Hyrcanian. On this [Roman] side, Maziris and Strator rivers; all [flowing down] from Caucasus. The following region is Margiane, famous for having so much sunshine; only in this area are vines cultivated. It is enclosed on all sides by lovely mountains 1,500 stades round, difficult to reach because of sandy desert extending for 120 miles, and itself situated opposite the region of Parthia.

47. Here Alexander had founded an Alexandria later destroyed by barbarians. Seleucus' son Antiochus re-established a Syrian [city] on the same site, with the Margus flowing through it and being channeled into lake Zotha; he had preferred that it be called Antiochia. In size, the city measures 70 stades round. Orodes settled Roman captives from Crassus' disaster here.

From its heights and through the Caucasus range as far as the Bactrians extend fierce Mardian people, a law to themselves. Peoples below this region: Orciani, Commori, Berdrigae, Pharmacotrophi, Chomarae, Choamani, Murrasiarae, Mandruani;

48. Mandrum and Chindrum rivers, and beyond them the Chorasmi, Gandari, Pariani, Zarangae, Arasmi, Marotiani, Arsi, Gaeli, Legi (called Cadusii by Greeks), Matiani, town Heraclea founded by Alexander but then destroyed and re-established by Antiochus who called it Achais, Dribices whose territory is bisected by Oxus river which rises in lake Oaxus, the Syrmatae, Oxyttagae, Moci, Bateni, Saraparae, Bactrians whose town Zariasta was later named Bactra after the river. This people occupies the far side of mount Paropanisus opposite the sources of the Indus and is enclosed by Ochus river.

49. Beyond are Sogdiani with town Panda and at the furthest limit of their territory Alexandria founded by Alexander the Great. Here there are altars set up by Hercules and Father Liber, also by Cyrus and Samiramis and Alexander, all of whom marked this as their endpoint in this part of the world enclosed by Jaxartes river, which Scythians call Silis and Alexander and his soldiers reckoned to be Tanais. Demodamas, the general of kings Seleucus and Antiochus and our main informant on these matters, crossed this river and erected altars to Apollo Didymaeus.

XIX Scythian peoples

50. Beyond are Scythian communities. Persians called all these peoples Sacae after the people nearest [to them], the ancients called them Aramii; Scythians themselves call the Persians Chorasari, and Caucasus mountain Croucasis, meaning 'white with snow'. The number of communities is beyond counting, and they live on equal footing with the Parthians. Their best-known ones: Sacae, Massagetae, Dahae, Essedones, Astacae, Rumnici, Pestici, Homodoti, Histi, Edones, Camae, Camacae, Euchatae, Cotieri, Authusiani, Psacae, Arimaspi, Antacati, Chroasai, Oetaei; here Napaei are said to have been wiped out by Palaei. Notable rivers in their territory, Mandragaeum and Caspasum.

51. In my opinion, because the number of peoples here is beyond counting and they are nomadic, there is more disagreement among writers about this area than any other. Not only did Alexander the Great report that water from this [Caspian] sea was fresh, but also Marcus Varro said that such [water] was brought to Pompey when he was on campaign nearby during the Mithridatic war – no doubt because the size of the rivers' inflow overcomes the salt.

52. [Varro] adds that scouts who went into Bactrian territory under Pompey's leadership showed that it takes seven days to go from India to Bactrus river which flows into the Oxus, and they were [next] conveyed from it across the Caspian to the Cyrus, and [then] it was no more than a five-day overland journey for Indian goods to be brought to Phasis on the Pontus. There are many islands in the entire [Caspian] sea, Zazata one especially well-known.

XX Places after the Eoan ocean. Seres

53. After Caspian sea and Scythian ocean, the route veers towards Eoan [ocean], with the coast's frontage facing east. Its first section after the Scythian cape is uninhabitable because of snow; the next part is uncultivated because of the peoples' savagery. Scythian Anthropophagi, who eat human bodies, live here. Nearby, in consequence, a vast wilderness where the numerous wild beasts are quite as monstrous as the humans. Then more Scythians and more beast-infested desert [extending] all the way to a mountain range called Tabis that overhangs the sea. And not until almost half-way along the length of its coast facing north-east is the region inhabited.

54. The first humans known there are Seres, famous for the wool in their forests; after soaking it in water they comb the white down off the leaves, so that our women have the double labor of unravelling the thread and weaving it back together: such complicated work is sought from such a faraway part of the globe so that ladies may appear diaphanous in public! Seres are in fact mild [people], but like wildlife they too shun contact with the rest of humanity and do not initiate commercial dealings.

55. Their first known river is Psitharas, next Cambari, third Lanos followed by cape Chryse, Cirnaba bay, Atianos river, and Attacori bay and people protected by sunny hills from every damaging breeze, the same climate as where Hyperboreans live. Amometus wrote a book exclusively about them, just as Hecataeus did about Hyperboreans. After the Attacori, the Thuni and Tocari people, and then Indian Casiri (who are cannibals) inland facing Scythians. Indian Nomads also wander in this direction. According to some reports, they have contacts with Ciconae and Brisari to the north.

XXI Indians

56. But from hereon there should be complete agreement about the peoples. The Hemodi mountains spring up and Indian peoples begin, bordering not just the Eoan sea but also the southern one which we called Indian. The part facing east extends in a straight line to a bend and the start of the Indian sea, totaling 1,875; from there it then curves itself southward for 2,475 (as Eratosthenes says) all the way to Indus river, which is the western border of India.

57. However, from the voyage of ships under sail several calculations of its entire length have been made as 40 days and nights, and from north to south 2,850 [miles]; Agrippa gave its length as 3,300, its width as 1,300. Posidonius measured it from north-east to south-east, placing it opposite Gallia (which he measured from north-west to south-west), all facing its west side. So he has shown beyond doubt that India's exposure to the west wind helps to make it healthy.

58. Here the sky looks different, the stars' risings are different; there are two summers a year, two harvests with a winter between them when the Etesiae blow, while during our midwinter the breezes here are light, the sea navigable. Anyone interested in identifying all its peoples and cities should realize that these are beyond counting. This has been made clear not only by Alexander the Great's armies and those of his successor kings Seleucus and Antiochus – who even sailed round to the Hyrcanian sea and Caspian, as did their fleet commander Patrocles – but also by other Greek authors such as Megasthenes and Dionysius (sent by Philadelphus for this very purpose), who stayed with Indian kings and reported on the strength of these peoples.

59. This is no place for precision, however, because the accounts are so varied and hard to credit. Alexander the Great's companions wrote that, in the region of India he conquered, there were 5,000 towns (none smaller than Cous), 9,000 peoples, and that India was one-third of the entire world and its communities beyond counting – no doubt a plausible claim, because the Indians are almost the only people who have never migrated from their own territory. The reckoning is that from Father Liber to Alexander the Great they had 153 kings over 6,451 years plus three months.

60. The immensity of their rivers is amazing; Alexander is said to have sailed at least 600 stades every day along Indus, and yet was not able to reach its mouth in under five months and a few days, even though this [river] is known to be smaller than Ganges. Our own Seneca, who also attempted a commentary about India, gave figures of 60 rivers, 118 peoples. It would be equally difficult to count the mountains. Interlocked are the Imavus, Hemodus, Paropanisus, and part of Caucasus from which the whole [chain] runs down into a vast plain as in Egypt.

61. But for the purpose of giving a comprehensible description of the country, we proceed in the footsteps of Alexander the Great. Diognetus and Baeton, the recorders of his journeys, wrote that it is as many miles as we said [6.44] from Caspian Gates to the Parthian [city] of Hecatompylos, then 575 from there to Alexandria of the Aries founded by the king, 199 to Prophthasia [city] of the Drangae, 565 to the town of the Arachosii, 175 to Ortospanum,

62. then 50 to Alexander's town (though different figures are found in some copies [of the record]); this city is said to be sited right below

Caucasus. From here it is 237 to Cophes river and the Indian town Peucolatis, 60 from there to Indus river and town Taxilla, 120 to the famous Hydaspes river, 390 to the equally well-known Hypasis which marked the end of Alexander's journeys, even though after crossing the river he dedicated altars on the opposite bank. The king's own letters also agree with these [figures].

63. From here Seleucus Nicator is the source for the remaining figures: 169 to Sydrus [river], the same to Jomanes river (in some copies the figure is five miles more), then 112½ to the Ganges, 569 to Rhodapha (but according to some this distance is 325), 167½ (or some say 265) to town Callinipaza, then 625 to the confluence of Jomanes and Ganges (most add 13½), 425 to town Palibothra, 637½ to Ganges mouth.

64. Peoples worth mentioning after the Hemodi mountains (one of its spurs is called Imaus, meaning 'snowy' in the local language): Isari, Cosiri, Izi and, all along the range, Chirotosagi and many peoples called Bragmanae, among them Mactocalingae; rivers – both navigable – Prinas, and Cainnas which flows into Ganges. By the sea, Calingae peoples and above them Mandaei; Malli with their mountain Mallus, and Ganges marking the end of this region.

XXII Ganges

65. Some say that [Ganges], like Nile, rises from obscure sources and irrigates the nearby land in the same way, while others have stated that it rises in Scythian mountains. Nineteen rivers flow into it. Navigable ones (besides those already mentioned [6.64]): Crenacca, Eramnomboua, Casuagus, Sonus. According to others, after its own source has immediately erupted with a great roar, it is tossed through rocks and falls, yet as soon as reaching gentle plains it lodges in some lake, and then flows placidly with a width of at least eight miles and normally 100 stades, and never less than 20 paces deep, its last people being Gangarides Calingae whose royal capital is called Pertalis.

66. The king has a standing army of 60,000 infantry, 1,000 cavalry, 700 elephants ready for war. The more moderate members of Indian communities have many different kinds of livelihoods: [some] till the soil, others undertake military service, others export local goods [or] import foreign ones, the best and wealthiest govern the state, serve as judges, assist kings.

A fifth class is devoted to wisdom, which they revere and almost turn into a religion, [men] who always end life of their own free will, dying on a pyre which they first light themselves. There is one [class] in addition, half-savage and preoccupied with the huge task – from which those mentioned above are barred – of hunting and domesticating elephants: they plough with these and travel on them, these are the livestock they are most familiar with, with them they make war and fight for the frontiers; strength and age and size determine which are recruited for campaigns.

67. In Ganges there is a very large island named Modogalinga where a single people lives. Beyond it are situated Modubae, Molindae, Uberae with their splendid town of the same name, Modressae, Praeti, Calissae, Sasuri, Passalae, Colebae, Orumcolae, Abali, Thalutae. Their king has 50,000 infantry, 4,000 cavalry, 4,000 elephants mobilized. Then come the more powerful Andarae people who have a great number of villages, 30 towns fortified with walls and towers, [altogether] providing the king with 100,000 infantry, 2,000 cavalry, 1,000 elephants. Dardae are the greatest producers of gold, as Setae are of silver.

68. But nearly everyone in India, and not just in this region, is outdistanced in power and prestige by the Prasi with their very extensive and very wealthy city Palibothra, hence some call the people themselves Palibothri and even the whole region beyond Ganges [Palibothra]. In their king's pay, day in day out, are 60,000 infantry, 30,000 cavalry, 8,000 elephants, from which the vastness of his resources can be gauged.

69. Further inland from them Monaedes and Suari who occupy mount Maleus, on which shadows fall to the north in winter and to the south in summer, over six-month periods. According to Baeton, the Septentriones [Great Bear constellation] appear in this region once a year, and only for 15 days. Megasthenes says that the same happens in many places in India. Indians call the south pole *dramasa*. Jomanes river runs through the Palibothri and into Ganges between the towns Methora and Chrisobora.

70. Turning to the zone south from Ganges, the populations are tanned by the sun's rays. They are in fact already colored, but not yet as burned as Aethiopians; the closer they come to the Indus, the swarthier they appear. Indus is immediately after the Prasian people in whose mountains there are said to be Pygmies. Artemidorus records the distance between the two rivers [Indus and Ganges] as 2,100.

XXIII Indus

71. Indus, called Sindus by locals, flows from an east-facing ridge of Caucasus mountains called Paropanisus; it takes in 19 rivers itself, the best-known ones being Hydaspes which brings four others, Cantaba which brings three, and Acesinus and Hypasis both themselves navigable. However, due to some restraint on the part of its waters, nowhere is [Indus] wider than 50 stades or deeper than 15 paces, [yet] it does form a very extensive island named Prasiane and another smaller one Patale.

72. By the most cautious accounts it is navigable for 1,240 miles and, after more or less trailing the sun westward, it flows into the ocean. I shall give figures for the coast to its [mouth] stage by stage just as I find them, despite the lack of agreement among [reports]: 625 from Ganges mouth to cape Calingon and Dandaguda town, 1,225 to Tropina, 750 to cape Perimula where India's best-known market is, 620 to Patala town on the island mentioned above [6.71 Patale].

73. Mountain peoples between [Indus] and Jomanes: Caesi, Caetriboni forest-dwellers, then Megallae whose king has 500 elephants and an uncertain number of infantry and cavalry, Chrysei, Parasangae, Asmagi whose area is infested by wild tigers. Their forces are 30,000 infantry, 300 elephants, 800 cavalry. They are surrounded by wilderness and a ring of mountains and encompassed by Indus. Dari and Surae 625 below the wilderness, and then wilderness again for 187; for the most part sand surrounds them just like sea around islands.

74. Below these deserts: Malthaecorae, Singae, Maroae, Rarungae, Moruni. These are free people living in mountains that stretch unbroken along the ocean coast; they have no kings and occupy the mountain slopes with many cities. Then Nareae hemmed in by mount Capitalia, the highest in India. People living on the other side of it work extensive gold and silver mines.

75. After them Oratae whose king has only ten elephants but substantial forces of infantry, Suarattaratae who (under a king) do not maintain elephants but rely on cavalry and infantry, Odonbaeorae, Sarabastrae with their beautiful city Thorax protected by swampy canals which crocodiles – insatiable consumers of human flesh – leave accessible only by bridge. Another of their towns that gains praise is Automula, located on the seashore where five rivers converge into one and with a famous market. Their king has 1,600 elephants, 150,000 infantry, 5,000 cavalry. The poorer king of the Charmae has 60 elephants and modest other forces.

76. After them, Pandae people, the only Indians ruled by women. The story is that just a single child of this sex was born to Hercules and was in consequence really dear to him and endowed with an outstanding kingdom; those who trace their ancestry back to her are rulers of 300 towns, 150,000 infantry, 500 elephants. After this group of 300 cities the Derangae, Posingae, Butae, Gogaraei, Umbrae, Nereae, Brangosi, Nobundae, Cocondae, Nesei, Palatitae, Salobriasae, Orostrae adjoining Patale island, from which it is said to be 1,925 to the Caspian Gates.

77. Then from here living along Indus, listed proceeding upstream: Mathoae, Bolingae, Gallitalutae, Dimuri, Megari, Ardabae, Mesae, Abisari, Silae; then 250 [miles] of desert to overcome. After them: Organagae, Abortae, Brasuertae, followed by as much desert as before. Then Sorofages, Arbae, Marogomatrae, Umbritae and Ceae comprising 12 tribes with two cities each; the Asini occupying three cities; their capital Bucephala named after king Alexander's horse and founded as its burial place.

78. Above them, mountain peoples below Caucasus: Sosaeadae, Sondrae and, after crossing Indus and proceeding downstream with it, Samarabiae, Sambraceni, Bisambritae, Orsi, Andiseni, Taxilae with their famous city. At this point the region slopes down to a plain, the overall name for which is Amenda, with four communities: Peucolitae, Arsagalitae, Geretae, Assoi. In fact most accounts do not set [India's] western border at Indus river, but add four satrapies, [those of] the Gedrosi, Arachotae, Arii, Paropanisidae, with Cophes river as the final boundary; others' preference is that the entire [region] belongs to Arii.

79. Moreover, most also assign to India Nysa city and mount Merus sacred to Father Liber, hence the origin of the myth that he was born from Jupiter's thigh; also [assigned there] are the Aspagani people who grow vines and laurel and boxwood and all the other fruits that thrive in

Graecia. There are memorable, almost mythical accounts of the soil's fertility and the kinds of crops and trees or beasts and birds and other animals that will be referred to at the relevant points in the rest of the work, as will the four satrapies in brief shortly [6.92]. But now our attention hurries on to Taprobane island.

80. There are, however, others before it: Patale which we have located [6.71] right at the Indus delta shaped like a triangle, 220 miles wide. Beyond Indus mouth Chryse and Argyre with abundant metals, I believe; that they have only gold and silver, as some have stated, I am loath to believe. Crocala 20 miles beyond them, and 12 further Bibaga flush with oysters and shellfish, then Coralliba eight beyond [Bibaga] just mentioned, and many insignificant ones.

XXIV Taprobane

81. Taprobane was long thought to be another world called Antichthones. But the time and achievements of Alexander the Great clearly demonstrated it to be an island. The commander of his fleet, Onesicritus, wrote that elephants bred there are larger and more aggressive than those in India; according to Megasthenes, it is divided by a river and the inhabitants are called Palaegoni, who produce more gold and impressive pearls than the Indians. Eratosthenes also reported dimensions of 7,000 stades long, 5,000 wide, and that there are no cities but 750 villages.

82. Beginning at Eoan sea, it stretches along India between east and west, and was once believed to be a 20-day sail from the Prasian people; but later, when ships of reeds and rigging from the Nile were used, our ships' voyage was reckoned to take seven days. The sea in between [Taprobane and India] is shallow, no more than six paces down, but in certain straits it is so deep that anchors do not reach the bottom. Consequently ships have prows at both ends so that there would be no need to turn them in the narrow channels; they load up to 3,000 amphoras. The stars are not observed for navigation;

83. Septentrio [Great Bear] is not visible. Birds are carried with them and are quite frequently released so as to follow their course in making for land. The sailing season is for no more than four months of the year. In particular the 100 days after the summer solstice are avoided, because then that sea is wintry.

84. Old accounts record this much. More accurate information reached us during the principate of Claudius when envoys even came from the island. It happened like this: a freedman of Annius Plocamus (who had gained a treasury contract for collecting Red sea tax) when sailing around Arabia was swept by north winds past Carmania and after 15 days brought to Hippuros, a harbor [in Taprobane]. Thanks to the king's hospitable kindness, he learned their language within six months, and then when questioned gave an account of the Romans and Caesar.

85. Among the marvels [the king] heard about, he admired [Roman] honesty, because the *denarii* in the money seized were of equal weight even though the different images on them pointed to issues by several [emperors]. This especially motivated him to seek [our] friendship, and he dispatched four envoys with Rachias their head. From them we discovered that there are 500 towns [and] a south-facing harbor located at the town Palaesimundum, the best-known one of all there and the king's own, with a population of 200,000.

86. Inland, Megisba marsh 375 miles round and encompassing islands that produce only grass. From it surge two rivers: Palaesimundus flows through three channels into the harbor next to the town of the same name, five stades at its narrowest, 15 at its widest; the other named Cydara flows north in the direction of India. The nearest cape in India is the one called Coliacum, a four-day sail passing half-way by Sol's island.

87. The sea there is very green in color as well as bushy with trees; rudders wear away their foliage. When with us, as if it were a new sky, [the envoys] were amazed by the Septentriones [Great Bear] and Pleiades, and they explained that where they live the moon is only visible above the earth between the eighth and sixteenth [days], and that the huge, brilliant star Canopus shines at night. But they were most amazed by the fact that their shadows fell towards our sky, not towards theirs, and that the sun rises on the left and sets on the right rather than the other way round.

88. They reported that the side of their island along south-east India extends for 10,000 stades, and that beyond the Hemodi mountains they also face the Seres, well-known to them through trade. Rachias' father traveled there, and on arrival they were met by Seres, men of distinctly superhuman size, with red hair, blue eyes, harsh-sounding voices, and no shared language for commercial transactions. The rest [of the report] matched our traders' account: that on the far bank of a river, next to what [Seres] were selling, goods were set for them to take if satisfied with the exchange. The worthiest context for detesting luxury is to imagine yourself right there and to consider what is being sought, by what means, and why.

89. But even Taprobane, despite Nature's removal of it beyond our world, is not free from our vices. Gold and silver are valued there too; a marble like tortoise-shell, pearls and gems are prestigious; altogether their mass of luxury beats ours by far. They said that they had greater wealth, but that we lived more opulently. None of them has a slave, nobody sleeps in once it is daytime or during the day, their buildings rise [only] a moderate height from the ground, [the price of] grain is never inflated, there are no courts or litigation, they worship Hercules, and the people elect a king on account of his great age, mildness and lack of children; he abdicates if he subsequently produces one, so as not to create a hereditary kingship.

90. The people assign him 30 'rectors', and only by their majority vote is anyone condemned to death; there is also an appeal to the people, for which 70 judges are chosen; if they free the accused, the 30 lose all respect, a really severe disgrace. The king dresses like Father Liber, the others like Arabians.

91. If the king commits any wrong, he is to be punished with death, but nobody executes him; instead they all shun him and refuse to exchange even a word with him. Holidays are spent hunting, tigers and elephants being considered the most popular prey; the land is farmed with care, vines are not cultivated, but there are plenty of fruit-trees. They also enjoy fishing, especially for turtles whose shells roof families' homes – ones big enough are found! A typical life-span is 100 years.

XXV Ariani and neighboring peoples

92. This is what we have learned about Taprobane. The four satrapies (which we held over to this point [see 6.79]) are situated thus: mountain regions follow the peoples nearest to the Indus. Capisene had a city, Capisa, which Cyrus destroyed. Arachosia, with a river and town of the same name (which some called Cufis), was founded by Samiramis.

Erymandus river flows past Arachosian Parabaeste. Next to them, to the south and towards a section of the Arachotae, are located Dexendrusi, Paropanisidae to the north, the town Cartana – later called Tetrogonis – beneath the Caucasus. This region is opposite Bactria, then the Arii with their town Alexandria named after its founder, Sydraci, Dangalae, Parapinae, Catages, Mazi; near Caucasus Cadrusi with a town founded by Alexander.

93. Below these everything is more on the flat. After Indus, Ariana region – scorched by high temperatures and surrounded by deserts, although with plenty of shade spreading between them – clusters cultivators around two rivers especially, Tonberum and Arosape; Artacoana town, Arius river flowing past an Alexandria founded by Alexander (a town covering 30 stades), and Artacabene much lovelier as well as older, fortified again by Antiochus, covering 50 stades;

94. Dorisdorsigi people; Pharnacotis, Ophradus rivers; Prophthasia, Zaraspadum town, the Drangae, Euergetae, Zarangae, Gedrusi, towns Peucolis, Lyphorta, Methorgum; desert; Manain river, Acutri people, Eorum river, Orbi people, the navigable river Pomanus bordering the Pandae, also the Cabirus bordering the Suari with a harbor at its mouth, Condigramma town, Cophes river; into it flow Saddaros, Parospus, Sodamus, all navigable.

95. Some wish Daritis to be part of Ariana, and they give the dimensions of both as 1,850 long and half as wide as India. Others have located the Gedrusi and Sires across 138 miles, and then Ichthyophagi [Fish-Eaters] Oritae – who speak their own language not that of Indians – across 200 miles; Alexander forbade all Ichthyophagi to subsist on fish. Then the Arbii people have been located across 200 miles, with desert beyond; then Carmania and Persis and Arabia.

XXVI Voyages to India

96. But before we pursue these places in detail, it is appropriate to point out what Onesicritus reported when he sailed with Alexander's fleet from India to the Persian interior (as most recently recounted by Juba), and then to describe the sea route discovered at that time as it is in use today.

In their log Onesicritus and Nearchus omit <all> the names of stopping-places as well as distances; they set off from Xylinepolis (founded by Alexander), but there is no proper explanation in the first place of which river it lay on or of [just] where it was.

97. However, they do record the following as worth a mention: a town founded by Nearchus during the voyage, and Arbium river which can handle ships; opposite, an island 70 stades away, an Alexandria founded by Leonnatus on Alexander's orders in this people's territory, Argenuus with its safe harbor, the navigable river Tonberum along which Pasirae live; then Ichthyophagi for such a long stretch that it takes 30 days to sail past them; an island called Solis as well as Nympharum Cubile, which is red and where no animal survives, although the reason is unclear;

98. Ori people; Hyctanis, a Carmanian river which has a harbor and produces gold. At this point, they [Onesicritus and Nearchus] noted that the Septentriones first appeared, and that Arcturus is not visible every night and never for a whole night. The Achaemenids ruled up to this point. Metals – copper and iron and arsenic and cinnabar – are mined. Next is the cape of Carmania where the crossing to the Macae, an Arabian people on the opposite coast, is 50 miles; three islands of which only Oracta, 25 miles from the mainland, is inhabited and has fresh water; four further islands in the bay off Persis (around these swam 20-cubit long sea-snakes that terrified the fleet),

99. Athotadrus island, as well as Gauratae where the Gyani people [live]; in the middle of Persian gulf, Hyperis river capable of handling cargo ships, Sitioganus river on which it is a seven-day voyage to Pasargadae, the navigable Phristimus river, an unnamed island, Granis river which can handle ships of average size (it flows through Susiane, with Deximontani who produce bitumen living on the right bank), Zarotis river with its mouth difficult for all but experts, two small islands; then a shallow passage as if through a marsh, but still penetrable along certain channels;

100. Euphrates mouth, a lake formed by the Eulaeus and Tigris next to Charax, then Susa by Tigris. There, after a three-month voyage, they found Alexander celebrating a festival, seven months after he had left them at Patala. Such was the route of Alexander's fleet.

Later, the most reliable route to be found seemed the one from Arabia's cape Syagrum to Patale by means of a west wind called Hippalus locally; it was estimated to be 1,332 miles.

101. In the subsequent period it was considered a shorter and safer route to make for the Indian port Zigerus from the same cape, and for a long time this was the route sailed, until a merchant found a shortcut and greed brought India closer: in fact the voyage is [now] made annually, with cohorts of archers on board because pirates have been a severe threat.

There will be no cause to apologize for describing the entire route from Egypt now that reliable information about it is available for the first time: the topic merits attention because in no year does India drain our empire of less than 50,000,000 *sesterces*, sending in return goods sold among us for 100 times their cost.

102. Two miles from Alexandria is the town Juliopolis; from there they sail on Nile 309 miles to Coptus, a trip that takes 12 days when the Etesiae are blowing. From Coptus the journey is by camel, with stations organized for taking water: the first, after 32 [miles], is called Hydreuma, the second in the mountains a day's journey, the third at another Hydreuma 85 from Coptus, the next in the mountains; then 184 from Coptus, Apollo's Hydreuma again in the mountains; then 236 from Coptus to Novum Hydreuma.

103. There is also another Hydreuma Vetus named Trogodyticum where a garrison two miles away on a branch-road keeps watch; the distance from Novum Hydreuma is seven. Then Berenice town with a harbor on Red sea, 257 miles from Coptus. But since the greater part of the journey is made at night because of the heat and the days are spent at rest-stops, the whole journey from Coptus to Berenice takes 12 days.

104. They begin to sail in midsummer before Canis rises or immediately after its rising, and after about 30 days they come to Arabian Ocelis or Cane, in an incense-producing region. There is also a third harbor called Muza which those sailing to India do not make for, unless they are traders of incense and Arabian perfumes. Inland, a town called Sapphar – a royal capital – and another [called] Save. When making for India, it is most practical to depart from Ocelis; from there with the Hippalus wind, it is a 40-day sail to Muziris, the first market in India. This has little appeal because of nearby pirates who occupy a place called Nitriae, and the goods on offer are limited; in addition, where ships lie [at anchor] is far from land, and cargoes are brought in and carried out by lighters. When I produced this work, Caelobothras was king there.

105. There is another more useful harbor belonging to the Naecyndes people which is called Becare. Pandion was ruling there, far from the market, in a distant town upcountry called Modura. The region from which they convey pepper to Becare in single tree-trunk lighters is called Cottonara. None of these names of peoples or harbors or towns is found in any earlier account, which is an indication of the changes occurring in the local situation.

106. They sail back from India at the beginning of the Egyptian month Tybis, our December, or at any rate before the sixth day of the Egyptian Mechir, that is, before our Ides of January [January 13]: hence it happens that they return within a single year. They sail from India with a southeast wind and, when they have entered Red sea, with the African or south wind. Now let us return to our main subject.

XXVII Carmania

107. Nearchus wrote that Carmania's coast extends 600[1] miles. From its start to Sabis river 100 miles; then a 25-mile stretch with vineyards and fields to Ananis river; the region is called Armuzia. Carmania's towns, Zetis and Alexandria.

XXVIII Persian gulf

Then also in this area the sea twice bursts into the land. What we call Red sea, Greeks call Erythrean after king Erythras or, as some say, because of the belief that it gains such a color from the sun's reflection, while others say [its name comes] from the sand and land, and others from the character of the water itself.

108. At any rate, it is divided into two bays. The one to the east called Persian is, according to Eratosthenes, 2,500 round; opposite is Arabia whose coastline is 1,500 long. Back on its other side it is surrounded by a bay named Arabian. Ocean flowing in here is called Azanian. [Some] have made the opening of Persian [gulf] five wide, others four. It has been established that from here to the innermost part of the bay in a straight line is nearly 1,125, and that it has the shape of a human head.

---

[1] Literally here, "12 × 50."

109. Onesicritus and Nearchus wrote that it is 1,700 from Indus river to Persian gulf and on to Babylon by the Euphrates marshes. In a corner of Carmania, Chelonophagi [Turtle-Eaters] who cover their huts with turtles' shells and feed on their flesh. They live right on the cape after Arabis river, their bodies hairy all over except for their heads, and their clothing fish-skins.

110. From their region towards India is said to be Caecandrus, a desert island 50 miles out to sea, and next to it, with a strait flowing between, Stoidis which produces pearls. After a cape, Harmozaei adjoin the Carmani. Some place Arbii between them for an entire 421-mile stretch of coast. Here Macedonians' harbor and Alexander's altars on a cape. Rivers Siccanas, then Dratinus and Salsum.

111. After it cape Themisteas; Aphrodisias, an inhabited island. Next Persis begins at Oratis river which divides it from Elymais. Opposite Persis the islands Psilos, Cassandra, Aracha – with a very high mountain – sacred to Neptune. Facing west, Persis itself takes up 550 miles of shoreline, rich to the point of luxury, with its name changed to Parthian long ago. Now for a brief account of their empire:

XXIX Parthians' kingdoms

112. There are 18 Parthian kingdoms in all; for this is how they divide their provinces around two seas, as we said [6.41], Red [sea] to the south and Hyrcanian to the north. Of these, the 11 designated 'upper' begin at the border of Armenia and the Caspian shoreline, and extend to the Scythians, with whom they live on equal terms. The remaining seven kingdoms are called 'lower'. With regard to Parthians, there has always been Parthyaea at the foot of the mountains very frequently referred to, which border all these peoples.

113. It has the Arii to the east, Carmania and the Ariani to the south, Median Pratitae to the west, Hyrcani to the north, with deserts around in every direction. More remote Parthians are called Nomads. On the near side of the desert to the west, their cities we have mentioned: Issatis and Calliope [6.44]; to the north-east Pyropum; to the south-east Maria; in the middle Hecatompylos, Arsace, Nisiaea a splendid region of Parthyene where there is Alexandropolis [named] after its founder.

114. At this point it is also necessary to draw attention to the position of the Medes, and to circle the shape of their country to Persian sea so that the rest should then be grasped more easily. For Media, extending to the west and lying diagonally across Parthia, blocks both kingdoms. Thus it has the Caspi and Parthians to the east, Sittacene and Susiane and Persis to the south, Adiabene to the west, Armenia to the north.

115. Persians have always lived by Red sea, which is why it is called Persian gulf. Ceribobus is the coastal region there; but the place where it reaches the Medes is called Climax Megale because of the tough mountain-climb by stages, with a narrow entrance to Persepolis, the kingdom's capital destroyed by Alexander. In addition, on the far border is a Laodicea founded by Antiochus.

116. Then to the east, Magi occupy Phrasargis fortress with Cyrus' tomb there. Ecbatana town is theirs too, transferred to the mountains by king Darius. Paraetaceni protrude between Parthians and Ariani. The 'lower' kingdoms are confined by these peoples and Euphrates. The rest we shall address after Mesopotamia, except for its tip and the Arabian peoples mentioned in the previous book [5.86–90].

XXX Mesopotamia

117. All Mesopotamia used to belong to Assyrians who, except for Babylon and Ninus, were dispersed in villages. Macedonians collected them into cities because of the richness of the soil. Besides the towns already mentioned it includes Seleucia, Laodicea, Artemita; and – in [the territory of] Arabian people called Orroei and Mardani – Antiochia founded by Nicanor, prefect of Mesopotamia, and called 'Arabian'.

118. Bordering these in the interior are Arabian Eldamari above whom, on Pallaconta river, Bura town, the Salmani and Arabian Masei; next to the Gordyaei are Azoni with Zerbis river dropping into Tigris; next to Azoni the Silices (mountain people), and Orontes with Gaugamela town west of them, as well as Suae on a cliff. Above the Silices, Sitrae with Lycus coming through from Armenia; south-east of the Sitrae Azochis town, then on plains the towns Diospege, Polytelia, Stratonicea, Anthemus.

119. Close to Euphrates, Nicephorium – already mentioned [5.86] – founded by Alexander's order because of its advantageous location.

Apamea in Zeugma has also been mentioned already [5.86]. Those proceeding east from it come to the fortified town Caphrena, once 70 stades wide and called a satraps' capital because tribute was collected there; nowadays it has been reduced to a citadel.

120. Thebata remains as it was, as does Oruros, the limit of the Roman empire during Pompey the Great's command and 50,200 [stades?] from Zeugma. There are those who relate that Euphrates was diverted by efforts of the prefect Gobares where we said it was split [5.90] to prevent Babylon being harmed by its raging current, and that every Assyrian called it Narmalchas, which means 'royal river'. Where it divides used to be a town Agranis, one of the largest, which Persians destroyed.

121. Babylon, capital of the Chaldaic peoples, was for a long time supremely famous among cities worldwide, and so the rest of Mesopotamia and Assyria was called Babylonia. It has two walls, 60 miles round, 200 feet high and 50 wide (there are three more digits [finger-widths] in their foot than in ours), and Euphrates flows through it with amazing structures on both sides. The temple of Jupiter Belus remains there still; he was the inventor of astronomy.

122. But it has reverted to wilderness after being depleted by nearby Seleucia, which Nicator founded for that very reason just short of the fortieth milestone at the confluence of the Euphrates canal and Tigris. [Seleucia] nonetheless has the further name Babylonia and today is free, with its own laws and a Macedonian culture. Its urban population is reported as 600,000, with walls laid out like an eagle spreading its wings and the most fertile fields in the entire east. To deplete it in turn, Parthians founded Ctesiphon in Chalonitis, near the third milestone from it, nowadays the capital of their kingdoms; when it was no longer thriving, king Vologesus recently founded another town, Vologesocerta, nearby.

123. Still today in Mesopotamia are towns [including] Hipparenum – also a center of Chaldaean learning, like Babylon – by a river which drops into the Narraga, hence the state's name. Persians destroyed the walls of the Hippareni. Orcheni too, a third center of Chaldaean learning, is located in the same area, facing south. After these, the Notitae and Orothophanitae and Gnesiochartae.

124. Nearchus and Onesicritus state that Euphrates is navigable for the 412 miles from Persian sea to Babylon, but according to later writers 440 to Seleucia, and according to Juba 175½ from Babylon to Charax. Some say that it flows beyond Babylon in a single channel for 87 before being dispersed into canals, and that its whole course is 1,200 miles. The variety of authors accounts for the inconsistent calculations, just as different Persian writers also give different measurements for the *schoenus* and *parasang*.

125. Where [Euphrates] ceases to give protection with its channel once its course reaches the border of Charax, it is immediately infested by raiders, the Attali, an Arabian people, beyond whom are Scenitae; but where it curves are Arabian Nomads as far as the Syrian desert, after which, as we have said [5.87], it turns south leaving Palmyra's desert behind.

126. From Mesopotamia's starting-point Seleucia is a 1,125-mile sail by the Euphrates; to sail by the Tigris from Red sea is 320, from Zeugma 724. Zeugma is 175 from Syrian Seleucia on our [Mediterranean] shore. This then is the width on land between the two seas, but that of the Parthian kingdom 944.

XXXI Tigris

There is also a Mesopotamian town called Digba on the Tigris bank near the confluence.

127. Something should be said about the Tigris itself too. It rises in the region of Greater Armenia from a very visible source on a plain. The place is called Elegosine and the river, where it flows rather slowly, Diglito. Once it accelerates, because of its speed it begins to be called Tigris, since this is the Median word for arrow. It flows into lake Aretissa, which floats all heavy objects thrown into it and gives off clouds of nitre. There is a single type of fish there, one that stays clear of the stream flowing through, just as the fish in it do not swim from Tigris into the lake.

128. But [Tigris] is propelled with both a different course and color, and after crossing [the lake] and reaching mount Taurus plunges into a cave, slides underground and bursts out on the other side. The place is called Zoaranda; that it is the same river is clear because it carries flotsam right through. Then it crosses another lake, called Thospites, and plunges again into underground channels, re-emerging after 22 miles around Nymphaeum. In Arrhene region it flows so close to the Arsanias that, according to Claudius Caesar, when both are in spate they merge and yet do not mix, since the less substantial Arsanias floats on top for a distance of nearly four [miles], and then – after the two separate – merges with Euphrates.

129. However, after Parthenia and Nicephorion – well-known rivers from Armenia – have become its tributaries, Tigris separates the Arabian Orroei and Adiabeni and forms Mesopotamia, as we have said [6.25]. After traversing the mountains of the Gordyaei around town Apamea in Mesene, 125 miles short of Seleucia Babylonia it is divided into two channels, one heading south to Seleucia while supplying Mesene with water, the other turning north behind the same people and cutting through the Cauchan plains. Once the waters have rejoined, it is called Pasitigris.

130. After that, Choaspes from Media becomes its tributary and, as we said [6.122], after being brought between Seleucia and Ctesiphon it discharges into Chaldaic lakes and expands them to a width of 62 miles; then, flowing in a vast channel, it is carried past Charax town on the right into Persian sea through a ten-mile-wide mouth. The mouths of the two rivers used to be 25 miles apart (or, as others say, seven), both navigable. But the Orcheni long ago blocked Euphrates, as did their neighbors for irrigating fields, and it only reaches the sea through Tigris.

131. The region next to Tigris is called Parapotamia. In it, as mentioned [6.129], is Mesene with its town Dabitha. Next to it is Chalonitis with Ctesiphon, [a region] wooded with not only palms but also olives and orchards. Mount Zagrus extends as far as this from Armenia, coming between the Medes and Adiabeni above Paraetacene and Persis. Chalonitis is 380 miles from Persis; some say that by the shortest routes it lies the same distance from Caspian sea as from Syria.

132. Between these peoples and Mesene is Sittacene, also called Arbelitis and Palaestine. Its town Sittace was founded by Greeks, with Sabdata to the east and to the west Antiochia between the two rivers Tigris and Tornadotus, also Apamea which Antiochus named after his mother; Tigris surrounds it and Archous splits it.

133. Below is Susiane and within it the Persians' ancient capital Susa founded by Darius, Hystaspes' son. It is 450 miles from Seleucia Babylonia, and the same distance from the Medes' Ecbatana across mount Carbantus. On Tigris' northern channel is the town Barbitace, 135 miles from Susa. Here are the only human beings who – remarkably – loathe gold, which they collect and bury so that no-one can use it. Bordering the Susiani to the east are raiders called Oxii, and 40 communities of thoroughly savage Mizaei.

134. Above them as well as being Parthian subjects are Mardi and Saitae spreading above Elymais, which in our account adjoins the coast of Persis [6.111]. Susa is 250 miles from Persian sea. When Alexander's fleet came here on the Pasitigris, there was a village called Aple on the Chaldaic lake, a 62½-mile sail from Susa. For the Susiani, the nearest people to the east are Cossiaei, and above Cossiaei to the north Massabatene at the foot of mount Cambalidus, which is a spur of the Caucasus with a very easy route to the Bactrians from there.

135. Susiane is separated from Elymais by Eulaeus river, which rises among Medes and after burial in an underground channel for a limited distance resurfaces and glides through Massabatene. It goes round Susa's citadel and the temple of Diana that those peoples highly venerate; there is also marked reverence for the river itself insofar as the kings drink from it exclusively, and thus convey it[s water] great distances. It gains as tributaries the rivers Hedyphon which goes past the Persians' Asylum [town], Aduna coming from the Susiani. Next to it Magoa town 15 miles from Charax; but some place [Magoa] at the extreme limit of Susiane very close to wilderness.

136. Below the Eulaeus, Elymais borders Persis on the coast, 240 miles from Oratis river to Charax. Its towns, Seleucia and Sostrate situated on mount Chasirus. As we mentioned previously [6.99], the coast in front of it – like Lesser Syrtis – is made inaccessible by mud since a mass of sediment is brought down by Brixa and Ortacia rivers, and Elymais itself is so marshy that no-one can reach Persis without detouring round it. It is also infested with snakes carried down by rivers. A particularly impassable part of it is called Characene after the Arabian town [Charax] that marks the end of these kingdoms; I shall discuss it [6.138] after first relating Marcus Agrippa's opinion.

137. For by his account Media and Parthia and Persis have Indus as their eastern boundary, Tigris to the west, Caucasian Taurus to the north, Red sea to the south, and their extent is 1,320 miles long, 840 wide; he said further that Mesopotamia itself is enclosed by Tigris to the east, Euphrates to the west, the Taurus to the north, Persian sea to the south, and that it is 800 miles long, 360 wide.

138. [The people of] Charax, a town situated deep within Persian gulf where Arabia (further called Eudaemon) projects out, live on a manmade hill two miles wide between the confluence of Tigris to the right and Eulaeus to the left. It was first founded by Alexander the Great for settlers brought from a royal city Durine (which then ceased to exist), and for disabled soldiers left there; he had ordered that it be called Alexandria, with one district Pellaeus – [named] after his own country – assigned specifically to Macedonians.

139. The rivers destroyed this town; later Antiochus, the fifth king, rebuilt it and named it after himself. After it was wrecked again, Spaosines, Sagdodonacus' son and king of the neighboring Arabians (but not Antiochus' satrap, as Juba said mistakenly), rebuilt it with protective breakwaters, gave it his own name, and secured the adjacent location for a length of six miles and width of a little less. In the past it used to be ten stades from the seashore – even [Agrippa's map in] Vipsania Porticus has it on the sea – but according to Juba 50 miles.

140. Nowadays Arabian envoys, as well as our own traders who have come from there, assert that it is 120 from the shore. Nowhere else has sediment brought down by rivers advanced further or more swiftly. More surprising is the fact that this has not been pushed back by the tide that comes up far beyond it.

141. It has not escaped me that this is the birthplace of Isidorus, the most recent author to describe the world, sent by Deified Augustus to the east to make a complete record when his elder son [Gaius Caesar] was going to Armenia to deal with the Parthians and Arabians; nor have I forgotten what we said at the beginning of this work that each author appears to be the most accurate regarding his own territory [3.1]. In this section, however, I opt to follow the Roman military and king Juba, who wrote books for the same Gaius Caesar about this same Arabian expedition.

XXXII Arabia

142. Arabia's size is greater than all other peoples. At its longest, as we have said [5.85], it descends from mount Amanus in the region of Cilicia and Commagene. Many of its peoples were brought there by Tigranes the Great, [some] of their own accord going to our [Mediterranean] sea and the Egyptian shore already discussed [5.65], and also Nubaei who penetrate as far as mount Libanus in central Syria; bordering them are Ramisi, then Teranei, then Patami.

143. The Arabia peninsula itself projects out between two seas, Red and Persian. Some contrivance of Nature surrounded it with sea and made it similar in shape and size to Italia, so that it even views the exact same part of the sky; it is fortunate to be situated thus. We have described its communities from our sea up to Palmyra's desert; now we shall complete the rest from there. As we said [5.65, 86; 6.125], hemming in the Nomads and those who harass the Chaldaeans are Scenitae who are themselves nomadic, but named after their tents which – made from goat's hair material – are pitched anywhere they like.

144. Next, Nabataeans who live in a town named Petra in a valley just under two miles wide and surrounded by impenetrable mountains with a river flowing through it. It is 600 from town Gaza on the shore of our sea, and 135 from Persian gulf. Two roads meet here, one for those on their way from Syria to Palmyra, the other for those coming from Gaza.

145. After Petra, Omani were settled all the way to Charax, with the once well-known towns Abaesamis and Soractia founded by Samiramis; nowadays these are wilderness. Next is a town on the Pasitigris bank named Forat, subject to the king of the Characeni; people from Petra convene here and sail on to Charax, a 12-mile trip with a favorable current. But those who sail from the Parthian kingdom come to Teredon village. Below the confluence of Euphrates and Tigris the left bank is occupied by Chaldaeans, the right by nomad Scenitae.

146. Some say that two other towns, far apart, are also passed when sailing on Tigris: Barbatia, then Dumatha, which is a ten-day sail from Petra. Our traders say that Apamea too is subject to the king of the Characeni, its location being where a Euphrates overflow and Tigris merge; so Parthians, when poised to invade, are blocked by the flood-barriers that have been built.

147. We shall now describe the coast from Charax, which was first explored for Epiphanes: the place where a Euphrates mouth used to be, Salsum river, cape Caldone, an estuary along 50 [miles] of coast more like a whirlpool rather than sea, Achenum river, desert for 100 miles as far as Icarus island, Capeus bay where Gaulopes and Gattaei live, Gerrhaicus bay, town Gerrha five miles wide; it has towers made from square salt-blocks.

148. From the seashore to Attene region 50, with Tylos island opposite it and the same number of miles from the seashore. It is best-known for its masses of pearls and its town of the same name, with a second smaller one 12½ miles from a cape. Large islands, to which no-one has gone, are said to be visible beyond it. It is 112½ miles round, further than that from Persis, and approached by a single narrow channel. Ascliae island, the Nochaeti, Zurazi, Borgodi, Catharraei, Nomades peoples, Cynos river.

149. Juba says there is no information about the sailing-route beyond it on the [far] side because of rocks, although he omits to mention Batrasavaves, a town of the Omani, and Omana which earlier [writers] claimed to be a well-known harbor of Carmania, also Homna and Attana, towns that nowadays our traders say are the busiest in Persian sea. After Canis river, according to Juba, a mountain which appears scorched, Epimaranitae peoples, then the Ichthyophagi, a desert island, Bathymi peoples <... >, Eblythaei mountains, Omoemus island, Mochorbae harbor, Etaxalos and Inchobriche islands, Cadaei people,

150. many islands without names, but also the well-known Isura, Rhinnea and one very nearby where there are stone stelae inscribed in strange lettering, Coboea harbor, the deserted Bragae islands, Taludaei people, Dabanegoris region, mount Orsa with its harbor, Duatas bay, many islands, mount Tricoryphos, Chardaleon region, Solanades and Cachina islands as well as some belonging to the Ichthyophagi. Then Clari, the Mamaean shore with its gold mines, Canauna region, Apitami and Casani peoples, Devade island, Coralis spring, Carphati, Alaea and Amnamethus islands, Darae people,

151. Chelonitis islands and many belonging to Ichthyophagi, deserted Odanda, Basa, many [islands] belonging to the Sabaei. Thanar and

Amnum rivers, Doricae islands, Daulotos and Dora springs, islands Pteros, Labatanis, Coboris, Sambrachate, and a mainland town with the same name. To the south, many islands, Camari being the largest, Musecros river, Laupas harbor, the Scenitae Sabaei, many islands and their emporium Acila (a starting-point for voyages to India),

152. Amithoscatta region, Damnia, the Greater and Lesser Mizi, Drymatina, the Macae with their cape 50 miles opposite Carmania. A remarkable event is said to have occurred here: Numenius, after being placed in charge of Mesene by king Antiochus when fighting Persians, defeated them here with his fleet and on the same day, after the tide turned, did so again with his cavalry, and then erected twin monuments, to Jupiter and Neptune, on the same spot.

153. Far out at sea lies Ogyris island, famous as the burial place of king Erythras; it is 125 miles from the mainland, 112½ round. Equally famous is a second island in Azanian sea, Dioscuridu, 280 from the tip of cape Syagrum.

The rest on the mainland, also to the south, the Autaridae an eight-day journey into the mountains, Larendani and Catabani people, Gebbanitae with several towns, but Nagia and Thomna (with 65 temples) the largest; this indicates its size.

154. A cape 50 away from where Trogodytae are on the mainland, Thoani, Actaei, Chatramotitae, Tonabaei, Antiadalaei, Lexianae, Agraei, Cerbani, Sabaei the best-known of the Arabians because of their incense; these peoples stretch out to both seas. Their towns on Red sea shore: Merme, Marma, Corolia, Sabbatha; inland towns: Nascus, Cardava, Carnus, and Thomala a perfume-marketing center.

155. Atramitae are one group of them, with their capital Sabota including 60 temples within its walls; yet the royal capital of them all is Marebbata. They occupy a bay 94 [miles round], filled with spice-producing islands. Bordering Atramitae inland are Minaei.

Also living along the sea are Aelamitae with their town of the same name, next to whom are Chaculatae, Sibi town which Greeks call Apate, the Arsi, Codani, Vadaei with their large town, Barasasaei, Lechieni, Sygaros island which does not admit dogs, so after being abandoned and wandering along the seashore they die.

156. An innermost bay where Laeanitae live, who give their name to it; their royal capital Agra and on the bay Laeana or, as others say, Aelana: we have written the name of the bay itself as Laeanitic, while others have written it as Aelanitic, Artemidorus as Alaenitic, Juba as Leanitic.

Arabia's perimeter from Charax to Laeana is said to amount to 4,765 miles, but Juba considers it a little less than 4,000. It is widest in the north between towns Heroum and Charax.

157. The rest of its inland should now be described. The ancients located the Thimanei next to Nabataeans, but nowadays there are the Taveni, Suelleni, Araceni, Arreni with a town where all business is conducted, the Hemnatae, Avalitae, towns Domata and Haegra, the Thamudaei, Baclanaza town, the Chariattaei, Toali, Phodaca town, the Minaei who consider themselves descendants of Creta's king Minos, the Carmei part of them; 14 miles [away] the town Mariba, [then] Paramalacum by no means an unimportant place, with the same to be said of Canon.

158. The Rhadamaei – also thought to be descended from Rhadamanthus, Minos' brother – the Homeritae with town Mesala, the Hamiroei, Gedranitae, Amphryaei, Lysanitae, Bachylitae, Samnaei, Amaitaei with towns Nessa and Chenneseris, the Zamareni with towns Sagiatta and Canthace, the Bacaschami with town Riphearina which is their word for barley, the Autaei, Ethravi, Cyrei with town Elmataeis, the Chodae with town Aiathuris 25 miles into the mountains, where there is Aenuscabales spring which means 'of camels';

159. Ampelome town a Milesian colony, Athrida town, the Calingi whose town Mariba means 'masters of all', towns Pallon and Murannimal next to a river through which Euphrates is thought to discharge, the Agraei and Ammoni peoples, Athenae town, the Caunaravi meaning 'richest in cattle', Chorranitae, Cesani, Choani. There also used to be the Greek towns Arethusa, Larisa, Chalcis, since destroyed in various wars.

160. To date, Aelius Gallus of the equestrian order is the one person to have led a Roman force into that land, because Augustus' son Gaius Caesar only glimpsed Arabia. Gallus destroyed towns not named by previous authors: Negrana, Nestus, Nesca, Magusum, Caminacum, Labaetia, Mariba just mentioned [6.159] (six [miles] round), also Caripeta which was the furthest he advanced.

161. He reported other discoveries: that the nomads' diet is the milk and flesh of wild animals, and that the rest extract wine (as Indians do) from palm-trees and oil from sesame. [He reported] that Homeritae have the greatest numbers, and Minaei lands fertile with lofty palm-groves as well as being rich in cattle; Cerbani and Agraei, and especially Chatramotitae, are outstanding warriors; Carrei have the most extensive and most fertile fields, Sabaei are the wealthiest because of the fertility of their perfume-producing forests, gold mines, irrigated fields, and high yields of honey and wax. We shall discuss perfume in its own book [Book 12*].

162. Arabians routinely wear turbans or do not cut their hair; they shave their beards but leave a moustache; some also leave [their beards] uncut. And as strange as it is to say, among their countless communities equal numbers live on trade or raiding. Altogether they are the richest peoples, since the greatest amounts of Romans' and Parthians' wealth remain in their hands, given that they sell what they take from sea and forest and buy nothing in return.

XXXIII Red Sea bay

163. We now follow the rest of the coast opposite Arabia. Timosthenes reckoned that the whole bay was a four-day sail long, two days wide, with 7½-mile straits; Eratosthenes [says] that it is 1,200 along either side from the mouth, Artemidorus that it is 1,750 along the Arabian side,

164. but 1,137½ miles along the Trogodytic side up to Ptolomais, Agrippa that it is 1,732 on either side. Most authors have reported the width as 475, while some say that the straits facing the south-east extend four, others seven, yet others 12.

165. It includes the following sites: after Laeanitic bay another bay which Arabians call Aeas, where the town Heroon is. Cambysu, where invalid soldiers were brought, also used to be between the Neli and Marchadae. Tyres people, a harbor of the Danei from which a navigable canal leads to Nile covering a distance of 62½ to the section called the Delta situated between the river and Red sea, [a project] conceived first of all by Sesostris king of Egypt, next by the Persian king Darius, then followed by Ptolomaeus who actually dug a canal 100 feet wide, 30 deep, 37 miles long as far as Fontes Amari.

166. Fear of a flood scared him from going further because it was discovered that Red sea is three cubits higher than the landmass of Egypt. Some do not give this explanation, but [say it was] so that Nile water, which provides the only drinkable supply, would not be spoiled by the influx of sea[-water]. Even so, the entire journey from Egyptian sea is regularly made by land along three routes: one from Pelusium through the desert, where the route is invisible without the guidance of rods fixed in the ground because breeze immediately covers footprints;

167. the second [begins] beyond mount Casius and after 60 miles rejoins the Pelusiac road (Arabian Autaei live here); the third from Gerrhum – [also] called Agipsum – [passing] by the same Arabians, nine shorter, but made rough by mountains and lacking water. All these routes lead to Arsinoe, founded by Ptolomaeus Philadelphus on Carandra bay and named after his sister; he was the first to explore the Trogodytic region, and he gave the name Ptolomaeus to the river which flows past Arsinoe.

168. Next is a small town Aenum – there are other writers who call it Philoterias – then there are Asaraei, a wild Arabian people intermarried with Trogodytae, islands Sapirine and Scytala, then desert as far as Myos Hormos where Tatnos spring is, mount Aeas, Iambe island, many harbors, Berenice town named after Philadelphus' mother and with a route to it from Coptus which we have described [6.103], Arabian Autaei and Gebadaei,

XXXIV Trogodytice

169. Trogodytice [region] called Midoe in the old days (Midioe by others), mount Pentedactylos, several islands [called] Stenae Thyrae, just as many [called] Halonesi, Cardamine, Topazos which gave its name to the gem. A bay crammed with islands: of these the ones called Maraeu have sources of water, while those of Eraton are arid; there used to be royal prefects for them. Inland the Candaei are called Ophiophagi because it is their habit to eat snakes; their region is more fertile than any other.

170. Even though Juba appears to have investigated it all with the greatest care, in this region he has omitted – unless a mistake has occurred in copying – a second Berenice also called Panchrysos, and a third called Epi Dires ['for Hunting'] remarkable for its location: for it sits on a neck [of land] projecting far out, where Red sea's passage puts Arabia 7½ miles away. Citis island here, itself also yielding topaz.

171. Further on, forests with the Ptolomais founded by Philadelphus for elephant-hunts – hence its further name 'for Hunting' – and lake Monoleus nearby. This is the region referred to in our second book [2.183] where, for 45 days before the solstice and the same number after it, shadows disappear in the sixth hour and for the remaining hours they fall to the south, but on other days to the north; yet, at the Berenice we mentioned first [6.103], on the very day of the solstice shadows completely disappear in the sixth hour and nothing else extraordinary happens, it being 602 miles from Ptolomais. This occurrence proved a powerful illustration and the source of prodigious insight because here the world was exposed and, with incontrovertible arguments from the shadows, Eratosthenes then grasped how he could proceed to measure the earth.

172. From here Azanian sea, a cape called Hippalum by some writers, lake Mandalum, Colocasitis island and out in open water many [others] full of turtles, Sace town, Daphnis' island, the town of the Adulites: it was founded by Egyptians' slaves who had fled from their masters.

173. Here Trogodytae, as well as Aethiopians, have their largest market; it is a five-day voyage from Ptolomais. A very great amount of ivory is conveyed here, as are rhinoceros horns, hippopotamus hides, turtle shells, apes, slaves. Above the Aethiopian Aroterae are islands called Aliaeu, as well as Bacchias and Antibacchias and Stratioton. After this on the coast of Aethiopia an unexplored bay – which surprises me, since traders investigate more remote places – and a cape with Cucios spring, which seafarers make for.

174. Beyond, Isis' harbor, a ten-day row from the town of the Adulites; Trogodytic myrrh is brought to it. Two islands off the harbor are called Pseudopylae, while there are also two in it called Pylae; on one of these, stone stelae with unknown lettering. Further on Avalitu bay, Diodorus' island and other deserted ones, as well as a deserted stretch of mainland, Gaza town, Mossylites cape and harbor from which cinnamon is exported. Sesostris led an army as far as this. Some place a single Aethiopian town, Barigaza, further down the shore.

175. Juba favors the view that the Atlantic sea begins after Mossylic cape, and that it is navigable with a north-west wind along his Mauretaniae up to Gades. His opinion in its entirety must not be omitted here: he argues that, in a straight line, it is 1,500 miles from the Indian cape called Lepte Acra (or Drepanum by others) past Exusta island to the Malichu ones, then 225 miles to the place called Scenei, then 150 to Adanu islands. Thus it is 1,875 miles to the open sea.

176. Everyone else's opinion has been that it is impossible to sail here because of the sun's heat. Moreover, cargoes are threatened from the islands by Arabians called Ascitae because they lay planks over a pair of oxhide bags and commit piracy with poisoned arrows. Juba also says that there are Trogodytic people called Therothoae – because they hunt with amazing speed, just as Ichthyophagi swim like sea-creatures – Bangeni, Zangenae, Thalibae, Saxinae, Sirechae, Daremae, Domazenes.

177. He says, moreover, that the communities living on Nile from Syene as far as Meroe are not Aethiopians but Arabians; also that Sol's town, which we have located in Egypt not far from Memphis [5.61], had Arabian founders. There are those, too, who take the further [Nile] bank from Aethiopia and attach it to Africa. But they settled the banks because of the water.[2] I leave everyone to form their own opinion, and proceed to set down the towns on both sides in the order they are recorded from Syene.

XXXV Aethiopia

178. And first on the Arabian side Catadupi people, then the Syenitae, towns Tacompsos which some call Thathice, Arama, Sesamos, Andura, Asarduma, Aidomacum with Arabeta and Bogghiana, Leuphitorga, Tautare, Emeae, Chindita, Noa, Goploa, Gistate, Megadale, Aremni, Nups, Direa, Patigga, Bacata, Dumana, Radata where a golden cat was worshiped like a god, Boron inland very near Mallos, Meroe. This is Bion's account.

179. Juba differs: fortified town Megatichos between Egypt and Aethiopia (called Mirsion by Arabians), then Tacompsos, Aramum, Sesamum, Pidema, Muda, Corambis next to a source of bitumen,

---

[2] This sentence may be a later addition.

Amodata, Prosda, Parenta, Mania, Tessata, Galles, Zoton, Grau Come, Emeum, the Pidibotae, Endondacometae (nomads living in tents), Cistaepe, Magadale, Parva Primis, Nups, Direlis, Patinga, Breves, Magasneos, Egasmala, Cramda, Denna, Cadeum, Athen, Nabatta, Alana, Macua, Scammos, Gora on an island; after these, Abale, Androcalis, Sere, Mallos, Agoce.

180. On the African side are said to be Tacompsos – either a second town with the same name or part of the one above [6.178, 179] – Mogore, Sea, Aedosa, Pelenariae, Pindis, Magassa, Buma, Lintuma, Spintum, Sidopt, Gensoe, Pindicitor, Agugo, Orsum, Suara, Maumarum, Urbim, Mulon (a town called Hypaton by Greeks), Pagoartas, Zamnes (where there begin to be elephants), Mambli, Berressa, Coetum. There also used to be a town opposite Meroe, Epis, but it was destroyed before Bion wrote.

181. These are the places recorded up to Meroe, but at this time hardly any of them exist on either side of the river. Recently, for certain, praetorian soldiers sent with a tribune by the *princeps* Nero to explore – when he was contemplating war in Aethiopia as well as elsewhere – reported a wilderness. However, the Roman military did penetrate even there during Deified Augustus' time, led by Publius Petronius who was himself also of the equestrian order and prefect of Egypt. He reduced their towns, and we shall mention only those which we found, in order: Pselcis, Primi, Bocchis, Forum Cambusis, Athena, Stadissis whose population is deafened by the din of Nile's fast flow. He also plundered Napata.

182. He advanced as far as 870 miles from Syene, but this Roman campaign did not make a desert there. Aethiopia, alternately in control and in subjection, had been worn out by wars with the Egyptians; it was famous and powerful even up to the Trojan wars, when Memnon was king. And it is clear from the stories about Andromeda that it controlled Syria and our [Mediterranean] shores in king Cepheus' time.

183. Likewise there have been various accounts of its dimensions too, first by Dalion who traveled far beyond Meroe, then by Aristocreon and Bion and Basilis, and even by the younger Simonides who lived in Meroe for five years while writing about Aethiopia. Timosthenes, prefect of Philadelphus' navies, reported that the journey from Syene to Meroe took 60 days, but he gave no distance for it; Eratosthenes gives it as

625 [miles], Artemidorus as 600. Sebosus says that the distance from Egypt's furthest [northern] point is 1,675 miles, although those just mentioned say 1,250.

184. But this whole controversy has recently been concluded because Nero's explorers have reported that it is 976 from Syene by these stages: 79 miles from Syene to Hiera Sycaminos, 72 from there to Tama in the Aethiopian Euonymites' region, 120 to Primi, 64 to Acina, 25 to Pitara, 106 to Tergedum. Gagaude island is in the middle of this area. From here onwards the birds called parrots were first seen; and after another [island], called Articula, the *sphingion* animal [ape], after Tergedum the *cynocephalos* [baboon]. From there 80 to Nabata, a small town and the only one of those mentioned [that still exists]. From it to Meroe island 360.

185. Only around Meroe was the foliage greener, and did something of forests appear as well as traces of rhinoceroses and elephants. Meroe town itself [was reported as] 70 miles from the initial landing on the island, and next to it – for those who go upstream by the righthand channel – is another island, Tadu, which forms a harbor.

186. The town [was reported] to have few buildings and a female ruler, Candace, a name used by queens in succession for many years past, with a sacred shrine of Hammon there too and sanctuaries all over the region. Otherwise, when Aethiopians were in overall control, this was a notably famous island. They said that it routinely fielded 250,000 armed men, supported 3,000 craftsmen. According to reports, still today Aethiopians have 45 kings.

187. The whole people was called Aetherian, then Atlantian, and finally Aethiopian after Aethiops, Vulcan's son. Little wonder that freakish likenesses of humans and animals are born around its furthest edges, since the liveliness of fire is the creator that forms their bodies and molds their shapes. It is definitely claimed that in the interior of the eastern part there are peoples without noses, the whole of their face being perfectly flat, while others have no upper lip, others no tongues.

188. One group, too, has a solid face, and without noses has only a single orifice through which to breathe, so ingests liquids through oat-stalks and subsists on its grains (oats grow wild there). Some replace speech by nods and gesturing with their limbs. Before [the time of] the Egyptian king Ptolomaeus Lathyrus some were ignorant of how to use fire. According to certain reports, there is a Pygmy people in the marshes where Nile rises. However, where we left off on the coast there is a chain of mountains which glows red as if on fire.

189. After Meroe, the whole region is bounded by Trogodytae and Red sea. Because it is a three-day journey from Napata to the Red sea shore, rainwater is saved for use in several places; the region in between them has great quantities of gold. Further on, Atabuli, an Aethiopian people, are in possession; then, opposite Meroe, Megabarri, called Adiabari by some. They control Apollo's town. Some of them are nomads who subsist on elephants.

190. Opposite them, on the African side, the Macrobii; again, after the Megabarri, Memnones and Dabelli and – a 20-day journey further on – Critensi. Beyond them, Dochi, then Gymnetes always nude, then Anderae, Mattitae, Mesagches; Hipsodores, black in color, daub their entire bodies with red ochre. Yet on the African side, Medimni, then nomads who subsist on baboons' milk, Alabi, Syrbotae who are said to be eight cubits tall.

191. Aristocreon relates that Tolle town is a five-day journey from Meroe on the Libyan side. From there 12 days to Esar, a town of Egyptians who fled from Psammetichus. He relates that the population was 300,000, and that the town Diaron on the Arabian side opposite was theirs. However, what he calls Esar, Bion calls Sape – a name meaning 'immigrants'. Their capital Sembobitis on an island, and a third [town] in Arabia, Sinat. Between the mountains and Nile are the Simbarri, Palugges, and on the mountains themselves many Asachae tribes. They are said to be a five-day journey from the sea; they live by hunting elephants. An island in Nile belonging to the Semberritae is ruled by a queen.

192. Eight days' journey from there Nubian Aethiopians, with their town Tenupsis located on Nile; the Sesambri in whose territory all four-footed animals, even elephants, have no ears. But on the African side Tonobari, Ptoemphani who have a dog for a king and infer its commands by its movements, Harusbi with a town located far from Nile, thereafter Archisarmi, Phalliges, Marigarri, Chasamari.

193. Bion says that there are also other towns on islands. After Sembobitis facing Meroe the whole journey takes 20 days; on the nearest island a Seberritan town ruled by a queen, and another [town], Asara; on the second [island] Darde town; on the third, called Medoe, Asel town; the fourth named Garroe with a town of the same name. Then along the banks towns Nautis, Madum, Demadatis, Secundum, Collocat, Secande, Navectabe with the Psegipta territory, Candragori, Araba, Summara.

194. Some say that the region above [that of] the Sirbites, where the mountains end, is occupied by coastal Aethiopians, the Nisicathae and Nisitae, whose names mean 'men with three or four eyes', not because they are like that, but because they are especially skilled at aiming their arrows. After them on the side of Nile that extends above Greater Syrtis and southern ocean, Dalion says there are Vacathi who only use rainwater, Cisori, Longopori a five-day journey from the Oecalices, Usibalchi, Isbeli, Perusii, Ballii, Cispii.

195. The rest is desert followed by an imaginary land. Facing west, Nigroe whose king has a single eye in his forehead, Agriophagi who subsist mainly on the flesh of panthers and lions, Pamphagi who devour everything, Anthropophagi whose diet is human flesh, Cynamolgi who have dogs' heads, Arthabatitae who walk on all fours and roam like wild animals, then Hesperioe, Perorsi, and those we said bordered Mauretania [5.10]. Quite a number of Aethiopians live only on locusts that have been smoked and salted to feed them for a year. These people do not live more than 40 years.

196. Agrippa reckoned that the Aethiopians' entire territory together with Red sea extends for a length of 2,170 miles and with Upper Egypt 1,296 width. Some have split up its length as follows: a 12-day voyage from Meroe to the Sirbites, 15 days from there to the Dabelli, a six-day journey from them to Aethiopic ocean. There is broad agreement among writers that the entire [distance] from the ocean to Meroe is 625 miles, then to Syene what we have stated [6.184].

197. Aethiopia lies from north-east to south-west. Along its centerline are thriving forests, mostly of ebony. At its mid-point a very high mountain glowing with perpetual fires rises from the sea. Greeks call it Theon

Ochema, and four days' voyage from it is the cape called Hesperu Ceras on the border of Africa next to the Aethiopian Hesperii. According to some reports, in this region there are also gentle hills covered with pleasant shade that belong to Aegipanes and Satyrs.

XXXVI Islands in the Aethiopic sea

198. Both Ephorus and Eudoxus and Timosthenes have said that there are several islands in this whole sea while, according to Clitarchus, king Alexander received a report about one so wealthy that its inhabitants bought horses with gold talents, and another on which a sacred mountain was found covered in woods where an amazingly attractive scent oozed from the trees. There is an island opposite Persian gulf and facing Aethiopia named Cerne, whose size and distance from the mainland are not agreed upon; only Aethiopian communities are said to be there.

199. According to Ephorus, because of the heat those sailing to it from Red sea are unable to proceed beyond certain columns – so some small islands are called. Polybius reported Cerne to be in the furthest part of Mauretania opposite mount Atlas, eight stades from the mainland. Nepos Cornelius says that it is almost opposite Carthago, one mile from the mainland, no more than two round. There is said to be another island too opposite mount Atlas, one itself also called Atlantis. From there, a two-day voyage past wilderness to the Aethiopian Hesperii and the cape we called Hesperu Ceras [6.197], from which the land's frontage first turns itself to the west and Atlantic sea.

200. There are stories that also opposite this cape are the Gorgades islands, once the Gorgons' home, a two-day sail from the mainland, according to Xenophon of Lampsacus. The Carthaginian general Hanno landed there and reported that the women had hairy bodies, while the men very nimbly got away. He deposited the skins of two of the Gorgades women in a temple of Juno as proof and as amazing sights. They were on display until Carthago was taken.

201. Still further on, the stories say, are two islands called Hesperides, although everything thereabouts is uncertain: Statius Sebosus reported that the route from Gorgons' islands sailing past Atlas to Hesperides islands was 40 days, with one day on to Hesperu Ceras.

Rumors about Mauretania's islands are just as vague. It is at least established that some opposite the Autololes were discovered by Juba, who began the dyeing of Gaetulic purple there.

XXXVII About the Fortunatae islands

202. Some think that beyond them are the Fortunatae [islands] and a few others. Again Sebosus gives their number and even calculates that Junonia is 750 miles from Gades, with Pluvialia and Capraria the same distance westward from there, and on Pluvialia rain providing the only water. From there the Fortunatae are 250 away opposite the left side of Mauretania towards the eighth hour of the sun [south-south-west]. Invallis [is so called] because of its convexity, Planasia because of its appearance, Invallis 300 miles round, with trees there growing to 140 feet tall.

203. Juba discovered the following about the Fortunatae: they too are positioned southward and a little to the west, 625 miles from the Purpurariae, provided the voyage is made for 250 above due-west, and then striking east for 375. The first is called Ombrios and lacks any trace of buildings; in its mountains it has a marsh, as well as trees like giant-fennel from which water is extracted – bitter from black ones, from whiter ones a pleasure to drink.

204. The second island is called Junonia, with a small shrine on it built of stone. Nearby a smaller [island] with the same name, then Capraria teeming with large lizards. Within sight of both is cloudy Ninguaria which gained its name from its perpetual snow.

205. Very near it, one called Canaria because it has so many dogs of massive build, two of which were brought back for Juba. Traces of buildings are visible here. Although [the islands] all have quantities of fruits and every kind of bird in plenty, [Canaria] is also filled with palm-groves that produce dates and pine-nuts; there is abundant honey too, as well as papyrus and catfish spawning in the rivers. Harm comes to [the islands] from rotting carcasses of sea-monsters that are constantly washed up there.

XXXVIII Earth's comparative dimensions

After this full description of the globe's land both inner and outer, the dimensions of its open waters should surely be summarized briefly.

206. Polybius said that the distance from Gaditan strait to Maeotis mouth is 3,437½ in a straight line, 1,250 from the same starting-point eastward to Sicilia, 375 to Creta, 187½ to Rhodos, the same to the Chelidoniae, 225 to Cyprus, 115 from there to Seleucia Pieria in Syria, a calculation which totals 2,340.

207. Agrippa reckons the same span from Gaditan strait to Issic bay to be 3,440 in a straight line. I suspect a numerical error here, since he also said that the route from Siculan strait to Alexandria is 1,350. However, the entire arc through the bays mentioned from the same starting-point up to lake Maeotis amounts to 15,509; Artemidorus adds 756 and also said that with Maeotis it is 17,390 miles.

208. These dimensions are taken by unarmed men who challenge fortune with peaceable daring.

We shall now compare the size of [earth's] parts as best we may, given the difficulty created by differences between authors. Even so, it will be most appropriate to state widths as well as lengths. So in that case the size of Europa is <... *missing text* ...> 8,714 [long]. The length of Africa – calculating the average of all the various figures being provided – comes to 3,698; the width of where it is inhabited is nowhere more than 250.

209. But since Agrippa made it 910 along its Cyrenaic side, including its known desert as far as the Garamantes, the entire dimension to be used for purposes of comparison amounts to 4,608. The length of Asia is acknowledged to be 5,013¾. Its width should reasonably be calculated from the Aethiopic sea to Alexandria situated by Nile so that – with the measurement running through Meroe and Syene – it is 1,875.

210. So evidently Europa is a little less than one and a half times bigger than Asia, and two and one-sixth times more extensive than Africa. If all these totals are combined, it will be strikingly clear that Europa is one-third and a little more than one-eighth of the entire earth, Asia a quarter and one-fourteenth, Africa one-fifth and over one-sixtieth.

XXXIX Division of the lands into parallels and equal shadows

211. To these we shall also add a single most meticulously subtle theory devised by Greeks, so that nothing may be missing from our overview of how the world is situated and what may be discovered about the regions noted, and to which stars each relates, or what the balance is between day and night, and where shadows have the same length and the world's curvature is equal. An account will therefore be offered of this [theory] too, and the whole earth will be laid out according to the divisions of the sky. The world has numerous sections, called 'circles' by our [Latin writers], but 'parallels' by Greeks.

212. At the start is the part of India turned southward. It extends as far as Arabia and those living on Red sea. Within it are the Gedrosi, Carmani, Persians, Elymaei, Parthyene, Aria, Susiane, Mesopotamia, Seleucia also called Babylonia, Arabia as far as Petra, Syria Coele, Pelusium, the lower parts of Egypt called Chora, Alexandria, coastal Africa, all Cyrenaica's towns, Thapsus, Hadrumetum, Clupea, Carthago, Utica, both Hippos, Numidia, both Mauretaniae, Atlantic sea, Hercules' pillars. In this arc of the sky at midday during the equinox, a seven-foot long pole – called a 'gnomon' – casts a shadow up to four feet long, but the longest night and day lasts for 14 equinoctial hours, whereas the shortest lasts for ten.

213. The following circle begins where India slopes west, and runs through the middle of the Parthians, Persepolis, the nearest part of Persis, nearer Arabia, Judaea, those living on mount Libanus; it encompasses Babylon, Idumaea, Samaria, Hierosolyma, Ascalon, Iope, Caesarea, Phoenice, Ptolomais, Sidon, Tyrus, Berytus, Botrys, Tripolis, Byblos, Antiochia, Laodicea, Seleucia, coastal Cilicia, southern Cyprus, Creta, Lilybaeum in Sicilia, and northern parts of Africa and Numidia. At the equinox a 35-foot pole makes a 24-foot long shadow; the longest day and night lasts almost 14 and two-fifths equinoctial hours.

214. The third circle begins where India is very close to [mount] Imavus. It stretches through the Caspian Gates, the nearest parts of Media, Cataonia, Cappadocia, Taurus, Amanus, Issus, Cilician Gates, Soloe, Tarsus, Cyprus, Pisidia, Pamphylia, Side, Lycaonia, Lycia, Patara, Xanthus, Caunus, Rhodos, Cous, Halicarnasus, Cnidus, Doris, Chius, Delos, middle of the Cyclades, Gythium, Malea, Argos, Laconica, Elis, Olympia, Peloponnesian Messania, Syracusae, Catina, middle of Sicilia, southern parts of Sardinia, Carteia, Gades. A 100-inch sundial makes a 74-inch shadow. The longest day lasts 14½ and one-thirtieth equinoctial hours.

215. Under the fourth circle are situated what is on the other side of [mount] Imavus, southern parts of Cappadocia, Galatia, Mysia, Sardis, Zmyrna, Sipylus, mount Tmolus, Lydia, Caria, Ionia, Trallis, Colophon, Ephesus, Miletus, Chius, Samus, Icarian sea, northern Cyclades, Athens, Megara, Corinthus, Sicyon, Achaia, Patrae, Isthmos, Epirus, northern parts of Sicilia, eastern parts of Gallia Narbonensis, coastal Hispania from Carthago Nova and on westward. Shadows of 16 feet are produced by a 21-foot gnomon. The longest day lasts 14 and two-thirds equinoctial hours.

216. The fifth section, beginning at the entrance to Caspian sea, contains Bactrians, Hiberia, Armenia, Mysia, Phrygia, Hellespont, Troas, Tenedus, Abydos, Scepsis, Ilium, mount Ida, Cyzicum, Lampsacus, Sinope, Amisum, Heraclea in Pontus, Paphlagonia, Lemnus, Imbrus, Thasus, Cassandria, Thessalia, Macedonia, Larisa, Amphipolis, Thessalonice, Pella, Edesus, Beroea, Pharsalia, Carystus, Boeotian Euboea, Chalcis, Delphi, Acarnania, Aetolia, Apollonia, Brundisium, Tarentum, Thuri, Locri, Regium, the Lucani, Neapolis, Puteoli, Tuscan sea, Corsica, Baliares, central Hispania. A seven-foot gnomon [casts] a six-foot shadow. At the most, a day lasts 15 equinoctial hours.

217. The sixth group – with the city of Rome in it – encompasses Caspian peoples, Caucasus, northern parts of Armenia, Apollonia on the Rhyndacus, Nicomedia, Nicaea, Calchadon, Byzantium, Lysimachea, Cherronesus, Melas bay, Abdera, Samothrace, Maronea, Aenus, Bessica, Thracia, Maedica, Paeonia, the Illyrians, Dyrrachium, Canusium, Apulia's borderlands, Campania, Etruria, Pisae, Luna, Luca, Genua, Liguria, Antipolis, Massilia, Narbo, Tarraco, central Hispania Tarraconensis, and then on through Lusitania. A nine-foot gnomon [casts] an eight-foot shadow. The longest day spans 15 and one-ninth equinoctial hours or – as Nigidius preferred – one-fifth.

218. The seventh division begins from Caspian sea's other coast and goes above Callatis, Bosporus, Borysthenes, Tomi, Thracia's back-country, the Triballi, the rest of Illyricum, Hadriatic sea, Aquileia, Altinum, Venetia, Vicetia, Patavium, Verona, Cremona, Ravenna, Ancona, Picenum, the Marsi, Paeligni, Sabini, Umbria, Ariminum, Bononia, Placentia, Mediolanum and everything from the Appenninus, as well as – across

Alpes – Gallia Aquitanica, Vienna, the Pyrenaei, Celtiberia. A 36-foot shadow [is cast] by a 35-foot pole, although in part of Venetia the shadow and gnomon match. The extreme is for a day to last 15 and three-fifths equinoctial hours.

219. To this point we have been discussing past writers' conclusions. The most conscientious of their successors have assigned the rest of the earth (beyond the above) into three sections. [The first extends] from the Tanais through lake Maeotis and the Sarmatians as far as the Borysthenes, and so on through the Daci and part of Germania and the ocean shore encompassing Galliae, with a 16-hour [longest day]. The second [extends] through the Hyperboreans and Britannia, with a 17-hour [longest day]. The last, the Scythian, [extends] from the Ripaean range to Thule where, as we have said [2.186, 4.104], there are alternating periods of continuous day and night.

220. The same [writers] have also placed two circles before where we made our start. The first is through Meroe island and the Ptolomais founded on Red sea for elephant-hunts, where the longest day would be 12 hours plus a half; the second passes through Egyptian Syene, with a 13-hour [longest day] there. And the same [writers] add half an hour [to the length of that day] for each circle up to the furthest.

And this is enough about earth.

****

Totals   Towns, 1,195
Communities, 576
Famous rivers, 115
Famous mountains, 38
Islands, 108
Extinct cities and peoples, 95
Facts and inquiries and observations, 2,214

[Roman] Sources

Marcus Agrippa. Marcus Varro. Varro Atacinus. Cornelius Nepos. Hyginus. Lucius Vetus. Mela Pomponius. Domitius Corbulo. Licinius Mucianus. Claudius Caesar. Arruntius. Sebosus. Fabricius Tuscus. Titus Livius the son. Seneca. Nigidius.

The East, India

Foreign Sources

King Juba. Hecataeus. Hellanicus. Damastes. Eudoxus. Dicaearchus. Baeton. Timosthenes. Patrocles. Demodamas. Clitarchus. Eratosthenes. Alexander the Great. Ephorus. Hipparchus. Panaetius. Callimachus. Artemidorus. Apollodorus. Agathocles. Polybius. Timaeus of Sicilia. Alexander Polyhistor. Isidorus. Amometus. Metrodorus. Posidonius. Onesicritus. Nearchus. Megasthenes. Diognetus. Aristocreon. Bion. Dalion. Simonides the Younger. Basilis. Xenophon of Lampsacus.

BOOKS 7 TO 37

Notable Geographical Passages

BOOK 7

XXVI Clemency, magnanimity

95. Truly, to state at this point all the names of Pompey the Great's victories and his triumphs is to reflect the Roman empire's glory, not just that of a single man – a match in their brilliance not only to Alexander the Great's feats, but also nearly to those of Hercules and Father Liber.

96. So, after recovering Sicilia – which first marked the start of his rise for the sake of the Republic among Sulla's group – then after subjugating the whole of Africa [province] and bringing it back under control, and at that stage acquiring the name Magnus [Great] as loot, [Pompey], a Roman *eques*, came back [to ride] in a triumphal chariot – something unprecedented [81 BCE]. Immediately he made his way over to the west where, with trophies erected in the Pyrenaei, he claimed that his victory brought back under control 876 towns from the Alpes to the borders of Hispania Further; with more decency he made no mention of Sertorius, and after quelling civil war which was rousing everywhere abroad, he again rode in a triumphal chariot as a Roman *eques* [71 BCE], having been a general that often rather than [merely] a soldier.

97. Subsequently, after being sent to every sea and then to the east, he brought back for his country the following banners, like the winners in sacred games – their crowns not for themselves, after all, but for crowning their countries. So he paid these honors to the city in the shrine of Minerva which he dedicated from spoils:

> Gnaeus Pompeius Magnus, Imperator, after completing 30 years of war, has scattered, routed, killed, accepted the surrender of 12,183,000 people, sunk or captured 846 ships, received the capitulation of 1,538 towns and

forts, conquered the lands from the Maeotae to Red sea; he duly makes this offering to Minerva.

98. This summarizes his achievements in the east. But the description of the triumph which he led three days before the Kalends of October [September 28th] in the consulship of Marcus Piso and Marcus Messala [62 BCE] was:

> After liberating the sea-coast from pirates and restoring command of the sea to the Roman people, he triumphed over Asia, Pontus, Armenia, Paphlagonia, Cappadocia, Cilicia, Syria, the Scythians, Judaeans, Albani, Hiberia, Creta island, the Basternae, and in addition over king Mithridates and Tigranes.

99. The supreme height of this glory (as he said himself in a public meeting when discussing his feats) was that he encountered Asia as the most distant province and returned it at the center of [Roman] territory. If, on the other hand, anyone may wish to review in a similar way the feats of Caesar – who appeared to be greater than [Pompey] – that really would mean enumerating the whole world, to which (we may agree) there is no end.

******

LVI About creativity, about the mind

191. Before ending [our discussion] of human nature, it seems appropriate to point out [various] inventions and who made them. <Mercury> introduced buying and selling, Father Liber <the harvesting of grapes>; he also invented that royal emblem the diadem, as well as the triumph; Ceres [discovered] grains, before which acorns were eaten; again, she [invented] milling and preparing [flour] in Attica (or some say in Sicilia): for this reason she was considered a goddess. She again was the first to issue laws, although others have thought it was Rhadamanthus.

192. I think that the Assyrians have always had an alphabet, but according to some (Gellius, for instance) Mercury discovered this while with the Egyptians, yet others opt for with the Syrians; there is agreement that Cadmus brought to Graecia from Phoenice a 16-letter alphabet, to which Palamedes during the Trojan war added the four characters H Y Φ

X, and after him the lyric poet Simonides as many again, Ψ Ξ Ω Θ; the impact of all these is recognized in our alphabet. To Aristotle, 18 letters were ancient, and his inclination is that two, Φ X, were added by Epicharmus rather than Palamedes.

193. Anticlides says that someone named Menon invented the alphabet in Egypt 15,000 years before Phoroneus, Graecia's most ancient king, and he attempts to prove this from monuments. By contrast, Epigenes, a serious authority of the first order, informs [us] that Babylonians' astronomical observations inscribed on baked bricks date back 720,000 years; Berosus and Critodemus give the shortest period for this, 490,000. Hence it appears that an alphabet has been in use from time immemorial. The Pelasgi brought it to Latium.

194. The brothers Euryalus and Hyperbius first set up brick-kilns and houses at Athens; previously, caves were used for homes. Gellius likes to think that Caelus' son Toxius was the inventor of mud-brick construction, an idea copied from swallows' nests. Cecrops named the first town after himself, Cecropia, which nowadays is Athens' citadel. Some favor Argos being founded earlier by king Phoroneus, and others Sicyon too, while according to Egyptians their Diospolis [dates to] very much earlier.

195. Agriopa's son Cinyra invented tiles and copper-mining, both on Cyprus island, as well as tongs, hammer, crowbar, anvil; Danaus, when brought from Egypt to [the region of] Graecia that was called Argos Dipsion, [invented] wells; Cadmus [discovered] quarries at Thebes or, according to Theophrastus, in Phoenice; Thrason walls; the Cyclopes towers according to Aristotle, or the Tirynthii according to Theophrastus;

196. Egyptians fabrics; Lydians at Sardis the dyeing of wool; Arachne's son Closter spindles for wool-working; Arachne linen and nets; Nicias of Megara the craft of fulling; the Boeotian Tychius shoemaking. Egyptians like to say that medicine is their discovery, but according to others it was made by Arabus, son of Babylone and Apollo; herbology and pharmacology were discovered by Chiron, son of Saturn and Philyra.

197. It was the Scythian Lydus who demonstrated how to smelt and work copper in Aristotle's view, but in Theophrastus' it was the Phrygian Delas; some think the Chalybes [discovered] bronze-working, others the Cyclopes; Hesiod thinks that those called the Dactyli Idaei on Creta [discovered] iron. The Athenian Erichthonius discovered silver – or Aeacus did, others say – as did the Phoenician Cadmus gold-mining and -smelting at mount Pangaeus or, according to others, Thoas or Aeacus in Panchaia, or Oceanus' son Sol, to whom Gellius also attributes the discovery of medicine from metals. Midacritus was the first to import lead from Cassiteris island.

198. The Cyclopes invented iron-working, the Athenian Coroebus pottery-work, the Scythian Anacharsis the wheel for it, or (others say) the Corinthian Hyperbius. Daedalus [invented] wood-working and along with it the saw, adze, plumb-line, drill, glue, isinglass; but the Samian Theodorus the square and level and lathe and lever; the Argive Phidon, or Palamedes (Gellius' preference), measures and weights; Cilix's son Pyrodes fire from flint, with Prometheus [discovering] how to keep it in a fennel stalk.

199. The Phryges four-wheeled vehicles, Carthaginians trade, the Athenian Eumolpus viticulture and arboriculture, Silenus' son Staphylus mixing wine with water, the Athenian Aristaeus olive oil and presses as well as mead, the Athenian Buzyges the ox and plough, or (say others) Triptolemus;

200. Egyptians the monarchic state, the people of Attica after Theseus the democratic one. The first tyrant was Phalaris at Agragas. Lacedaemonians invented slavery. Capital trials were first held on the Areopagus.

Africans first went into battle against Egyptians with clubs they call *phalangae*. Proetus and Acrisius invented shields when fighting one another (or it was Athamas' son Chalcus), the Messenian Midias the breastplate, Lacedaemonians the helmet, sword, spear, the Cares greaves and plumes.

201. Some say that Jupiter's son Scythes invented the bow and arrow, others that arrows were invented by Perseus' son Perses; the Aetoli lances, Mars' son Aetolus the javelin with throwing-strap, Tyrrenus the spears used by *velites* [light-armed troops], the Amazon Penthesilea the *pilum* [another type of spear], Pisaeus the axe, the Cretans hunting-spears and the *scorpio* as well as other pieces of artillery, Syrians the catapult, Phoenicians the *ballista* [artillery-piece] and sling, Tyrrenus' son Pisaeus the bronze trumpet, the Clazomenian Artemon the *testudo* [mobile screen used in sieges],

202. Epius the horse (nowadays called the ram) at Troy among other machines for attacking walls, Bellerophon horse-riding, Pelethronius horse-reins and saddles, the Thessali called Centauri living along mount Pelium fighting from horseback. Phrygian people were the first to harness two horses to a chariot, Erichthonius four of them. In the Trojan war Palamedes invented army formations, giving signals, password-tokens, watches; also during it Sinon signaling from lookout posts, Lycaon truces, Theseus treaties.

203. Car (from whom Caria is named) [invented] predicting the future from birds, Orpheus added doing the same from other animals, Delphus from internal organs, Amphiaraus from fire, the Theban Tiresias from birds' organs, Amphictyon by interpreting visions and dreams. Libya's son Atlans [invented] astrology (others say it was Egyptians, others still Assyrians), the Milesian Anaximander the use of a globe in it [astrology], Hellen's son Aeolus a grasp of the winds.

204. Amphion [invented] music, Mercury's son Pan the pipe and flute, Midas in Phrygia the slanted pipe, Marsyas among the same people the twin-pipe, Amphion Lydian rhythms, the Thracian Thamyras Dorian ones, the Phrygian Marsyas Phrygian ones, Amphion the lyre (others say it was Orpheus, others still Linus). Terpander was the first to play seven strings after adding three to the first four; Simonides added an eighth, Timotheus a ninth. Thamyris first played the lyre without vocal accompaniment, Amphion (others say Linus) did so with singing. Terpander composed songs to accompany the lyre. The Troezenian Ardalus introduced voice accompaniment to pipes with song. The Curetes taught the dance-in-armor, Pyrrus the *pyrriche* [war-dance], both in Creta.

205. We owe epic verse to the Pythian oracle. The origin of poetry is a major question; there is proof of its existence before the Trojan war. The Syrian Pherecydes began the practice of prose oratory during king Cyrus' time, the Milesian Cadmus did so for history, Lycaon for gymnastic games in Arcadia, Acastus for funeral ones at Iolcus, subsequently Theseus at the Isthmos, Hercules at Olympia, Pythius athletic ones, the Lydian Gyges playing ball; painting [was invented] by Egyptians, and in Graecia by Daedalus' relative Euchir (as Aristotle favors) or by the Athenian Polygnotus (as Theophrastus favors).

206. Danaus was the first to come to Graecia from Egypt by ship; previously, sailing had been between islands in Red sea on rafts invented by king Erythras. People can be found who think that the Mysians and Trojans contrived them earlier on the Hellespont, when they crossed it to engage the Thracians. Even nowadays in Britannic ocean wicker ones are made with hides sewn all around; on Nile there are ones made from papyrus and rushes and reeds.

BOOK 8

LXXXIII Creatures possibly in one location but not another

225. It is remarkable that the natural order has not only bestowed different creatures on different lands, but has also barred some from certain places within the same area. Dormice are found in only a part of Italia's Mesia forest. In Lycia, gazelles do not cross the mountains near Sexi, nor do wild asses the border dividing Cappadocia from Cilicia. Stags in the Hellespont do not migrate into unknown territory, and those around Arginusa do not go past mount Elaphus, even though on the mountain they have split ears.

226. On Poroselene island weasels do not cross a road. Similarly in Boeotia moles that burrow through entire fields nearby in Orchomenus run from the very soil when brought to Lebadea. We have seen bed-coverings made from their pelts: so not even respect for religion deters luxury from exploiting these creatures with supernatural properties. On Ithaca any hares brought in expire at the very edge of the shoreline, as do rabbits on Ebusus, [even though] nearby Hispania and the Baliares are swarming with them.

227. In Cyrene the frogs used to be mute, and even after ones that make sounds were brought over from the mainland, the former type hangs on. Even nowadays those on Seriphus island are mute, although they croak when brought elsewhere; this is also said to be the case on Thessalia's lake Sicandrus. In Italia a bite from a spider mouse is poisonous – a type that the region beyond Appeninnus lacks. Also, wherever these ones are, they die once they have crossed a wheel-rut. There are no wolves on Macedonia's mount Olympus nor on Creta island.

228. In fact there are no foxes or bears there, and no vicious creature at all except for a venomous spider. We shall discuss this type in the appropriate place, among spiders [11.79*; 18.156*]. It is more remarkable that on the same island there are no stags except in the region of the Cydoneatae; the same goes for boars and partridges, hedgehogs, while in Africa there are no boars or stags or goats or bears.

LXXXIV Which [creatures] may only harm outsiders and where; which may [harm] only locals and where

229. Moreover, some creatures that do no harm to native [species] kill invasive ones: such are the small snakes at Myrinthe that are said to be earth-born. Similarly serpents in Syria, especially around the banks of Euphrates, do not touch sleeping Syrians or, even if they have bitten them after being trampled upon, no harm from them is felt; [but] they are dangerous to any other peoples, keen to kill them with excruciating pain. For this reason Syrians also do not kill them. By contrast, Aristotle says that on Caria's mount Latmus visitors are not harmed by scorpions, [but] natives are killed.

BOOK 9

XVII Which fish are the largest

44. Tunny are exceptionally large. We have encountered one weighing 15 talents, with its tail two cubits and a palm wide. And from certain rivers come [fish] just as large, the catfish in Nile, pike in Rhenus, and in Padus sturgeon fattening from laziness sometimes up to 1,000 pounds; it is caught with a hook on a chain, and only dragged out by a team of oxen. Yet a very small [fish] called the anchovy kills it by biting into a certain vein in its throat, an assault made with remarkable passion.

45. The catfish, wherever it is, roams and assaults every creature, often drowning horses as they swim. In appearance very like a sea-pig, it is dragged out by double ox-teams from Germania's Moenus river especially, and by mattocks from Danuvius. In Borysthenes too there are accounts of an exceptionally large [fish] with no bones or spine inside it and very soft flesh.

46. In India's Ganges are ones called *platanistae* that have a dolphin's snout and tail and measure a hefty 16 cubits. Here too Statius Sebosus refers to the far from slight marvel of worms that have two gills each of six cubits, are dark blue and take their name from their appearance.

Evidently they have great enough strength to drag off elephants coming down to drink by gripping the trunks with their teeth.

******

49. Every type of fish grows especially fast, particularly in Pontus; this is because of the quantity of rivers conveying fresh water into it. The day-by-day growth of the one called *amia* is perceptible. Along with tunny, this [fish] and the *pelamydes* enter Pontus in schools for fresher feeding-grounds, each one with its own leader, and ahead of them all the mackerel, which is yellow-colored in the water, but like the rest when out of it. [Mackerel] fill Hispania's fishponds, but tunny do not go there.

50. Yet no creature harmful to fish enters the Pontus except seals and small dolphins. Tunny enter along the right bank, exit along the left. This is assumed to happen because they see more with their right eye, both [eyes] being naturally dull. In the channel of Thracian Bosporus joining Propontis to Euxinus [Pontus], in those very narrows of the strait separating Europa and Asia, is a remarkably dazzling rock that shines from the depths of the sea to its surface near Calchadon on Asia's side. Panicked by the sudden sight, [the tunny] always rush in a throng to make for the opposite promontory of Byzantium,

51. called for this reason Aureum Cornu [Golden Horn]. So Byzantium gains the entire catch, and at Calchadon there is a great dearth because of the 1,000-pace channel flowing between them. But [the tunny] wait for a north wind to blow in order to exit the Pontus with a favorable current, and only those entering Byzantium's port are caught. They do not rove around in winter; wherever it grips them, they hibernate there until the equinox. For quite a number of hours and miles, and visible from the helm, in a wonderfully delightful way they often accompany ships proceeding under sail, and are not even frightened by being made the repeated target of harpoons. Some people call tunnies that act like this pilot-fish.

52. Many spend the summer in Propontis and do not enter Pontus; the same goes for the sole, while turbots do enter; the *sepia* is not there, while the cuttle-fish is found; among rock-fish, the *turdus* and *merula* are absent, as are shellfish, although there are plenty of oysters. But all of them spend the winter in the Aegean. Of those that enter Pontus, only the *trichiae* do not return – it will be best to use the Greek names for most of these [fish] since different regions have different names for the same ones –

53. but these alone enter Hister river [Danuvius] and by its underground passages float down from it into Hadriatic sea. As a result, they are observed there descending, yet never on the move from the sea.

The period for catching tunny is from Pleiades' rising to Arcturus' setting.[1] For the remaining wintertime they hide in the lowest eddies unless lured by some warmth or the full moon. They even fatten up so large that they are liable to burst. Their lifespan is two years at its longest.

BOOK 10

XXXVII Memnonides

74. Sources say that every year birds fly from Aethiopia to Ilium and fight at Memnon's tomb, hence their name Memnonides [Memnon's daughters]. Cremutius relates his own discovery that they do the same every fifth year in Aethiopia around Memnon's palace.

XXXVIII Meleagrides

The Meleagrides in Boeotia fight in much the same way. This is a type of hen from Africa, hunch-backed, flecked with varied plumage. They are the least likely of foreign birds to be served at dinners because of their unpalatable taste; but Meleager's tomb has made them famous.

XXXIX Seleucides

75. The arrival of birds called Seleucides is what inhabitants of mount Cadmus beg for in their prayers to Jupiter when locusts are devastating their crops. Where the [birds] come from is not known, nor where they go to, as they are never seen except when their protection is needed.

XL The ibis

Egyptians also pray for their ibis to counter the arrival of snakes, and Eleans call upon the god Myiagrus when a swarm of flies brings disease; these die the moment a sacrifice has been made to this god.

---

[1] Between May and November.

XLI Birds possibly in one location but not another

76. Yet on the matter of birds hiding away, night-owls also are said to remain concealed for a few days – a kind of bird that does not exist on Creta and, if imported there, even dies. For this too is an instance of remarkable differentiation on Nature's part. Some [things] she bars from some places, such as types of crops or bushes, and creatures also; for one not to be born in a place this is no surprise, but it is astonishing for one imported to die. Why this hostility to the well-being of a single type, or what is this ill-will on the part of Nature? What are the territorial limits prescribed for birds?

77. There are no eagles on Rhodos. In Transpadana Italia by the Alpes, the lake called Larius with its delightful wooded countryside is a place to which storks do not venture; the same applies to jays and jackdaws which come no nearer than the eighth milestone from it, despite there otherwise being vast flocks of them in the neighboring region of the Insubres (this bird's unique passion for theft – of silver and gold especially – is remarkable). There are said to be no Mars' woodpeckers in the territory of Tarentum.

78. Recently there have begun to be sightings of magpies coming from the Appenninus towards the city [Rome] (rare though they still are) – the ones called 'varied' conspicuous for their long tail. A distinguishing feature of theirs is that they moult every year at the time when turnips are planted. Partridges in Attica do not fly across the border with Boeotia, nor does any bird [fly] to the temple dedicated to Achilles on the island in Pontus where he is buried. In the Fidenae area near the city [Rome] storks make neither chicks nor nests. But every year a mass of pigeons flies out of the sea to the Volaterrae area.

79. In Rome neither flies nor dogs enter Hercules' temple in the forum Boarium. There are many similar [examples] beyond these, which I consistently take care to omit when treating this species or that, with concern not to become tedious, especially when (as Theophrastus says) even in Asia there are invasive doves and peacocks and deer, and in Cyrenaica croaking frogs.

******

LXVII Foreign birds: phalerides, pheasants, numidicae

132. We have heard of strange kinds of birds in Germania's Hercynian forest, ones with plumage that shines like fires at night. As for other birds there, nothing of note arises beyond the fact that their remoteness confers fame.

The phalerides in Parthian Seleucia and in Asia are the most celebrated water-birds, as again are the pheasants in Colchis – they lower and raise twin 'ears' from their plumage – so too the numidicae [guinea-fowl] in the African region Numidia; all these are now found in Italia.

LXVIII Foreign birds: flamingos, partridges, cormorants, choughs, ptarmigans

133. That deepest sump of all spendthrifts, Apicius, claimed the flamingo's tongue to be exceptionally tasty. The Ionian partridge is especially famous and otherwise vocal, but it goes dumb when captured; once classed among rare birds, it is now in Gallia and Hispania. It is caught around the Alpes too, where there are also cormorants, a bird unique to Baliares islands, as are to the Alpes the chough – black with a yellow beak – and the specially appetizing ptarmigan whose name [*lagopus*] comes from its feet with their shaggy hair like that of a hare; otherwise it is white and the size of a dove.

134. It is not easy to nurture outside that region, since it does not adapt to a tamed lifestyle and its body very quickly grows weak. There is also another [bird] with the same name – different from the quail only in size – that makes a most delicious meal with saffron sauce. A prefect of the Alpes, Egnatius Calvinus, has reported personally sighting there the ibis too, unique though it is to Egypt.

LXIX 'New' birds, *vibiones*

135. 'New' birds also entered Italia during the civil wars across the Padus around Bedriacum [69 CE] – and are still so called – resembling thrushes, a little smaller than doves in size and distinctly tasty. Baliares islands export the purple gallinule, an even finer type of this bird than the one mentioned above [10.129*]. There too are the buzzard, a kind of hawk much appreciated by diners, as well as the *vibio*, the name given to the lesser crane.

BOOK 12

XXX About the frankincense region

51. Cinnamon – very closely related in type [to cardamon] – should come next, were it not appropriate to draw attention first to the riches of Arabia and to why it has gained the names Felix [fortunate] and Beata [blessed]. What it chiefly has then are frankincense and myrrh, the latter in common with the Trogodytae too. Frankincense is nowhere but Arabia, and not even throughout Arabia.

52. Round about its middle is Atramitae, a Sabaean district with the capital of the kingdom Sabota on a lofty mountain, an eight-stage journey away from their frankincense-producing region called Sariba; Greeks say this means 'a secret'. It faces north-east, with rocks blocking the way in every direction and to its right the sea made inaccessible by crags. Its milk-white soil is reported to be partly red.

53. The woodlands are 20 *schoeni* long and half as wide. According to Eratosthenes' calculation, a *schoenus* is 40 stades in length – thus five miles – but some have reckoned 32 stades to the *schoenus*. The hills there rise high, and trees that spring up of their own accord spread down to the plains too. It is agreed that the soil is full of clay with springs few and alkaline.

54. Also bordering it are the Minaei in another district, through which [frankincense] is conveyed out along a single narrow track. These [people] were the first to engage in the frankincense trade and they continue to be the main ones involved, hence from them it is also called Minaean. Them aside, the sight of a frankincense tree means nothing to other Arabians, nor does it even to all of them, given their claim that no more than 3,000 families maintain for themselves an inherited right [to the trade]. As a result, members of these families are called sacred, and they are not to be contaminated by any contact with women or funerals when they are tapping into the trees or gathering from them, a religious scruple that therefore raises the price of the product. Some say that the woodlands' frankincense is the common property of these people, others that it is divided each year by rotation.

XXXI About trees that produce frankincense

55. There is no agreement about the appearance of the tree itself. We Romans have operated in Arabia, and our forces have penetrated into a large part of it. Even Augustus' son Gaius Caesar sought glory there and yet, so far as I know, there is no description of these trees by anyone in Latin. Accounts by Greeks differ:

56. some state that it has the leaf of a pear-tree, only smaller and with a greenish color; others that it resembles the mastic-tree with a reddish leaf; several that it is a turpentine-tree as witnessed by king Antigonus when presented with a plant. King Juba – in the volumes he wrote for Augustus' son Gaius Caesar who was so inspired by Arabia's renown – says that it has a twisted trunk with branches very like those of the Pontic maple, and oozes sap like the almond, and that such [trees] planted on the initiative of the ruling Ptolemies are to be seen in Carmania and in Egypt.

57. It is agreed that it has the bark of the laurel-tree, and some have stated that its leaves are similar too. That was certainly the case with a tree in Sardis that even kings of Asia took the trouble to cultivate. Envoys who had come from Arabia in my time have cast further doubt on everything; hence our justifiable surprise, because even slips from frankincense-trees reaching us make it possible to believe that the mother-tree too is smooth, and puts out its branches from a trunk with no knots.

******

63. After being gathered, frankincense is conveyed by camel to Sabota, where just one gate is left open for the purpose. The kings have made it a capital offence to deviate from the route. Here priests exact one-tenth of it (by volume, not weight) for the god called Sabis, and only after that is it legal to trade it. Public expenditure is covered as a result, because the god is even kind enough to feed guests for a fixed number of days. [The frankincense] can only be exported through [the territory of] the Gebbanitae, and so a tax is paid to their king too.

64. Their capital Thomna is 2,437½ miles from Gaza, a town in Judaea on our shore; this [distance] is divided into a 65-stage journey for the camels. There are fixed shares given to priests and the kings' clerks. But in addition to these, guards too and escorts and gatekeepers and servants pilfer.

65. Anyway, wherever the route goes, they pay here for water, there for fodder or lodging, as well as various tolls, so that the expenses for every camel as far as our shore amount to 688 *denarii*, and there is a further payment to our empire's customs officials. As a result, a pound of the best frankincense costs six *denarii*, second grade five *denarii*, third three *denarii*. Once within our empire, it is adulterated with drops of very similar white resin, but this is detected by the means already mentioned [12.36*]. It is tested for whiteness, heft, brittleness, flammability to ensure that it burns instantly, also that it is impervious to being bitten and instead crumbles into powder.

******

XLI About Arabia's good fortune

82. Even though Arabia is without cinnamon or cassia, it is called Felix [Fortunate], a spurious and unwanted designation crediting this favor to the gods above, despite the debt to those below being greater. It is humans' [taste for] luxury that has produced this 'good fortune', even in death when they burn for the deceased what had been understood to be for gods.

83. Experts in these matters claim that [Arabia] does not grow as much in a year as *princeps* Nero burned on his [wife] Poppaea's final day [65 CE]. Reckon, then, the number of funerals worldwide every year, and how much of what is offered to gods drop by drop is heaped up in a pile to honor corpses. Yet [the gods] were no less well disposed to petitioners offering salted flour; in fact they were clearly kinder to them.

84. But even now Arabia's sea is more fortunate, since from it pearls are sent, and by the lowest estimate India and the Seres and this peninsula draw 100,000,000 sesterces from our empire annually: it is this much that our pleasures and women cost. What fraction of this, let me ask, is offered to gods, even those of the Underworld?

******

XLIX *Hammoniacum*

107. We have left lands facing ocean for those curving around our seas.
So [the part of] Africa lying below Aethiopia oozes a sap called *hammoniacum* in its sands; the oracle of Hammon is even named from

this, and it is produced near there from a tree called *metopon*, with the character of resin or gum. There are two kinds of it: *thrauston* which is like male frankincense and especially valued, and the other, called *phyrama*, rich and resinous. It is adulterated with sand which gives the impression of having stuck to it while growing; so it is the smallest, purest lumps that are valued. The best cost 40 *asses* per pound.

L *Sphagnos*

108. The *sphagnos* [moss] that is most valued is in Cyrenaica province below these locations; others call it *bryon*. The Cyprian kind is graded second, the Phoenician third. It is said to grow in Egypt too, and in Gallia for certain. I would have no doubt about this because there are grey tufts with this name on trees (oaks especially, we observe), ones very notable for their smell. Top rating goes to the whitest and widest, second to the reddish, none to the black. Also, [mosses] that grow on islands and rocks are rejected, as are all those which lack their distinct smell and have that of a palm-tree.

LI *Cypros*

109. Egypt has the *cypros* [henna-tree], with the leaves of a jujube-tree and coriander-like seeds, white and fragrant. These are cooked in oil and then crushed to become what is called *cypros*. It costs five *denarii* per pound. The Canopic area on the banks of Nile produces the best, Ascalon in Judaea second, Cyprus island third. Its smell is quite pleasant. By all accounts this tree is what in Italia would be called the *ligustrum* [privet].

BOOK 14

VIII Fifty fine wines

59. Who would doubt that some kinds of wine are more pleasing than others, or that one of a pair from the same vintage may outclass the other in quality because of the jar or some random accident? This is why everyone is self-appointed judge of top quality.

60. Julia Augusta credited her 86 years of life to wine from Pucinum, which she consumed exclusively. It is produced on a bay of Hadriatic sea, not far from the source of the Timavus, on a rocky hill where sea-breeze ripens a few casks. No other wine is reckoned to match its curative effect; I should imagine that it is what Greeks lavished remarkable praise upon, calling it Praetetian from a Hadriatic bay.

61. Deified Augustus, as well as most subsequent emperors, preferred Setinian to all others, because experience proved that its taste did not cause unexpected harmful indigestion; it originates above Forum Appi. Previously, the highest reputation for excellence was held by Caecuban [grown] in pine-grove marshes on the bay of Amynclae, but no longer because of cultivators' neglect and spatial constraints, but even more because of the ship-canal from Baiae's lake as far as Ostia which Nero had begun.

62. The Falernian region used to have the second level of distinction and within it Faustinian especially, the product of painstaking cultivation; it is [now] deteriorating because of a drive for quantity instead of quality. The Falernian region begins at the Campanian bridge on the left in the direction of Sulla's colony Urbana (recently attached to Capua); the Faustinian [begins] about four miles from Caedicium village, this village being six miles away from Sinuessa. No other wine is more highly esteemed nowadays; it is the only wine to be set ablaze with flame.

63. There are three types: dry, sweet, light. Some distinguish them as follows: Caucinian growing on the highest hills, Faustinian midway, Falernian at the bottom. The point must be made, however, that despite their fame none of these comes from grapes with an appealing taste.

64. Third prize is more or less gained by Alba's vines close to the city [of Rome], extra-sweet and occasionally dry, also by those of Surrentum grown only on trellises, very much recommended for convalescents because it is light and salubrious. Tiberius Caesar used to call it a doctors' plot to confer distinction on Surrentine when it was just a kind of high-class vinegar. To his successor Gaius Caesar it was flat wine of distinction. Rival wines are Massic ones and those from mount Gaurus facing Puteoli and Baiae.

65. Statanan wines (on the Falernum border) undoubtedly did attain top-quality status and made it plain that any terroir has its day, both better times and worse. The Calenian ones adjacent typically used to be rated higher, as well as the Fundanian cultivated both on trellises and up trees, and others from the city's [Rome] environs such as Veliternian, Privernatian. One cultivated at Signia is classed as a drug because of its extreme acidity which makes it an effective remedy for diarrhea.

66. Mamertine ones grown around Messana in Sicilia have gained fourth place in the running thanks to Deified Julius [Caesar] (his letters reveal that their superior status was first bestowed by him); among them, there is special praise for Potulanan – so called after the man who developed them, and nearest to Italia. Respect is also paid to Tauromenitanan – again from Sicilia – which after bottling commonly stands in for Mamertine.

67. Among the rest are Praetutian from the upper [Hadriatic] sea and those grown at Ancona, as well as what are called palm wines because they originated by accident from a single palm; inland, Caesenatine and Maecenatian; also Raetic in the Verona region, rated by Vergil [*Georgics* 2.95–96] as inferior only to Falernian; then from the [Hadriatic] sea's inmost bay Hadrian, while from the lower [Tuscan sea] Latiniensian, Graviscan, Statoniensian.

68. Luna wins the prize for Etruria, Genua for Liguria and, for between Pyrenaei and Alpes, Massilia with its pair of tastes since it produces a richer type called full-bodied for blending other wines. Respect for Baeterrae's is limited to the Galliae. As for the rest grown in Narbonensis there is no statement to be made, because they are a factory product tainted with smoke and (though I wish it were otherwise) harmful herbs and potions. In fact dealers even fake the taste and color with aloe.

69. [Wines] in Italia further from Ausonian sea gain glory too, Tarentine and Servitian and those grown at Consentia and Tempsa and Babbia, also Lucanian with its Thurine being foremost. Yet most famous of all these are the Lagaran – grown not far from Grumentum – thanks to Messala Potitus' recovery. Recently, by effort or by chance, Campania has aroused respect with new names, praised for Trebellic at the fourth milestone from Neapolis, Cauline close to Capua, and in its territory Trebulan (generally always rated as being one for the masses), and Trifoline.

70. Pompeii's reach the peak of their maturity in ten years, and gain nothing from further aging. Also, and unsuspected, is the nasty headache they cause through to the sixth hour [noon] on the next day. Unless I am mistaken, what is clear from those examples is how place of origin and soil matter, not the grape, and it would be pointless to press on enumerating the types, when the same vine is liable to fare differently in different places.

71. In the Hispaniae, Laeetan are honored for their high yield, yet for refinement Tarraconensian and Lauronensian and Baliaric from the islands [of that name] rival Italia's topmost rated. I am well aware that most people will consider there to have been many omissions here when each has their own favorite and, no matter where one goes, the same story surfaces:

72. that one of Deified Augustus' freedmen, most experienced in discernment and palate, when assessing wine for his banquet said to the host about the local wine that its taste was indeed new to him and not in the remarkable category, but that Caesar would drink nothing else. Far be it from me to deny that other wines also deserve to be well-known, but those mentioned here are ones whose ratings have been agreed upon for a long time.

IX Thirty-eight fine overseas wines

73. Now in a similar way I will discuss overseas ones. The most famous in post-Homeric times (as we treated above [14.53*]) have been those of Thasus and Chius, in Chius the one called Ariusium. To these the authority of Erasistratus, a most eminent doctor, added Lesbian about 450 years after the city of Rome's foundation [around 300 BCE]. Nowadays Clazomenian – after being seasoned with rather less seawater – is the all-out favorite.

74. Lesbian tastes naturally of the sea. Tmolitan is not a wine with appeal for drinking as such but, with its sweetness blended into other wines, their dryness gains charm and at the same time maturity, since they instantly seem more aged. Ranked next after these are Sicyonian, Cyprian, Telmesic, Tripolitic, Berytian, Tyrian, Sebennytic.

75. This [last] is grown in Egypt from three kinds of grapes very famous there: Thasius, aethalus, peuce. Ranked below: Hippodamantian, Mystic, cantharitan, Cnidian protropum, Catacecaumenitan, Petritan, Myconian.

Mesogitan has been found to cause headaches, and Ephesian is not healthy either because it is seasoned with sea water and grape syrup. Apamenan is said to be particularly useful for making mead, as is Praetutian in Italia, since this is the distinctive character of such types: the fact is that sweet wines do not blend.

76. Also now deteriorated is Protagian, which Asclepius' followers had ranked only just below Italian ones. In his book advising king Ptolomaeus on which wines to drink (at a time when Italian ones were still unknown), the doctor Apollodorus praised Naspercenitan in Pontus, then Oretic, Oeneatian, Leucadian, Ambraciot, and above all the rest Peparethian although he did say that its reputation suffered because it was no pleasure to drink until six years old.

BOOK 16

I Peoples who are without trees

2. As we have said, in the east there are indeed several peoples next to ocean deprived in this way [lack of trees]. In the north too, as we have witnessed, there are Chauci termed Greater and Lesser. There, twice in the course of each day and night, with its vast sweep ocean is stirred to spread over a huge expanse, covering where the natural order's never-ending dispute is situated along with doubts about whether it belongs to land or to sea.

3. There, a miserable people occupy high knolls or platforms erected by hand above the highest tide level in their experience. Their cottages set up thus, they are like voyagers with water covering everything around, but like castaways when it has receded, and around their huts they hunt the fish retreating with the sea. They have no livestock and, unlike their neighbors, gain no sustenance from milk; their situation does not call for them even to contend with wild beasts, because all undergrowth is far removed.

4. With sedge and marsh rushes they braid ropes to create fish-nets; mud that they have grabbed with their hands they dry more with wind than with sun; their food and their own frames chilled by the north wind they warm with turf. Their sole drink is rainwater stored in pits in their hallway at home. And these peoples, should they be subdued today by the Roman people, claim that they are becoming slaves! So, really, fortune spares many to punish them.

II Tree marvels in the northern region

5. Another marvel from the forests: they completely cover the rest of Germania and with their shade they add to the cold, but the tallest are not far from the Chauci mentioned above [16.2], especially around two lakes. Even the shores are occupied by very fast-growing oaks which, when undercut by waves or felled by gusts, detach with them vast islands in their tangle of roots. So balanced, they float along upright, often frightening our fleets with their tackle of huge branches; when – as if on purpose – driven by waves against vessels anchored for the night, the only recourse these have is to begin a naval battle against trees!

6. In this same northern region the vast Hercynian hardwood forest, untouched by time and as old as the world, is a marvel in a class of its own with its near-immortal status. Leaving aside [much] that should not be credited, it is established that hillocks are thrown up by the pressure of roots colliding with one another. Alternatively, where no earth has followed, arches vying with one another right up to the branches bend over like gates open [wide enough] to let cavalry squadrons through. All [the trees] are mostly acorn-bearing types, which Romans perpetually honor.

******

LXV Reeds, twenty-eight types of bamboo

159. With reeds eastern communities make war; by means of reeds with a feather attached they hasten death; to reeds they add points that cause harm because with their barbs they cannot be removed, and if the weapon is broken in the wound then another one is created from it. With these weapons they obscure the sun itself. Hence they very much prefer clear days, with a hatred of wind and rain which require that there be peace between them.

160. A fairly methodical count, too, taken of the Aethiopians, Egyptians, Arabians, Indians, Scythians, Bactrians, the numerous Sarmatian and eastern peoples and all the Parthians' kingdoms would show that close to half the entire world's population lives subdued by the reed.

161. Outstanding use of it on Creta has made warriors from there famous. But in this too, as in other respects, Italia has been the victor, since the reeds in Rhenus, a river of Bononia, are unsurpassed for arrows; they contain the most pith, as well as having the [right] weight for flight and a balance that steadies them even in the face of gusts; Belgic ones are without such qualities. All Cretic ones also have these advantages, although there is a preference for Indian ones which to some seem a different type, especially when with tips added they offer a substitute for spears.

162. Indian bamboo has the girth of a tree in fact, such as we routinely see in temples. According to Indians, there is also a distinction between the male and the female, the male trunk being denser, the female more capacious. Believe it or not, a single stretch between knots can even serve as a boat. It grows especially around Acesines river.

******

LXXXVIII Trees planted by Agamemnon. Trees from the first year of the Trojan War. Trees at Troy named after Ilium older than the Trojan War

238. According to some sources, a plane-tree at Delphi was planted by Agamemnon's hand, and another in a grove at Caphya in Arcadia. Today there are trees on Protesilaus' tomb by Hellespont opposite the city of the Ilienses [Trojans]; in every period since his, once these have grown tall enough to glimpse Ilium, they dry out and then revive again. The oaks on Ilus' tomb by the city, however, are said to have been planted at the time when it began to be called Ilium.

LXXXIX The same at Argos. Ones planted by Hercules. Ones planted by Apollo. A tree older than Athens

239. The olive tree to which Argus tied Io when she was transformed into a bull is said to survive still. On this side of Heraclea in Pontus are altars to Jupiter with the name *Stratios* [in Greek], and two oak trees there planted by Hercules. In the same region is the harbor of Amycus, famous for king Bebryx's murder. Since the day of his death his tomb has been covered by a laurel tree called 'insane' because, if anything lopped from it is brought on board ship, quarrels occur until it is thrown away.

240. We have mentioned [5.106] Aulocrene region traversed by the route from Apamea to Phrygia. A plane-tree is pointed out there from which Marsyas hung after being defeated by Apollo; even at that time it was chosen because of its size. Also, a palm-tree from this god's time is to be seen on Delos, and at Olympia a wild olive from which Hercules' first crown was made – still piously tended today. At Athens, too, the olive-tree produced by Minerva in the contest is said to be still alive.

BOOK 18

LVII The constellations' pattern day by day and signal on earth of agricultural work to be done

210. There is a two-step process: first of all to gain a rule from the sky, then to investigate this with the evidence. Above all [to be taken into account] is the convex shape of the world [i.e. universe] and the different perspectives from the globe's lands, with the same constellation revealing itself to different peoples at different times, so that it does not make the same impact everywhere on the same days. Additional difficulty has been caused by authors taking observations in different locations, or then even putting forward different findings for the same locations. However, there have been three groups: Chaldaean, Egyptian, Greek.

211. To these the dictator Caesar has added ours as a fourth, restoring individual years to the sun's orbit in consultation with his scientific expert Sosigenes; and this system was itself corrected after a mistake had been discovered, with the consequence that no intercalations were made for 12 straight years because the year, after previously moving ahead of the constellations, had begun to fall behind them.

212. Sosigenes himself too in his three treatises, although more careful than the others, nonetheless did not stop expressing doubts even while correcting himself. The authors with whom we prefaced this volume have covered these issues, although it is rare for the opinion of any one of them to agree with another. This is less surprising among the rest, who had the excuse of being in different areas.

213. Of those who differed in the same region, we shall offer one example of a disagreement: Hesiod – there even exists a work of astronomy with his name on it – stated that the morning setting of the Vergiliae occurred at the end of the fall equinox, Thales on the 25th day after the equinox, Anaximander on the 29th, Euctemon <on the 44th, Eudoxus> on the 48th.

214. We follow Caesar's observation above all; this will be Italia's system. However, we shall also state others' preferences, since we are explaining not a single land, but the entirety of Nature, ordered not by authors – this is too long-winded – but by regions. Readers should just bear in mind that for the sake of brevity when Attica is mentioned, they should at the same time understand the Cyclades islands; when Macedonia, then also Magnesia, Thracia;

215. when Egypt, then also Phoenice, Cyprus, Cilicia; when Boeotia, then also Locris, Phocis, and always the areas adjacent; when Hellespont, then also Chersonesus and the mainland as far as mount Athos; when Ionia, then also Asia and Asia's islands; when Peloponnesus, then also Achaia and the lands situated west of it.

216. Chaldaeans will indicate Assyria and Babylonia. It will be no surprise that Africa, the Hispaniae and Galliae receive no mention, since nobody who has written about the constellations' risings has made observations there. Even so, the calculation for those lands too will not be a difficult one using the arrangement of circles which we made in Book Six [6.211–220]; with this it is feasible to comprehend the astronomical relationship not just of peoples, but also of individual cities.

BOOK 19

I Nature and remarkable features of flax

2. Many [authors] have made gardening their next [topic]. To us, it does not seem the right point to move on to this at once, and we are surprised that some – eager for the satisfaction of knowledge or a reputation for learning by this means – have passed over such a lot without making any reference to so much that grows of its own accord or with cultivation, especially when in plenty of instances the respect to be gained in terms of their market-price and practical value is higher even than for cereals. And to begin with what possesses acknowledged value and has filled not only every land but also the seas, flax is grown from seed and cannot be termed either cereal or vegetable. But in which part of life is it not to be found?

3. And what is more amazing than the existence of a plant which moves Egypt so close to Italia[2] that Galerius reached Alexandria from Sicilia's strait in seven days, Balbillus – likewise a prefect [of Egypt] – in six, the next summer Valerius Marianus, a senator of praetorian rank, in nine days from Puteoli with the lightest breeze?

4. [Or] the existence of a plant that brings Gades from Hercules' pillars to Ostia in seven days and Hispania Nearer in four, Narbonensis province in three, Africa in two, as happened to Gaius Flavius legate of the proconsul Vibius Crispus with just the slightest breeze? How daring life is, how full of crime! So something is sown in order to grasp winds and storms, and we are superior to being conveyed by waves alone,

5. with sails larger than ships nowadays not enough; yet although one entire tree may suffice for the breadth of a yard-arm, still further sails upon sails are added above it, spread in the prow especially and others at the stern, and death is risked in so many ways. In short, the means of carrying the world to and fro is created from such a small seed, one with a thin stalk that stays close to the ground, and is not woven spontaneously but broken and crushed and beaten to the softness of wool – injury from which it attains the utmost daring!

6. There is no means of adequately cursing the inventor (mentioned by us in the appropriate place [7.209*]) who, far from being satisfied with people dying on land, had them also perish without a grave. In fact in the previous book [18.326–339*] we kept offering advance warning to be on the lookout for rain and wind in order to protect crops and food. How amazing it is to have something planted by human hand, and likewise harvested by human skill, that longs for sea-breezes. Moreover – just to be aware of the favor shown us by the Poenae [goddesses of revenge] – nothing grows more easily; and, so that we may recognize this happens in defiance of Nature, it burns fields and even damages the soil further.

II Seventeen outstanding varieties

7. It is sown mainly in sandy [soils] and in a single ploughing; nor is anything else faster-growing. Planted in spring, it is uprooted in summer, and for this additional reason harms the ground. Even so, you might

---

[2] Because sails are made from it.

forgive Egypt for planting it so as to import goods from Arabia and India. In that case aren't the Galliae also assessed on this output? Isn't it enough that they are blocked from the sea by mountains, and that on the ocean side an actual so-called void stands in their way?

8. Cadurci, Caleti, Ruteni, Bituriges and Morini (considered to be the remotest humans) – in fact really all the Galliae – weave sails, as indeed the enemy across Rhenus also do now, their women being unaware of any finer clothing. This point brings to mind Marcus Varro's account of the Serrani family's custom that the females should not wear linen clothing.

9. In Germania, moreover, [women] are even sent below to do this work underground. It is much the same too in the Aliana region of Italia between Padus and Ticinus rivers, where the linen is ranked third in Europa down from that of Saetabis [in first place]; those of Retovium near Alia and of Faventia on Via Aemilia are rated second. For luster, Faventia's is preferred to Alia's, which is never beetled. Retovium's stands out for its fineness and thread-count, with a luster to match Faventia's but not beetled, which pleases some people but puts off others. The thread's strength is nearly more uniform than that of a spider's web, and it twangs, should you be interested in testing it with your teeth; as a result, it costs twice what others do.

10. And, after these, Nearer Hispania's linen has a remarkable sheen because of the character of the stream – flowing by Tarraco – in which it is dressed; here, where linen was first produced, its fineness is astonishing too. It was not so long ago that, again from Hispania, Zoelic [linen] – especially useful for hunting-nets – came to Italia. This state is one in Gallaecia, near the ocean. For catching fish and fowl, [linen] from Cumae in Campania also has special distinction.

11. The same material is [used] too for hunting-nets – for we use linen to set no lesser traps for all beasts as for our own selves – but Cumae's nets slice off a boar's bristles and even outclass an iron blade; and before now we have seen [netting] so fine that it and its bolt-rope would pass through a person's ring, and one man may carry sufficient mass of it to encircle a grove. Nor is this what is most amazing: each of its threads consists of 150 strands, just as was recently [made for] Julius Lupus, who died when prefect of Egypt.

12. This might amaze anyone unaware that for the corselet once the property of a king of Egypt called Amasis, and [now] in the temple of Minerva at Lindus on the Rhodians' island, each individual thread comprises 365 threads [i.e. strands]; Mucianus, three times consul, just lately stated that he tested this, and that now only small remnants of [the corselet] survive because of the damage caused by those testing it.

13. Italia still has respect for Paelignian linen also, but it is only used by fullers. Its luster and similarity to wool are unmatched; so too, what the Cadurci produce for mattresses has an outstanding reputation. They and the flock [for stuffing them] are both inventions of the Galliae. Even so, in Italia the term *stramentum* [straw] continues in current use still [for bedclothes].

14. Egyptian linen is least robust, but most profitable. The four kinds there are named after the region in which each grows: Tanitic, Pelusiac, Butic, Tentyritic. The upper part of Egypt facing towards Arabia produces a bush called *gossypion* by some, but *xylon* by most, and so linen made from it is *xylina*. It is a small [bush], and from it hangs a fruit like a bearded nut with silky fiber inside, the down of which is spun. Along with its luster it has no match for softness or ease of combing. Clothing made from it has very great appeal to priests in Egypt.

15. The fourth kind is also called *othoninum*. It is made from a type of marsh-reed, though only from its tufts. From broom Asia makes thread that is particularly strong for fishing-nets because the plant is soaked for ten days; Aethiopians and Indians [make it] from apples, Arabians from gourds grown on trees, as we have said [12.38*].

BOOK 27

I [Medicinal plants]

1. As I deal with the present topic [medicinal plants], I certainly have growing admiration for the distant past, and the greater the quantity of plants waiting to be described, the greater correspondingly is the awe I feel for the care that the ancients took over their discoveries and for their generosity in reporting them. Without doubt, it would seem possible by this means for even the natural order's own bounty to have been outclassed, if the discoveries were a matter of human effort.

2. But it now appears that they were really the work of gods, or at any rate divinely inspired, even when a human made the discovery, and that the same universal parent both produced these [plants] and publicized them, there being no greater miracle of life, if we are willing to admit the truth. Think of the Scythian plant from the Maeotis marshes, and euphorbea from mount Atlas and from beyond Hercules' pillars where the natural order itself fades, in another direction britannica from islands lying beyond [earth's] landmass in the ocean, and similarly aethiopis from a zone scorched by celestial bodies,

3. others too from elsewhere being carried here and there throughout the world for mankind's well-being. All this is due to the immense majesty of Roman peace, which exhibits in turn not only humans with different lands and peoples between them, but also mountains and ranges soaring to the clouds and what they produce, including plants too. I beg that this bounty from the gods may never end, because by this means they seem to have made a gift of Romans as a twin source of life, so to speak, for the human condition!

BOOK 31

II Different kinds of water

4. Everywhere and in the great majority of lands a profusion of [waters] gushes out, here cold, there hot, there again both (as among the Tarbelli, an Aquitanic people, and in the Pyrenaei mountains just a slight distance away), there warm and cool. They offer aid against disease and they surge for humans' benefit exclusively among all creatures, enlarging the number of gods by various names and founding cities such as Puteoli in Campania, Statiellae in Liguria, Sextiae in Narbonensis province. However, the generosity with which they do this in Baian bay is unsurpassed, as is the range of their assistance here:

5. some have the power of sulfur, some of alum, some of salt, some of soda, some of bitumen, some are even a mix of acid and alkali; in some instances it is their steam that has value, this being powerful enough to heat baths and even to make cold water in bath-tubs boil. Waters called Posidianae in the Baian region – named after a freedman of Claudius Caesar – even cook food. In the sea itself also, waters that used to belong to Licinius Crassus give off steam: hence something good for health rises right amid the waves.

III Types of waters beneficial to the eyes

6. According to type, some waters benefit muscles or feet or hips, others sprains or fractures, they empty bowels, clean wounds; they heal the head and especially the ears, and Ciceronian [waters] the eyes in fact. It is worth remembering that there is a villa situated on the shore stretching from lake Avernus to Puteoli, famous for its portico and grove which Marcus Cicero used to call the Academy after the Athenian model; his volumes of the same name were composed here and here also he had set up memorials to himself, as if indeed he had not already done that worldwide as well.

******

IV Types producing fertility. Waters curing insanity

8. ... In the same region of Campania Sinuessa's waters are said to put an end to infertility in women and insanity in men,

V Types for gallstones

9. those on Aenaria island are a cure for gallstones, as is the cold one called Acidula four miles from Teanum Sidicinum, also one in the Stabiae region called Dimidia, and one from Acidulus spring in the Venafrum region. Those who drink from lake Velinus feel the same effects, and the same applies to the Syrian spring next to mount Taurus according to Marcus Varro, as well as to Gallus river in Phrygia according to Callimachus. But here it is essential to drink only a limited amount in case an onset of madness occurs, as happens in Aethiopia (Ctesias writes) to those who have drunk from the Red spring.

VI Types for wounds

10. Lukewarm as they are, Albula's waters near Rome heal wounds, but the really freezing ones of Cutilia in Sabine territory assail bodies with a sort of suction that could seem almost a bite; these ones are best for treating the stomach, sinews, whole body.

VII Types to protect the fetus

A spring at Thespiae causes women to conceive immediately, as does Elatum river in Arcadia, but Linus spring, also in Arcadia, protects the fetus and prevents a miscarriage. By contrast, a river in Pyrrha called Aphrodisium causes infertility.

VIII Types to remove acne

11. According to Varro, acne is removed by lake Alphius, and because of this disorder the ex-praetor Titius had a face like a marble statue. Cydnus river in Cilicia cures gout, as a letter from Cassius of Parma to Marcus Antonius shows. On the other hand, waters are to blame for everyone in Troezen feeling they have foot problems.

12. The state of the Tungri in Gallia has a remarkable spring sparkling with very many bubbles and tastes like iron, although there is awareness of this only after drinking. This [spring] purges the body and shakes off tertian fevers as well as disorders caused by gallstones. The same water, when placed over a fire, becomes agitated and eventually turns red. The Leucogaei springs between Puteoli and Neapolis cure eyes and wounds. Cicero claimed in his *Wonders* [book] that pack animals' hooves are toughened only by Reate's marshes.

IX Ones for coloring fleeces

13. Eudicus says that there are two springs in Hestiaeotis, Cerona which turns sheep that drink from it black, and Nelea which turns them white; by drinking from both, they become piebald. According to Theophrastus, oxen and cattle are made white by the Crathis at Thuri, black by the Sybaris,

X Ones for humans

14. and in fact humans feel that difference too: those who drink from Sybaris are darker and tougher and have curls; those who drink from Crathis are fair, gentler and have straight hair. Also, he says, people in Macedonia who may want their [livestock] born fair lead them to the

Haliacmon, while those wanting black or dark lead them to the Axius. [Theophrastus] again says that in certain places everything generated is dark, including crops (as among the Messapi), and that umber mice live and throng by a certain spring at Lusi in Arcadia. At Erythrae the Aleon river causes body-hair to grow.

XI Ones for memory, ones for forgetfulness

15. In Boeotia by [a sanctuary of] the god Trophonius next to Hercynna river there is a pair of springs, one inducing memory, the other forgetfulness – hence the names they have acquired.

XII Ones for heightened senses, ones for sloth, ones for melodious voices

In Cilicia at Cescum town there flows Nous stream, from which – says Marcus Varro – drinkers develop more discriminating instincts, but on Chia island is a spring which makes them dimwitted, [while] one at Zama in Africa gives them melodious voices.

XIII Ones to induce distaste for wine, ones for drunkenness

16. Eudoxus says that distaste for wine comes to those who have drunk from Clitor's lake, and drunkenness (says Theopompus) from springs we have mentioned [2.230]. According to Mucianus, wine flows from Father Liber's spring on Andrus during the regular seven-day [ceremonies] for that god, yet should it be removed out of sight of the temple, its taste changes into water.

XIV Ones to substitute for oil

17. According to Polyclitus, Liparis river near Soli in Cilicia is oily; Theophrastus says the same about a spring in Aethiopia with the same name, and according to Lycos there is a spring among the Indian Oratae with water that fuels lamps. The same is said [about a spring] in Ecbatana. According to Theopompus, the people of Scotusa have a lake able to heal wounds.

XV Salty and bitter ones

18. According to Juba, lake Insanus among the Trogodytae is named for its nefarious power, becoming bitter and salty and then fresh three times during the day and as many at night; it is swarming with white snakes 20 cubits long. He also says that a spring in Arabia ejects with such force that it spews out anything plunged into it regardless of weight.

XVI Ones that toss out rocks, induce laughter or tears, or allegedly cure love

19. According to Theophrastus, a spring of Marsyas in Phrygia near Celaenae town tosses out rocks. Not far from it are two springs called Claeon and Gelon after the meaning of their Greek names ['Crying' and 'Laughing']. A spring at Cyzicum is called Cupid's; Mucianus believes that those drinking from it no longer feel love.

XVII Ones that, once drawn, stay hot for three days

20. There is a hot spring at Crannon which almost boils, and when [its water] is added to wine in a container it keeps the drink hot for three days. There are also hot springs at Mattiacum across the Rhenus in Germania; a cupful from them stays hot for three days, with the water actually forming pumice around the rim.

XVIII–XX Water marvels

XVIII Ones in which everything is liable to sink, in which nothing [sinks]

21. Anyone who might think that some of these accounts are untrustworthy should realize that in no part of Nature are there greater marvels, although around the beginning of this work I recorded many at length [2.224 – 241]. Ctesias says that there is a pool among Indians called Sila in which nothing floats, everything sinks. According to Caelius, in our own [lake] Avernus even leaves go down; according to Varro, birds that have flown there die.

22. By contrast, everything floats in Africa's lake Apuscidamus, nothing sinks; it is the same in a spring at Phintias in Sicilia, according to Apion, and among the Medes in a lake and well of Saturn. But the spring of Apollo Surius at Myra in Lycia has the habit of shifting to somewhere else in its vicinity – which is some kind of omen; amazingly, its fish shift with it. Locals seek responses from them with food, which they grab if positive, but push away with their tails if negative.

23. Olcas river in Bithynia laps against Brietium – this is the name of a temple and a god – and those who break oaths cannot withstand its surge, which burns like a flame. In Cantabria too the sources of the Tamaris are considered prophetic. There are three of them, eight feet apart, combining into one channel to make an immense river.

24. Each one separately dries up for 12 days, sometimes 20, with not even a trace of water, although near them there is a spring brimming all the time. It is a dreadful outlook for those wanting to watch [the three] flow if they do not – as recently for the propraetorian legate Larcius Licinius: he died seven days later. In Judaea a stream runs dry every sabbath.

XIX Deadly waters, poisonous fish

25. Otherwise some marvels are dangerous. Ctesias writes that there is a spring in Armenia with black fish that when eaten instantly cause death. I have heard this also about fish around the source of the Danuvius, until it comes to a spring located near its channel which is the limit for this kind of fish; so, according to reports, the head of that river is taken to be here. It is said that the same occurs too in a Nymphs' pool in Lydia.

26. At Pheneum in Arcadia there flows out of rocks a watercourse called Styx which, as we have stated [2.231], kills instantly; but (says Theophrastus) in it are small – and likewise deadly – fish, something not found in other types of poisonous spring.

27. Theopompus says that there are also waters which kill at Cychri in Thracia, Lycos that at Leontini are ones [which kill] on the third day after being drunk, Varro that a spring four feet wide on [mount] Soracte [does likewise]. At sunrise it overflows as if boiling, [and] birds that have sipped from it lie dead nearby. For a further deceptive feature is that some such waters make an attractive sight, as at Nonacris in Arcadia where they have absolutely no off-putting property. The water there is considered dangerous because it is extremely cold, given that as it flows it turns itself into stone.

28. Quite the opposite occurs around Tempe in Thessalia, since [the spring's] venom terrorizes everyone and even bronze and iron are said to be corroded by that water. It flows for a short distance, as we have indicated [4.31], and amazingly roots of a wild carob-tree always in purple flower are said to entwine its source. Some unique plant thrives on the brim of the spring as well. In Macedonia, not far from the poet Euripides' tomb, two streams converge, one being an excellent tonic to drink, the other deadly.

XX Ones that may turn into stone or make stone

29. A spring at Perperene turns whatever land it soaks to stone, as do hot waters at Aedepsos on Euboea because wherever their stream flows the rocks grow taller. Garlands tossed into a spring at Eurymenae are turned to stone. At Colossae there is a river where bricks thrown in are pulled out as stone. In a mine on Scyrus whatever trees the river wets are turned into rock along with their branches.

30. Drops, too, that drip in the caves at Corycus harden as stone; <...> at Mieza in Macedonia [stone stalactites] actually even hang down from ceilings, although in one place at Corinthus [drops only harden] after falling, while in certain caves this happens at both ends [i.e. top and bottom] and columns are formed, as in a great cavern at Phausia in the Rhodian Cherronesus where these even have different colors. To my mind, this is enough illustration for now.

BOOK 32

VI Fishes' mental abilities. Fishes' remarkable qualities

15. Trebius Niger says that the xiphias, or swordfish, has a pointed bill, and that ships pierced by it sink; also that in the ocean off the place in Mauretania called Cottae, not far from Lixus river, flying fish leap out of the water in such great numbers that they sink ships.

VII Where fish may eat out of the hand

16. Fish at several imperial residences eat from the hand, but – as related with admiration by the ancients – in ponds not in fishpools, as at Helorus fortress in Sicilia not far from Syracusae, likewise in Jupiter's spring at Labrayndus where eels also wear earrings attached to them, just like fish on Chius next to the shrine of the Senes ['Aged'], as well as in Mesopotamia at Chabura spring which we have mentioned [31.37*].

VIII Where there may be responses from fish. Where they may recognize a voice

17. At Myra in Lycia at Apollo's spring called Curius, after three pipe-calls fish come to give signs; ones that tear apart the flesh tossed in serve to delight inquirers, while those that push it away with their tails cause dismay [cf. 31.22]. At Hieropolis in Syria [fish] in Venus' lake obey the voices of the temple staff; decorated in gold, they come when called, squirming to be tickled, their mouths open in wait for hands to be put in them. At Hercules' rock in the Stabiae area of Campania black-tails grab bread thrown in the sea, but keep clear of any food attached to a hook.

IX Where fish may be pungent, where salty, where freshwater, where not mute. Also their feeling, or lack of feeling, for locations

18. This is not the end of these marvels either, because the fish are pungent at Pele island and at Clazomenae, opposite Sicilia's rock and at Lepcis in Africa and at Euboea and Dyrrachium, then again so salty around Cephallania and Ampelos, Paros and Delos' rocks that they could be considered pickled, [although] not salty at that same island's harbor. Undoubtedly their food is what makes the difference.

19. Apion says that the largest fish is the porcus, which Lacedaemonians call orthagoriscus because it grunts when caught. A further, all the more remarkable feature of this type of fish is its occurrence in various locations: one case to hand that arises is Beneventum in Italia in that there – as is well known – picklings of all types are redone.

******

XI Coral

21. Indians value coral as highly as we do Indian pearls (which I have treated adequately in their proper place [9.104*]); it is a matter of the worth that different peoples attach to them. [Coral] even originates in Red sea – although a darker type – as well as in the Persian where it is called lace; the most prized occurs in Gallic bay around Stoechades islands and in Siculan [bay] around Aeoliae [islands] and Drepana. It is also formed at Graviscae and off Neapolis in Campania, an especially red kind, while the type at Erythrae is soft and therefore the cheapest.

BOOK 37

XI About amber

30. Items of amber hold the next place in [order of] luxuries, though to date only for women, and altogether their valuation matches that of jewels. There are some reasonable uses for what has been covered above: crystal for cold drinks, myrrhine ware for both [hot and cold]; so far, not even luxury has been able to devise a use for amber.

31. Here is a chance to expose the vanity of Greeks; readers should be fair-minded and patient, since it benefits the human condition to know that not everything reported by [Greeks] must be admired. After Phaethon was struck by a thunderbolt, grief transformed his sisters into poplar trees, and every year by Eridanus river (which we call Padus) they used to shed tears of amber, calling it electrum, since their name for the sun was Elector [shining one] – a story told by plenty of poets, the first, I understand, being Aeschylus, Philoxenus, Euripides, Nicander, Satyrus; but it is false, as Italia plainly attests.

32. Those who took more care stated that there were islands in Hadriatic sea – the Electridae, to which the Padus brought down [amber]. [But] there definitely have never been any islands there of that name, nor indeed are there any sited so that Padus' flow could convey anything to them. Aeschylus in fact said that the Eridanus is in Hiberia – in other words, in Hispania – and that this very [river] is called Rhodanus, while according to Euripides again and Apollonius, the Rhodanus and Padus merge on the shore of the Hadriatic; such great ignorance of the world makes it easier to forgive their ignorance of amber.

33. More cautious but equally mistaken [Greeks] have said that there are trees standing on inaccessible rocks at the farthest limit of Hadriatic bay which shed this gum when Canis rises. Theophrastus says that it is dug up in Liguria, while according to Chares Phaethon really died in Aethiopia on *Hammon's island* [in Greek], and that this is where his shrine and oracle are, and where amber originates. According to Philemon, it is a fossil extracted at two locations in Scythia: at one, it is transparent, with the color of wax, and called electrum; at the other, tawny and called sualiternicum.

34. Demostratus calls it lyncurium and says it is the product of wild lynxes' urine – the tawny and fiery one from males, the duller and transparent from females; others call it langurium, and say there are beasts in Italia called languri. Zenothemis' name for these is langes, and he claims they are native to the Padus region, while according to Sudines a tree that produces [amber] in Liguria is called lynca.

35. Metrodorus' opinion too was the same. Sotacus believed that in Britannia it seeped from rocks which he called electridae, Pytheas that [it came from] the Gutones, a people in Germania living by an estuary of ocean called Metuonis 6,000 stades wide; Abalus island is a day's sail from it; [amber] is brought here by spring currents and is frozen seawater's excretion. Locals use it instead of wood for kindling, and sell it to the Teutoni nearby.

36. Timaeus also believed this, but he called the island Basilia. Philemon denied that electrum gives off a flame. Nicias wanted to explain it as fluid from sunrays: around sunset, on being directed to earth with more vigor, they leave a rich perspiration there which ocean tides then deposit on the Germans' shores. A similar process occurs in Egypt – where it is called sacal – likewise in India, and with even more appeal to Indians than incense;

37. in Syria too women make whorls from it and call it 'harpax' because it snatches leaves, straw and fringes of clothing. Theochrestus says that it is driven onto Pyrenaean capes when ocean surges, as Xenocrates – who wrote about this most recently and is still living – also believed. Asarubas says that near Atlantic sea is lake Cephisis called Electrum by the Mauri; when this lake is well heated by the sun, floating amber emerges from the mud.

38. Mnaseas [mentions] a place in Africa called Sicyon, and Crathis river flowing into the ocean from a lake where birds he calls meleagrides and penelopes live. Here [amber] is produced in the same way as described above. According to Theomenes, near Great[er] Syrtis is the garden of the Hesperides and Electrum lagoon, where there are poplar trees from whose tops [amber] falls into the lagoon to be collected by girls of the Hesperides.

39. Ctesias says that among the Indians there is Hypobarus river, whose name indicates that it delivers all good things; it flows from the north into the eastern ocean next to a forested mountain whose trees bear electrum. These trees are called psitthachorae, a name which means 'cloying charm'. Mithridates says that there is an island off the shores of Carmania called Serita forested with a type of cedar from which [amber] flows down onto the rocks.

40. Xenocrates says that in Italia it is called not only sucinum, but also thium, or sacrium by Scythians, since it is produced there too; others think it is generated in Numidia from mud. Above all of these is the tragic poet Sophocles – quite a surprise, since his dramas are so solemn, not to mention his famous life, aristocratic Athenian birth and achievements and leadership of an army: he said [amber] originates beyond India from the tears of meleagrides, [the birds] who weep for Meleager.

41. Isn't it amazing that he believed this, or hoped he would be able to convince others of it? How could such childishness and naiveté be found, believing in birds' yearly laments, or heavy tears, or birds that would go from Graecia where Meleager died to weep among the Indians? Really? But don't poets come up with just as much fiction? But for anyone to have talked seriously about something imported every day, and not in short supply, which [itself] exposes the lies told about it – this is the utmost insult to mankind, one that takes insufferable liberties with lies!

42. It is certain that [amber] originates on islands in the northern ocean and that it is called glaesum by the Germans, and that in consequence we called one of these islands Glaesaria when Germanicus Caesar was conducting naval operations there – the one referred to as Austeravia by barbarians. [Amber] is produced from sap seeping out of a type of pine tree, like the gum in cherry trees and the resin in pines that erupts with a full flow. Hardened by frost or weather or by the sea after a rising tide has snatched it from the islands, for certain it is driven onto the shore so smoothly that it gives the impression of floating instead of settling on the sea-bottom.

43. Our ancestors also believed that it was a tree sap, and so called it sucinum. That the kind of tree is a pine is proven by its piney odor when rubbed, and because when lit it burns and smells like pinewood. Germans mostly bring it to the province of Pannonia, and from there it was Veneti (called Eneti by Greeks) who first made it famous, given their proximity to Pannonia and their involvements around Hadriatic sea.

44. The tale's link with the Padus has an obvious explanation: today peasant women in Transpadana wear amber in the form of necklaces, mostly as ornaments but also for medicinal purposes; it is believed to counteract tonsillitis and throat ailments, since the water near the Alpes is harmful to the human throat in various ways. Only recently has it been established that the distance from the shore of Germania (from where it is imported) to Carnuntum in Pannonia is nearly 600 miles.

45. There is a Roman *eques* still living who was sent by Julianus to procure it when he was in charge of gladiatorial games given by *princeps* Nero. He traversed both that trade-route and the shores, and brought back such a quantity that the nets for keeping off wild beasts which protect the front row were knotted with amber, while for one day the arms and biers and all the equipment – each day's parade being different – were made of amber.

46. The heaviest lump of it he brought weighed 13 pounds. It definitely also originates in India. Archelaus, who was king in Cappadocia, says that it was imported from there in a rough state – with pine-bark still sticking to it – and was finished by being boiled in a suckling pig's fat. Proof that it is initially liquid drops comes from the fact that sundry objects are visible inside, like ants and gnats and lizards; undoubtedly they became stuck when the sap was fresh, and remained immured as it hardened.

******

LXXVII Nature's works compared by region. Its products compared by price

201. Now that Nature's works have been covered in their entirety, some ranking of the actual products and regions will be appropriate.

So then, on the entire globe, wherever the vault of heaven turns, loveliest with everything that merits the leading position in Nature is Italia: leader and co-parent of the world with her men and women, commanders and soldiers, slaves, artistic excellence, outstanding talent and – with its location and healthy and mild climate – easy access for all peoples, shores with harbors, favorable winds. All this [Italia] has because of a position jutting towards the most useful part midway between east and west, ample water resources, healthy woodland, broken mountains, unthreatening wildlife, fertile soil, rich pasturage.

202. Nowhere is superior in what should be life's essentials: crops, wine, olive oil, fleeces, flax, cloth, bulls; even at training courses elsewhere her horses are the first preference. So long as mining was legal, her gold, silver, copper, iron were second to nowhere else on earth. Nowadays, while still pregnant with these, as an entire dowry she bestows juices and crops and fruits with a variety of tastes.

203. After [Italia], excluding India's marvels, I would rank Hispania next (at least wherever it is bordered by sea), because although part of it is barren, really where it is productive it thrives with crops, olive oil, wine, horses, and mines of all kinds. Thus far Gallia rates equal. But Hispania wins out because of its deserts' esparto-grass and its selenite with such exquisite shades too, its energetic efforts, training of slaves, people's physical toughness with such feisty spirit.

204. Among products themselves, however, the most expensive of those originating in the sea are pearls; of those above ground, crystals; in it, diamonds, emeralds, gems, myrrhine ware[3]; emerging from the ground, cochineal berries, silphium, nard from leaves, Seric clothing [silk], citron wood from its tree, cinnamon, cassia [perfume], amomum [spice] from bushes, from trees or bushes amber sap, balsam, myrrh, frankincense, from roots [aromatic] costum. Among what has the capacity to breathe, the most expensive product from land animals is elephants' tusks, from

---

[3] See 37.30 above.

sea [creatures] turtle-shells. From their coats, the hides that the Seres dye, and the hair of Arabian nanny-goats we call ladanum. Among what is amphibious, purple dyes from shellfish. There is nothing special to note about the nature of birds except warrior crests and the fat of geese from Commagene. It is not to be overlooked that gold – over which humans all go crazy – barely gains tenth place on the scale of expense, while in fact silver (used to purchase gold) is hardly twentieth.

205. Farewell, Nature parent of everything, and be favorably disposed to me as the only one of the Quirites [Romans] to have discussed you in all your aspects.

APPENDIX I

Titles and Technical Terms

consul	Two were the joint chief annual magistrates of the Roman Republic; the office was maintained under the rule of the emperors. A year was dated by the names of the consuls.
conventus	District (comprising numerous communities and peoples) in a Roman province, with a center which the governor would visit periodically to adjudicate cases and handle administrative business.
dictator	Special magistrate appointed – when Rome was a republic – to handle pressing matters with sweeping authority superior to that of regular magistrates.
eques (pl. *equites*)	Wealthiest, most privileged Roman citizens other than senators.
haruspices	Etruscan, and later Roman, priests noted for determining the will of the gods by inspection of animal entrails.
imperator	Originally, a title for a successful Roman general; later used of the emperor.
nome	District in Egypt, administered from its chief town.
prefecture	Roman district for administrative and judicial purposes.
princeps	During the Republic, an informal term for a 'leading figure', later applied to the emperor alone.
proconsul	Ex-consul who at the end of his term of office accepted an assignment (typically the governorship of a province) that continued his authority.
satrapy	In the Persian and Parthian empires, a large region governed by a satrap.
strategia	District controlled by the military.
tetrarchy	Administrative district, especially in the eastern Roman empire, sometimes under a semi-independent ruler.

APPENDIX 2

Units of Measurement

Area	
iugerum	240 by 120 Roman feet (approximately ⅗ of an acre)
Distance	

Because he draws from many different sources, Pliny mentions a variety of distance units. Conversion is always difficult, but he does offer some guidance.

cubit	Measurement from the elbow to the end of the longest finger
digit	Pliny uses *digitus* (6.121) – meaning breadth of a finger and understood to be a sixteenth part of a foot
foot (Babylonian)	According to Pliny (6.121), it is three finger widths (*digiti*) longer than the Roman foot
foot (Roman)	16 finger widths (*digiti*) or 12 inches (*unciae*); slightly shorter than the modern foot
inch	One *uncia* equals a twelfth part of a foot
mile	*mille passus* or *milia passuum*: 1,000 paces, hence 5,000 Roman feet (slightly less than one modern mile)
pace	Five Roman feet, as calculated by Pliny (2.85)
parasang	Mentioned once (6.124), but no distance is given in this unit
schoenus	Pliny (5.63) claims it equals 30 stades, but observes that Persians (6.124) give a different length and that Eratosthenes (12.53) equates it to 40 stades, while others (12.53) say 32
stade	Pliny (2.85) claims it equals 125 paces, so 625 Roman feet
Monetary	
as (pl. *asses*)	Roman bronze coin
denarius	Roman silver coin worth four *sesterces* or 16 *asses*
sesterce	Roman bronze coin worth four *asses*

	Weight
pound	equals 12 ounces; a Roman ounce is slightly less than a modern one
talent	weight and value varied; according to Pliny, Marcus Varro said an Egyptian talent amounted to 80 pounds of gold (33.52*), while an Attic one equaled 6,000 denarii (35.136*)

APPENDIX 3

Latin Editions Translated (Books 2 to 6)

Book, Sections	Budé	Sammlung Tusculum
2	Beaujeu, Jean. *Pline L'Ancien. Histoire Naturelle. Livre* II. Paris: Belles Lettres, 1950	
3	Zehnacker, Hubert. *Pline L'Ancien. Histoire Naturelle. Livre* III. Paris: Belles Lettres, 2004	
4	Zehnacker, H., and Alain Silberman. *Pline L'Ancien. Histoire Naturelle. Livre* IV. Paris: Belles Lettres, 2015	
5.1–46	Desanges, Jehan. *Pline L'Ancien. Histoire Naturelle. Livre* V, *1–46 (L'Afrique du Nord)*. Paris: Belles Lettres, 1980	
5.47–151		Winkler, Gerhard, and Roderich König. *C. Plinius Secundus d. Ä., Naturkunde. Lateinisch-deutsch. Buch* V. Munich: Artemis & Winkler, 1993
6.1–45		Brodersen, Kai. *C. Plinius Secundus d. Ä., Naturkunde. Lateinisch-deutsch. Buch* VI. Munich: Artemis & Winkler, 1996
6.46–106	André, Jacques, and Jean Filliozat. *Pline L'Ancien. Histoire Naturelle. Livre* VI *2e partie (L'Asie centrale et orientale, l'Inde)*. Paris: Belles Lettres, 1980	

Book, Sections	Budé	Sammlung Tusculum
6.107–162		Brodersen, Kai. *C. Plinius Secundus d. Ä., Naturkunde. Buch* VI (as above)
6.163–220	Desanges, J. *Pline L'Ancien. Histoire Naturelle. Livre* VI *4e partie. (L'Asie africaine sauf l'Égypte, les dimensions et les climats du monde habité)*. Paris: Belles Lettres, 2008	

Modern Works Cited

Beagon, Mary. 1992. *Roman Nature: The Thought of Pliny the Elder.* Oxford: Oxford University Press.
    2005. *The Elder Pliny on the Human Animal: Natural History Book 7. Translation with introduction and historical commentary.* Oxford: Oxford University Press.
Bowen, Alan C. and Francesca Rochberg (eds.) 2020. *Hellenistic Astronomy: The Science in Its Contexts.* Leiden: Brill.
Cornell, Tim J. (ed.) 2014. *The Fragments of the Roman Historians.* Oxford: Oxford University Press.
Doody, Aude. 2010. *Pliny's Encyclopedia: The Reception of the Natural History.* Cambridge: Cambridge University Press.
    2015. "Pliny the Elder." In *Oxford Bibliographies Online: Classics.*
Dueck, Daniela. 2012. *Geography in Classical Antiquity.* Cambridge: Cambridge University Press.
Gillespie, Stuart. 2001. *Shakespeare's Books: A Dictionary of Shakespeare Sources.* London: Athlone Press.
Healy, John F. 1987. "The Language and Style of Pliny the Elder." In *Filologia e Forme Letterarie: Studi Offerti a Francesco della Corte,* iv, ed. Sandro Boldini, 1–24. Urbino: Università degli Studi.
    1991. *Pliny the Elder, Natural History: A Selection.* London: Penguin.
    1999. *Pliny the Elder on Science and Technology.* Oxford: Oxford University Press.
Hiatt, Alfred. 2020. *Dislocations: Maps, Classical Tradition, and Spatial Play in the European Middle Ages.* Toronto: Pontifical Institute of Mediaeval Studies.
Ilyushechkina, Ekaterina, Alexander Podosinov and Alexius Belousov (eds.) 2021. *Plini Secundi: Naturalis Historia Tomus* I *Libri* 1–11. Moscow: Academia Pozharskiana.
Irby, Georgia L. (ed.) 2016. *A Companion to Science, Technology, and Medicine in Ancient Greece and Rome,* 2 vols. Malden, MA: Wiley Blackwell.
König, Jason, and Tim Whitmarsh (eds.) 2007. *Ordering Knowledge in the Roman Empire.* Cambridge: Cambridge University Press.
Lozovsky, Natalia. 2000. *"The Earth Is Our Book": Geographical Knowledge in the Latin West, ca. 400–1000.* Ann Arbor: University of Michigan Press.

McHam, Sarah Blake. 2013. *Pliny and the Artistic Culture of the Italian Renaissance: The Legacy of the* Natural History. New Haven, CT: Yale University Press.

Murphy, Trevor. 2004. *Pliny the Elder's* Natural History: *The Empire in the Encyclopedia.* Oxford: Oxford University Press.

Pinkster, Harm. 2005. "The Language of Pliny the Elder." In *Aspects of the Language of Latin Prose*, ed. Tobias Reinhardt, Michael Lapidge and James N. Adams, 239–256. Oxford: Oxford University Press.

Rackham, Harris (ed. and trans.) 1938. *Pliny: Natural History.* Volume I, *Libri* I–II. Loeb Classical Library. Cambridge, MA: Harvard University Press.

———. 1942. *Pliny: Natural History.* Volume II, *Libri* III–VII. Loeb Classical Library. Cambridge, MA: Harvard University Press.

Reeve, Michael. 2007. "The Editing of Pliny's Natural History." *Revue d'histoire des textes* 2: 107–179.

Riggsby, Andrew. 2019. *Mosaics of Knowledge: Representing Information in the Roman World.* New York: Oxford University Press.

Roller, Duane W. 2015. *Ancient Geography: The Discovery of the World in Ancient Greece and Rome.* London: I.B. Tauris.

Talbert, Richard J. A. (ed.) 2000. *Barrington Atlas of the Greek and Roman World.* Princeton: Princeton University Press.

———. 2020. "An English Translation of Pliny's Geographical Books for the 21st Century." *Shagi/Steps* 6.1: 214–228.

Travillian, Tyler T. 2015. *Pliny the Elder: Book* VII *(with Book* VIII *1–34).* London: Bloomsbury Academic.

Index

The character of Pliny's geographical books makes a full index overwhelming. A concise, selective one is offered here, with particular attention to the following (cited by book and paragraph):

- names of individuals mentioned (human and divine)
- features and themes of concern to Pliny (for any that occur often, only more significant instances)
- notable placenames, especially those in Table of Contents (Book 1)

A complete list of placenames with citations is included in the database available at https://isaw.nyu.edu/research/pliny-the-elder. For a full index, see Karl Bayer and Kai Brodersen (eds.), *C. Plinius Secundus d. Ä., Naturkunde. Gesamtregister.* Munich: Artemis & Winkler, 2004.

Acarnania 4.5–6
Achaia/Achaeans 4.12–32
Achilles 4.83, 93; 5.125
  temple of 10.78
Acilius, Manius 2.98, 147
Adiabene 6.41
Aedemon 5.11
Aegipanes 5.7, 44, 46; 6.197
Aemilianus, Cornelius Scipio 5.9, 25
Aemilius Paulus 2.53; 3.138; 4.39
Aeneas 3.82
Aeolis 5.121–125
Aeolus 3.92–94
Aeschylus 37.31–32
Aesculapius 4.18
Aethiopia 2.184, 189; 5.43–64; 6.173–197
Aethiops 6.187
Aetolia 4.6–8
Afranius 2.170
Africa (continent) 2.173; 3.3; 5.1, 6, 40, 47–48; 6.208–210
Africa (province) 5.23–25; 7.96
Africanus, *see* Aemilianus
Agamemnon 16.238
Agathocles Bk.4–6 source
Aglaosthenes Bk.4 source; 4.66
Agrippa, Marcus Vipsanius Bk.3–6 source; 3.8, 16–17, 37, 86, 96, 150; 4.45, 60, 77–83,
  91, 98–99, 102, 105, 118; 5.9, 40; 6.3, 37–39, 57, 136, 139, 164, 196, 207, 209
air/breath 2.10–12, 33, 102–153, 162, 166, 192, 197, 200–201, 218–223
Ajax 5.125
Albania 6.29, 38–39
Alexander [the Great] Bk.6 source; 2.168, 180–181, 185; 3.57; 4.75; 5.62, 76, 107, 116–118, 134, 142; 6.40–41, 45–51, 58–62, 77, 81, 92–97, 100, 110, 115, 119, 134, 138, 198; 7.95
Alexander I of Epirus 3.98
Alexander, Cornelius 3.124
Alexander Polyhistor Bk.3–6 source
Alexandria (in Egypt) 2.178; 5.62, 128
Alpes 3.132–138
alphabet 5.67; 6.174; 7.191–193
Amasis 5.60; 19.12
amber 3.152; 4.94, 97, 103; 37.30–46
Amometus Bk.6 source; 6.55
Amphitus 6.16
anatomy (including hair), especially exceptional 2.189; 4.88, 95; 5.46; 6.35, 88, 109, 162, 187–188, 200; 31.14
Anaxagoras 2.149–150
Anaximander Bk. 2, 4 source; 2.31, 187, 191; 4.58; 7.203; 18.213
Anaximenes 2.187

Index

Andromeda 5.69, 128
Angerona 3.65
Annius Milo, Titus 2.147
Annius Plocamus 6.84
Antaeus 5.2–3
Antichthones/Antipodes 2.161–166; 6.81
Anticlides Bk.4 source; 4.67; 7.193
Antigenes Bk.5 source
Antigonus 12.56
Antiochus 2.167; 6.47–49, 58, 93, 115, 132, 139, 152
Antonius, Marcus 2.98–99; 31.11
Apicius 10.133
Apion 31.22; 32.19
Apollo 4.5, 66, 91–92; 5.60, 106, 116; 6.189; 7.196; 16.240
   cave of 2.232
   Didymaeus 6.49
   Clarius 2.232
   oracle of 4.7; 5.112
Apollodorus Bk.4, 6 source; 14.76
Apollonius 37.32
Appenninus 3.48
apsides 2.63–73, 79
Arabia 5.65, 68, 85–90; 6.108, 142–162; 12.51–57, 63–65, 82–84; 19.7
Arcadia 4.20; 31.27
Archelaus 37.46
Archimedes Bk.2 source
Arcturus, *see* constellations
Argo (ship) 3.128
Argus 16.239
Ariani 6.92–95
Aries, *see* constellations
Arimphaei 6.34–35
Aristarchus Bk.5 source
Aristides Bk.4 source; 4.64, 70
Aristocreon Bk. 5–6 source; 5.59; 6.183, 191
Aristocritus Bk. 4–5 source; 5.135
Aristotle Bk. 2, 5 source; 2.150, 220; 4.65–66, 70; 5.135; 7.192, 195, 197, 205; 8.229
Armenia (Greater and Lesser) 6.25–28, 40
Arruntius Bk. 3, 5–6 source
Artemidorus Bk.2–6 source; 2.242–246; 4.77, 121; 5.40, 47, 59, 129; 6.36, 70, 156, 163, 183, 207
Asarubas 37.37
Asclepius 14.76
Asia (continent) 2.173, 205; 3.3; 4.49, 75, 78, 87, 90; 5.47, 141; 6.1, 33, 209–210; 9.50
Asia (province) 2.200; 7.99
Asopis 4.13
Aspagani 6.79
Asphaltites lake 2.226; 5.71–73
astrology 2.28–29, 54–55
astronomy 2.95; 5.67; 6.87, 98, 121; 18.210–216
Astynomus Bk. 4–5 source; 5.129

Atargatis, *see* Derceto
Ateius, Lucius Bk. 3–4 source
Ateius Capito Bk. 3–4 source
Athens 4.24–25
Atilius Regulus, Gaius 3.138
Atlans 7.203
Atlantes, sub-human 5.45
Atlantic ocean/sea 2.205; 3.3–5; 4.119–120; 5.40; 6.175
Atlas 2.31
Atlas mt. 5.5–16
Attica 4.23–24; 18.214
Aufidius 6.27
Augustus Bk. 3–4 source; 2.24, 93–94, 98, 167–168, 178; 3.17, 46, 49, 62, 136; 4.5, 111; 5.2, 5, 20–21; 6.141, 160, 181; 12.55–56; 14.61, 72

Babylon 6.121–122
Baetica 3.6–17; 4.116–119
Baeton Bk. 5–6 source; 6.61, 69
Balbillus 19.3
Balbus, Cornelius 5.36
Baliares 3.76–79
bamboo 16.162
Basilis Bk. 6 source; 6.183
Bebryx 16.239
Bellerophon 7.202
Berosus 7.193
Bion Bk. 6 source; 6.178, 180, 183, 191, 193
birds 2.196, 207; 4.47; 10.74–79, 132–135
Bithynia 5.148–149
bitumen 2.226, 235–237; 5.72; 6.99, 179; 31.5
Bocchus 5.19
Boeotia 4.8, 25–26; 18.215
Bosporus, Cimmerian 6.1–3, 18
Bosporus, Thracian 9.50–51
breakwaters, *see* environment
breath, *see* air
bridges, *see* environment
Britannia 2.187; 3.119; 4.102–104; 6.219
bronze 2.137, 233; 7.197, 201; 31.28
Bucephalus (Alexander's horse) 6.77
buildings, *see* environment

Cadmus 7.192, 195, 197
Cadmus (the Milesian) 7.205
Caecilius, Gaius 2.100
Caecina Bk. 2 source
Caelius Antipater Bk. 2–3 source; 2.169; 3.132; 31.21
Caelobothras 6.104
Caelus 7.194
Calchas 3.104
calendar, *see* time
Calidonia 4.102
Caligula (emperor) 4.10; 5.2, 11; 14.64

Callicrates Bk. 3, 5 source
Callidemus Bk. 4 source; 4.64
Callimachus Bk. 4–6 source; 3.139, 152; 4.52, 65, 69–70, 73; 5.28; 31.9
Calliphanes Bk. 3, 5 source
Calpurnius Piso, Marcus 7.98
camels 5.72; 6.102, 158; 12.63–65
Camillus 3.125
Campania 2.180; 3.38, 60, 70; 14.62, 69
canals, *see* environment
Cancer, *see* constellations
Candace 6.186
Cappadocia 6.8–9, 23–24; 8.225
Capricorn, *see* constellations
Caria 5.103–109; 7.203
Carmania 6.107
Casius mt. 5.80
Caspian Gates 6.40, 43–45
Caspian sea 6.17, 26–40, 46–49
Cassius of Parma 31.11
Castor(es) 2.101; 6.16
Catiline 2.137
Cato the Elder Bk. 3–4 source; 3.51, 98, 114, 116, 124–125, 130, 133–134
Cato the Younger 5.24
Cato, Lucius 3.70
cattle 4.120; 6.2, 159–161; 31.13
Caucasian Gates 6.30–31, 40
Caucasus mts. 6.15, 28, 50, 60, 71
Cepheus 6.182
Ceres 3.60; 7.191
Ceto 5.69
Chabrias 5.68
Charax 6.138–139
Chares 37.33
Chauci 16.2–6
Chione 5.136
Chius 5.136; 14.73
Cilicia 5.80, 91–93; 6.24; 18.215
'circles', *see* zones
civilization/culture (or lack of) 3.31, 39; 5.44–46, 73; 6.35, 53, 55; 16.4; 27.3; 37.201
Claudius (emperor) Bk. 5–6 source; 2.92, 99; 3.119; 5.2, 11, 20, 58, 63, 75; 6.8, 17, 27, 31, 84, 128; 31.5
Claudius Marcellus, Marcus 3.131
Cleobulus Bk. 4–5 source; 5.136
Cleostratus 2.31
climate/weather 2.105–153, 189–190, 245; 3.41; 4.58, 89–90; 5.15, 65; 6.33–34, 55, 103, 176, 199; 16.5; 37.201
Clitarchus Bk. 6 source; 3.57; 6.36, 198
cloth/clothing/wool 2.211, 228; 3.41; 4.62; 5.14; 6.54, 109, 162, 190; 7.196; 19.5, 8, 13–14; 37.37, 44, 202

cloudbursts 2.131–134
Coeranus Bk. 2 source
coinage, Roman 6.85
colors of planets 2.79
comets, *see* shooting stars
commerce/trade 3.21, 41; 4.66; 5.34, 60, 63, 144; 6.15, 52–54, 82, 88, 99–101, 104, 140, 146, 149, 162, 173, 176; 7.199; 12.51–57, 63–65; 19.2; 37.45, 201–205
constellations 2.7–9, 31, 48–88, 106–110, 122–124, 177, 184, 188; 18.210–216
copper 2.158; 3.30; 4.64; 6.98; 7.195–197; 37.202
coral 32.21
Corcyra 4.52–53
Cornelii Scipiones (Scipios) 2.241, 3.21
Cornelius Nepos Bk. 2–6 source; 2.169–170; 3.4, 125–127, 132; 4.77; 5.4; 6.5, 31, 199
Cornelius Orfitus 2.99
Cornelius Scipio 3.9
Cornelius Sulla, Lucius 2.144; 3.70, 80; 7.96; 14.62
Corsica 3.80–83
Cous 5.134; 6.59
Crassus, *see* Licinius
Crates Bk. 4 source; 4.58
Cremutius 10.74
Creta 4.57–62; 6.157; 16.161
Critodemus Bk. 2 source; 7.193
crocodiles 5.9–10, 51; 6.75
Ctesias Bk. 2 source; 2.236; 31.9, 21, 25; 37.39
Curio Bk. 3 source
Cybele 5.147
Cyclades 4.65–67; 18.214
Cyclopes 7.195–198
Cyprus 5.129–131; 7.195; 18.215
Cyrenaica 5.31–38
Cyrus 6.49, 92, 116; 7.205

Daedalus 3.102; 7.198, 205
Dalion Bk. 6 source; 6.183, 194
Damastes Bk. 4–6 source
Danae 3.56
Danuvius/Hister river 3.127–128, 146, 149; 4.41–44, 79, 82; 9.53
Dardanus 3.63
Darius I 4.76; 6.41, 116, 133, 165
day, length of 2.186–187
Delmatia 3.141–145
Delos 4.65–66
Delphi 2.208; 3.120
Demetrius I 4.10
Democritus Bk. 2 source; 2.14
Demodamas Bk. 6 source; 6.49
Demostratus 37.34
Derceto/Atargatis 5.81

Diana 5.115
Dicaearchus Bk. 2, 4–6 source; 2.162
diet, especially exceptional 2.21; 4.95–97; 5.15, 45; 6.35, 53–55, 75, 109, 161, 169, 189–190, 195; 7.191; 16.3–4
Dinochares 5.62
Diodorus Bk. 3, 5 source
Diognetus Bk. 6 source; 6.61
Diomedes 3.103–104, 120, 141, 151; 4.42
Dionysius Bk. 4–5 source; 4.64; 5.134; 6.58
Dionysius the Elder 3.95
Dionysius (II) 2.222
Dionysodorus 2.248
distance, difficulty of measuring accurately 3.16; 4.91, 98; 6.124
dogs 2.107; 3.152; 5.15; 6.2, 155, 192, 205; 10.79
Dolabella, Publius 2.99
Domitius, Gnaeus 2.99
Domitius Corbulo Bk. 5–6 source; 2.180; 5.83; 6.23
Dosiades Bk. 4 source; 4.58
dwellings, *see* environment

earth 2.154–177, 210–212, 242–248; 6.1, 205–220
earthquakes 2.191–206
echoes 2.115
eclipses 2.42–58, 98, 180, 195
eclipticts 2.68
Egnatius Calvinus 10.134
Egypt 3.121; 5.1, 39–40, 43, 47–65; 6.166, 178–196, 220; 14.75; 18.215; 19.7, 12–14
elements (*see also* air, earth, fire, water) 2.10–11, 162
elephants 2.183; 5.5, 15, 18, 26; 6.66–68, 73–76, 81, 91, 171, 180, 185, 189–192, 220; 9.46; 37.204
Elymais 6.135–136
empire
    Carthaginian 5.76
    Macedonian 4.33, 39; 6.58–59, 81
    Parthian 5.88; 6.111–116
    Roman 3.56, 110; 4.98; 5.2, 29, 36, 76, 88; 6.101, 120, 160; 7.95–99; 12.65; 16.4; 27.3
environment, abuse of 2.148, 157–159; *see also* landscape changes
environment, built
    breakwaters/flood barriers 6.138–140, 146
    bridges 3.101; 4.4, 63, 75–76; 5.41, 85–86, 128; 6.13, 75; 14.62; land-bridges 5.76, 142
    buildings/dwellings 2.196–199; 3.54, 67; 5.34; 6.89–91, 121, 153, 186, 203–205; 7.194; 16.2–3
    canals/channels 3.34, 95, 119–120; 4.10, 84; 5.23, 116; 6.130, 165; 14.61
    city plans 3.66; 5.62, 148; 6.47, 121–122

    gates, city 3.66; 5.60; marking passes 6.30, 40, 43–44
    roads 2.199; 3.66–67; 6.103, 144, 167; 19.9
    walls 2.193, 197, 235; 3.66–67; 4.24, 43, 48, 85; 5.77; 6.67, 121–123, 155; 7.195, 202
    wells 2.58, 191, 197, 219; 4.120; 5.34, 38, 57; 7.195; 31.22
envoys 2.238; 6.84–88, 140; 12.57
Ephorus Bk. 4–6 source; 4.64, 120; 5.136; 6.198–199
Epigenes Bk. 2 source; 7.193
Epiphanes 6.147
Epirus 3.145; 4.1–4, 32
eques/equites 2.199; 5.11–12; 6.160, 181; 7.96; 37.45
equinox 2.50, 108, 124–125, 151, 176, 182, 188, 213–215; 4.89, 110; 6.212–213; 9.51; 18.213
Erasistratus 14.73
Eratosthenes Bk. 2, 4–6 source; 2.185, 247; 3.75; 5.39–41, 47, 127, 132; 6.3, 36, 56, 81, 108, 163, 171, 183; 12.53
Erythras 6.107, 153
Esseni 5.73
Etruscan scholarship Bk. 2 source; 2.138, 143, 199
Euboea 4.63–64
Euclid Bk. 2 source
Euctemon 18.213
Eudicus 31.13
Eudoxus Bk. 2, 4–6 source; 2.130, 169; 6.198; 18.213; 31.16
Euphrates river 5.83–90; 6.25, 120–130, 146
Euripides 31.28; 37.31–32
Europa 3.5, 74, 97, 145; 4.1, 49–52, 75–78, 87, 90, 94, 102, 121–122; 5.141; 6.1, 208–210; 9.50; 19.9
exploration (especially by Roman army) 2.168; 4.97, 102; 5.9–11, 14–15, 51; 6.40, 141, 160, 167, 173, 181–184; 12.55; 37.42

Fabianus Bk. 2 source; 2.121, 224
Fabius, Quintus 2.98
Fabricius Tuscus Bk. 3–4, 6 source
Fannius, Gaius 2.99
fertility, agricultural 3.41, 60, 117; 4.90; 5.24, 58, 113; 6.79, 91–93, 161; 7.199; 19.7; 37.202
fire 2.10, 162, 234–241; 4.66; 6.187–188; 7.198
fish 2.109, 203, 226–227, 231, 237; 3.60; 5.51; 6.91, 95, 109, 127, 205; 9.44–53; 16.3; 19.10; 31.22, 25–26; 32.15–19
Flavius, Gaius 19.4
flax 19.2–15
Fonteius 2.180
Fortunatae islands 4.119; 6.202–205
Fortune (goddess) 2.22
frankincense 12.51–57, 63–65, 107; 37.204
freedmen 5.11; 6.84; 14.72; 31.5

Gades 3.7; 4.119–121; 5.76
Galatia 5.146–147
Galba (emperor) 3.37
Galerius 19.3
Galli 3.57
Gallia 3.31; 4.105; 19.7–8; 37.203
  Aquitanica 4.105, 108–109
  Belgica 4.105–106; 16.161
  Lugdunensis 4.105, 107
Gallus, Aelius 6.160
games 2.93–94; 4.14, 18; 7.97, 205; 37.45
Ganges river 6.63–70
gates, *see* environment
Gellianus Bk. 3 source; 3.108
Gellius 7.192, 194, 197–198
gems, *see* jewelry
Germania 4.96–101; 16.5–6; 19.9
Germanicus 2.96; 5.2; 37.42
gladiators 2.96; 37.45
Glaesaria 4.97; 37.42
glass-making 5.75
gnomon(ics) 2.182–187; 6.210–220
Gobares 6.120
God/gods 2.14–27, 154, 192
gold 2.137, 158–159; 3.30; 4.112, 115; 5.3; 6.30, 67, 74, 80–81, 89, 98, 133, 150, 161, 178, 189, 198; 7.197; 10.77; 32.17; 37.202–204
Graecia 4.1, 23–28
Graecus 4.28
Great Mother 2.208
greed, *see* luxury
Greeks, especially criticism of 2.248; 3.42, 122, 152; 4.1, 4; 5.4; 37.31–33
Gutones 37.35
Gyges 7.205

hair, *see* anatomy
halcyon days 2.125
Hammon 6.186
  oracle of 5.31, 50
Hannibal 3.103; 5.148
Hanno Bk. 5 source; 2.169; 5.8; 6.200
Hecataeus Bk. 4–6 source; 4.94; 6.55
Hecuba 4.49
Hellanicus Bk. 4–6 source
Hellen 4.28; 7.203
Hellespont 2.205; 4.49, 75–76, 92; 5.127, 141–144; 6.1–2; 18.215
Heraclides Bk. 4 source; 4.70
Hercules 3.4, 8, 34, 47, 84, 123, 134; 4.25, 39, 120; 5.2–3, 7, 46, 50, 61; 6.49, 76, 89, 212; 7.95, 205; 16.239–240
  pillars of 19.4; 27.2
  temple of 10.79
  rock of 32.17

Herennius, Marcus 2.137
Herod I 5.69
Herodotus Bk. 2, 5 source; 2.201; 5.57, 68
Hesiod 7.197; 18.213
Hesperides 5.3, 31, 46; 37.38
Hesperis 4.58
Hiberia 6.29; Gates 6.40
Hibernia 4.102–103
Hierosolyma 5.70–73
Himantopodes 5.44, 46
Himilco Bk. 5 source
Hipparchus Bk. 2, 5–6 source; 2.53, 57, 95, 188, 247
Hispania 3.6–30; 4.118; 14.71; 37.203
Hispania Further, *see* Baetica
Hispania Nearer/Tarraconensis 3.18–29; 4.110–112; 19.10
Hister river, *see* Danuvius
Histria 3.129
Homer 2.13, 119, 201; 3.57, 82, 96; 4.13, 28, 31, 52, 69; 5.43, 53, 124, 141–143; 14.73
horses 2.90; 3.123; 4.42, 116; 6.77, 198; 7.202; 37.202–203
hours, equinoctial 6.212–220
Hyginus Bk. 3–6 source
Hyperboreans 4.89–91; 6.34, 55, 219
Hyrcanian sea, *see* Caspian sea
Hystaspes 6.133

Ilium, *see* Troas
Illyricum 3.139–140, 150
Ilus, tomb of 16.238
India/Indians 2.170; 6.56–95; 16.161–162; 19.7
  voyages to 6.96–106
Indus river 6.60, 71–80
inscriptions 2.154; 3.18, 129, 136; 6.150, 174; 7.97
inventions 7.191–206; 27.1–2
Io 16.239
Ionia 5.112–120
iron 2.112, 147, 157, 211; 3.30, 81; 4.112; 6.98; 7.197–198; 31.12, 28; 37.202
Isaurica 5.94
Isidorus Bk. 2–6 source; 2.242, 245–246; 4.9, 102, 121; 5.40, 47, 127, 129, 132, 135–136, 140, 150; 6.141
islands 2.202–204, 209
Italia 2.136, 199; 3.31, 38–75, 95–128, 138; 4.122; 6.143; 14.59–70; 16.161; 18.214; 19.9, 13; 37.201–202
Ithaca 4.55
ivory 5.12; 6.173

Jason 3.70; 6.38
jewelry/gems 2.21, 98, 158, 207; 5.34, 37; 6.89, 169; 19.11; 32.16; 37.30, 204

Juba Bk. 5–6 source; 5.16, 20, 51, 59; 6.96, 124, 139–141, 149, 156, 170, 175, 179, 201–205; 12.56; 31.18
Judaea 5.70–73
Julia Augusta 14.60
Julianus 37.45
Julius, Lucius 2.98
Julius, Sextus 2.199
Julius Aquila Bk. 2 source
Julius Caesar, Gaius (dictator) 2.92–94, 98; 5.128; 6.9; 7.99; 14.66; 18.211, 214
Julius Caesar, Gaius (adopted son of Augustus) 2.168; 4.10; 6.141, 160; 12.55–56
Julius Lupus 19.11
Junius Silanus 2.202
Juno 2.16; 3.70; 4.120; 6.200
   Lacinia, altar of 2.240
   temple of 2.144
Jupiter 2.19, 82, 138, 140; 4.52; 5.60, 68; 6.79, 152; 7.201; 10.75; *see also* planets
   Belus 6.121
   Elicius 2.140
   Feretrius 2.140
   Hammon, oracle of 5.49
   Stator 2.140
   Stratios, altars of 16.239
   spring of 32.16
   Tonans 2.140

Laelius Balbus 2.202
landscape changes 2.201–206; 3.16, 57, 86, 108, 112; 4.62–63, 85; 5.42, 114–115, 134; 6.138–140, 146; *see also* environment
languages 3.39
   Aethiopia 5.53
   Celtici 3.13
   Colica 6.15
   Gallic 3.122
   Greek 2.10; 3.50; 4.18; 5.136
   in Hemodi mts. 6.64
   Ichthyophagi 6.95
   Liguria 3.122
   Mauretania 5.13
   Media 6.127
   Punic 4.120
   Scythia 4.94
   Serae 6.88
   Taprobane 6.84
   (un)pronounceable 3.7, 39, 139; 5.1, 45; 6.188
   Volsci 3.59
Lares 3.66
Lathyrus, *see* Ptolomaeus
Latium 3.53–59, 68–70
lead 2.158, 233; 3.30; 4.112, 119; 7.197

Leonnatus 6.97
Lepidus 3.116
Lepidus, Marcus 2.99
Lesbos 5.139; 14.73–74
Leucothea 3.83
Liber/Father Liber 2.231; 3.8, 60; 4.25, 39; 5.74; 6.49, 59, 79, 90; 7.95, 191; 31.16
Libra, *see* constellations
Liburnia 3.129, 139–140
Libya 7.203
Licinius 3.116
Licinius Crassus, Marcus 2.147; 5.86; 6.47; 31.5
Licinius Larcius 31.24
Licinius Lucullus, Lucius 2.235
Licinius Mucianus Bk. 2–6 source; 2.231; 3.59; 4.66, 77; 5.50, 83, 128, 132; 19.12; 31.16, 19
lightning, *see* thunderbolts
linen 7.196; 19.8–14
Liparus 3.93
Livia (wife of Augustus), *see* Julia Augusta
Livius 3.116
Livius, Titus (Livy) Bk. 2–3 source; 3.4, 132
Livius the son Bk. 5–6 source
Lucullus, Marcus 4.92
Lusitania 3.6, 13; 4.113–118
luxury/greed 2.25, 118, 125, 158, 174–175, 207; 4.89; 5.12; 6.54, 88–89, 101, 111, 133; 8.226; 10.133; 12.82–84; 37.30, 203
Lycaon 7.202, 205
Lycaonia 5.95
Lycia 5.100–102
Lycos 31.17, 27
Lydia 5.110–111
Lydus 7.197

Macedonia 3.145; 4.33–39, 50; 18.214
Maeander river 2.201; 5.113
Maeotis 6.19–22
Magnesia 4.32; 18.214
'maps'
   of Agrippa 3.17
   drawings of landscape 6.40
marble 3.30, 141; 4.64, 67; 5.22, 136, 151; 6.89; 31.11
Marcellus, Gaius 2.147
Marcellus, Marcus 2.53
Marcia 2.137
Marcius, Lucius 2.241
Marcius, Lucius (consul 91 BCE) 2.199
Marcius, Quintus 2.99
Margiane 6.46
Marius, Gaius 2.100, 148; 3.34, 80
Mars 6.32; 7.201; 10.77; *see also* planets
Marsyas 3.108; 7.204; 16.240
marvels, earth's 2.208, 241

Mauretania 5.1–21, 51–52; 6.198–201
Medea 3.151
Media 6.43, 114
medicine/poison 2.155–159; 5.16; 7.196–197; 27.1–3; 31.4–19; 37.44
Megasthenes Bk. 5–6 source; 6.58, 81
Meleager 10.74; 37.40–41
Memnon 5.60; 6.182; 10.74
Menaechmus Bk. 4 source; 4.64
Menon 7.193
Mercury 5.61; 7.191–192, 204; *see also* planets
Meroe 6.185–186
Mesopotamia 5.66, 86; 6.25, 117–126
Messala, Marcus 7.98
Messala Potitus 14.69
Messapos 3.99
Metellus Celer, Quintus 2.170
Metrodorus Bk. 3–6 source; 3.122; 5.136; 37.35
Midacritus 7.197
Midas (in Phrygia) 7.204
Midias 7.200
Minerva 2.210; 4.72; 16.240
   shrine of 7.97
   temple at Lindus of 19.12
mines/mining 2.158, 207; 3.30, 81, 138; 4.112; 6.30, 74, 80, 98, 150, 161; 7.195; 37.202–203
Minos 6.157–158
Mithridates (The Great) 2.209; 6.7; 7.98; 37.39
Mithridates (king of Bosporus) 6.17
Mnaseas 37.38
Moeris lake 5.50
Moesia 3.149–150
moon 2.32–88, 99, 212–223
Mucius, Quintus 2.99
Muses 4.25
music 2.6, 84, 93, 148, 209; 5.7; 7.204
Myiagrus 10.75
Myrsilus Bk. 4–5 source; 3.85; 4.65
Mysia 5.142–144

Narbonensis 2.121; 3.31–37; 14.68
Nature 2.3, 13, 25, 27, 41, 52–53, 76–77, 101, 104, 113, 116, 121, 141–142, 145, 154, 160–162, 166, 190, 202, 206, 208, 223, 233, 236; 3.40, 132; 4.22, 29, 88; 5.88; 6.1–2, 30, 89, 143; 10.76; 19.6; 37.201–205
navigation 2.167–170, 179–181, 218; 5.67; 19.3–6
Nearchus Bk. 6 source; 6.96–98, 107, 109, 124
Nechepso Bk. 2 source; 2.88
Neptune 4.18; 5.150; 6.111, 152
Nero (emperor) 2.92, 199, 232; 4.10, 22; 6.40, 181, 184; 12.83; 14.61; 37.45
nets 7.196; 19.11; 37.45
Nicander 37.31
Nicanor 6.117

Nicias 2.54
Nicias of Megara 7.196; 37.36
Nicodorus 3.58
Nigidius Bk. 6 source; 6.217
Nile river 3.121; 5.47–64; 6.102, 165–166, 177, 181, 188
Nomads 4.83–84, 88; 5.22, 72; 6.38, 51, 55, 113, 125, 143–145, 148, 161, 179, 189–190
Norici 3.146–147; 4.98
Numa 2.140
Numenius 6.152
Numidia 5.22, 25
Nymphodorus Bk. 3, 5 source

ocean 2.164–174, 217–219, 242–245; 6.1, 33–35
Octavius 2.92
Octavius, Gnaeus 2.100
Oenotrus 3.99
oil/olives 2.108, 199, 226, 234; 3.41, 60; 5.3; 6.131; 7.199; 16.239–240; 37.202–203
Omanades 5.94
omens 2.24, 82, 89–94, 97, 101, 113, 137–150, 208, 222, 232; 3.55; 5.58; 7.203; 31.22–24; 32.17
Onesicritus Bk. 2, 6 source; 2.183, 185; 6.81, 96, 98, 109, 124
Opimius, Lucius 2.98
oracle of
   Branchidae 5.112
   Colophon 5.116
   Delphi 2.208; 4.7; 7.205
   Ephesus 5.115
   Hammon 5.31, 49–50; 12.107; 37.33
   Zeus Dodone 4.2
orbits 2.30–46, 56; 6.211; *see also* apsides
Orodes 6.47
Orpheus 4.41; 7.203–204
Osiris 5.60

Padus river 2.229; 3.49, 115–132
Palamedes 7.192, 198, 202
Palibothra 6.68
Palmyra 5.88–89; 6.143–144
Pamphylia 5.96
Pan 3.8; 7.204
Panaetius Bk. 5–6 source
Pandion 6.105
Pannonia 3.147–148
Paphlagonia 6.5–7
Papirius, Gnaeus 2.100
'parallels', *see* zones
Parthia 5.88; 6.41, 44, 50, 111–126, 134, 146, 162; 16.160
Patale island 6.71–72
Patrocles Bk. 6 source; 6.58
Paulus, Lucius 2.147

Index

pearls 6.81, 89, 110, 148; 12.84; 32.21; 37.204
Peloponnesus 4.9–22
Perses 7.201
Perseus 3.56; 5.7; 7.201
Perseus (Macedonian king) 2.53; 3.114
Persian gulf 6.107–111
Petosiris Bk. 2 source; 2.88
Petra 5.87–89; 6.144–146
Petronius, Publius 6.181
Peucetius 3.99
Phaethon 3.117; 37.31, 33
Pherecydes 2.191
Philemon Bk. 4 source; 4.95; 37.33, 36
Philip of Macedon 2.97
Philistides Bk. 4 source; 4.58
Philistus 4.120
Philonides (Alexander's courier) 2.181
Philonides Bk. 4–5 source; 5.129
Philoxenus 37.31
Phoenice/Phoenicians 5.67, 75–78; 18.215
Phrygia 5.144–145
Pindar 2.54
piracy/pirates 2.117, 125, 181; 3.101, 152; 5.38; 6.101, 104, 125, 176; 7.98
Pisidia 5.94
Piso, Lucius Bk. 2–3 source; 2.140; 3.131
Plancus, Marcus 2.99
planets/stars 2.12, 28–84, 213–214
Poenae 19.6
poison, *see* medicine
Pollux 2.101; 6.16
Polybius Bk. 4–6 source; 3.75; 4.77, 119–122; 5.9–10, 26, 40; 6.199, 206
Polydorus 4.43
Polygnotus 7.205
Pompeius, Gnaeus 3.70
Pompey (Magnus) 2.92; 3.12, 18, 101, 138; 5.58, 68; 6.51–52, 120; 7.95–99
Pomponius Mela Bk. 3–6 source
Pontus (Black Sea) 2.206, 219; 4.75–79, 92–93; 5.47; 6.1–4, 32; 9.49–52
Pontus (region) 2.126–127
Poppaea 12.83
Porcius, Gaius 2.98, 147
Porcius, Marcus 2.99
Porsina 2.140
Posidonius Bk. 2, 4–6 source; 2.85, 6.57
Postumius, Spurius 2.99
pottery 3.82; 7.198
Prometheus 7.198
Protesilaus 4.49
tomb of, 16.238
Psammetichus 6.191
Ptolemy, Juba's son 5.11, 16
Ptolomaeus Ceraunus 6.31
Ptolomaeus Lathyrus 2.169; 6.188

Ptolomaeus Philadelphus 6.58, 165–183; 14.76
purple dye 2.109; 5.12, 76; 6.201; 37.204
Pyrenaei 3.6, 18, 22, 29–32; 4.105, 109–118; 7.96; 14.68; 31.4
Pyrene 3.8
Pyrrandrus Bk. 5 source
Pyrrhus 3.101
Pythagoras 2.37, 83–84, 191
Pythagoreans Bk. 2 source
Pytheas Bk. 2, 4 source; 2.187, 217; 4.95, 102; 37.35

Rachias 6.85, 88
Raetus 3.133
rainbows 2.150–153
recession/retrogradation (of planets) 2.69–77
Red sea 5.65; 6.107, 163–168, 196
reeds 5.44; 6.82; 7.206; 16.159–162; 19.15
Rhadamanthus 6.158
Rhenus river (Germania) 4.101
Rhenus river (Italia) 16.161
Rhodos 5.132–133
Ripaean mts. 4.78; 5.98; 6.15, 34
rivers 2.166; 3.54; 6.60
 intermixed 2.224; 4.31; 6.128
 underground 2.225; 9.53
roads, *see* environment
Rome 2.16, 93, 144, 180–182, 200; 3.38–40, 65–67; 4.20; 10.79
 compared to Taprobane 6.84–91
Romulus 3.66
Rutilius, Publius 2.98

Sabis 12.63
Sagdodonacus 6.139
salt 2.222, 233–234; 5.34; 6.43, 51, 147, 195; 12.83; 31.5, 18; 32.18
Samiramis 6.8, 49
Samus 5.135
Sardinia 3.83–85
Sarmatia 2.246; 4.80–81, 91; 6.19, 40
Saturn 3.8, 19; 7.196; *see also* planets
 lake of 31.22
Satyrs 5.7, 44–46; 6.197
Satyrus 37.31
Scaurus 2.144
scholarship, Pliny's
 criticism of others 2.4, 14, 43, 73, 82, 85–88, 110, 161, 164, 214, 248; 3.17, 42, 128, 152; 4.83, 114–115; 5.4, 12; 6.40, 51, 124, 170, 219; 19.2; 37.31–33
 methodology 2.8, 62, 241; 3.1–2, 39–46, 122; 4.98–99; 6.23, 141
 praise of others 2.54, 62, 95, 117–118, 164; 3.17, 57–58; 5.16; 6.170; 27.1
Scribonius, Gaius 2.100

Scythia 2.135, 167; 4.80–86, 91; 6.20, 33–38, 50–55, 112
sea 2.201–206, 222–224
Sebosus Bk. 2–3, 5–6 source
Seleucus Nicator 2.167; 5.86, 127; 6.31, 43, 47–49, 58, 63, 122
senate/senator(s) 2.238; 3.131, 138; 5.11–12; 19.3
Seneca Bk. 6 source; 6.60
Septentrio(nes), *see* constellations
Serapion Bk. 2, 4–5 source
Seres 6.53–55, 88; 37.204
Sergius Plautus Bk. 2 source
Serrani 19.8
Sertorius 7.96
Servius Tullius 2.241
Sesostris 6.165, 174
Sestius 4.111
shadows 2.47–48, 50–51, 182–187; 5.56; 6.87, 171, 211–218
ships 2.101, 132, 164; 3.54; 4.10; 5.59; 6.104; 7.206; 19.3–6
shooting stars/comets 2.89–101
Sibylline books 3.123
Sicilia 2.204; 3.86–94
Silanus 2.100
Silenus 4.120
silver 2.90, 137, 158; 3.30–31; 4.112; 6.67, 74, 80, 89; 7.197; 10.77; 37.202, 204
Simonides (poet) 7.192, 204
Simonides Bk. 6 source, 6.183
Sirens 3.62, 85
sky 2.12–13, 147–153, 172; 6.58
slavery/slaves 6.89, 172; 7.200; 16.4; 37.201
snakes 2.155, 159; 3.78; 5.15–16, 45; 6.43, 98, 136, 169; 8.229; 10.75; 31.18
snow 4.88
Social War 2.199, 238
Sol 7.197
Sol Indiges 3.56
solstice 2.50, 81, 108, 119–122, 151, 176–188, 215, 229; 4.73, 89, 104; 5.57; 6.83, 171
Sophocles 37.40
Sosigenes Bk. 2 source; 2.39, 18.211–212
Sotacus 37.35
Sotades Bk. 5 source
Spaosines 6.139
speed of voyages 19.3–4
Sporades islands 4.68–74
springs 2.224–234; 31.10–29
Staphylus Bk. 4–5 source; 5.134
stars, *see* planets
statio (of planets) 2.70–77
Statius Sebosus 6.183, 201–202; 9.46
stones, fallen from sky 2.149–150
storms 2.131–134

Strato 5.69
Sudines 37.34
Suetonius Paulinus Bk. 5 source; 5.14–15
suicide 2.27, 156; 4.89; 6.66
Sulpicius, Servius 2.147
Sulpicius Gallus Bk. 2 source; 2.53, 83
Summanus 2.138
sun 2.12–13, 35–52, 58, 81, 222; 5.56–57
sundials, *see* time
Susa 6.133–134
Syria 5.66–70, 79–82
Syrtes 5.26–30

Tanais river 3.3, 5; 4.121; 6.19–20, 49
Taprobane 6.81–91
Tarquinius, Lucius 3.67, 70
Tarquitius Bk. 2 source
Taurus mt. 5.85, 97–99
Telchius 6.16
Tentyris 5.60
Teos 5.138
Tereus 4.47
territorial limits for
  animals 8.225–228
  birds 10.76–79
Thales 2.53; 18.213
Theochrestus 37.37
Theodorus 7.198
Theomenes 37.38
Theophrastus Bk. 3 source; 3.57–58; 7.195, 197, 205; 10.79; 31.13–14, 17, 19, 26; 37.33
Theopompus Bk. 2–3 source; 2.237; 3.57, 98; 4.2; 31.16–17, 27
Theseus 7.200, 202, 205
Thessalia 4.28–32
Thracia 4.40–50; 18.214
Thrasyllus Bk. 2 source
Thucydides Bk. 3–4 source; 3.86
Thule 2.187; 4.104
thunderbolts/lightning 2.82, 112–113, 135–146
Tiberis river 3.53–56, 62, 109
Tiberius (emperor) 2.200; 3.82; 14.64
tides 2.212–220, 229; 3.151; 5.26; 6.17, 140, 152; 16.3; 37.36, 42
Tigranes the Great 6.142; 7.98
Tigris river 6.25, 126–146
Timaeus Bk. 2, 4, 6 source; 2.38; 3.85; 4.94, 104, 120; 37.36
Timaeus the mathematician Bk. 5 source; 5.55–56
Timagenes Bk. 3 source; 3.132
time/sundials 2.6, 35, 45, 181–188, 213; 4.14; 6.171, 212–220; 18.210–216
Timosthenes Bk. 4–6 source; 5.47, 129; 6.15, 163, 183, 198
Timotheus 7.204

tin 4.104, 112
Tingitana 5.17–21
Titius 31.11
Titus (emperor) Bk. 2 source; 2.57, 89; 3.66
trade, *see* commerce
Trebius Niger 32.15
trees 2.240; 3.122; 4.9; 5.14, 33, 73; 6.91, 161, 202–203; 12.54–57, 107–109; 16.2–6, 238–240; 19.5; 37.30–46, 204
trench separating Africa Nova and Vetus 5.25
Triarius 6.9
triumph 5.36–37; 7.95–99, 191
    triumphal records Bk. 5 source
Troas/Troy 5.124–127; 16.238
Trogodytice 2.178, 185; 6.165–177, 189
trophies of
    Augustus 3.136–137
    Pompey 3.18; 7.96
Tubero, Quintus Bk. 2 source
Tuditanus 3.129
Tullius Cicero, Marcus 31.6, 12
Tullius Hostilius 2.140
Tullius Tiro Bk. 2 source
tunny 9.50–53
Turanius Gracilis Bk. 3 source
turtles 6.89, 91, 109, 172–173; 37.204
Typhon 2.91
Tyrus 5.76

Ulysses 3.85; 4.53
universe 2.1–9, 28–104, 160; 18.210

Vacuna 3.109
Valerianus Bk. 3 source; 3.108
Valerius, Lucius 2.100
Valerius Antias, Titus Bk. 2–3 source; 2.241; 3.70
Valerius Marianus 19.3
Valerius Soranus 3.65
Vannius 4.81
Varro, Marcus Bk. 2–6 source; 2.8; 3.8, 45, 95, 101, 109, 142; 4.62, 66, 77–78, 115; 6.38, 51; 19.8; 31.9, 11, 15, 21, 27

Varro Atacinus Bk. 3–6 source
Venus 2.93; 3.22; 5.60, 92; *see also* planets
Verrius Flaccus Bk. 3 source
Vergil 14.67
Vespasian (emperor) 2.18, 57; 3.30, 66; 5.20, 38, 69
Vettius Marcellus 2.199
Vetus, Lucius Bk. 3–6 source
Vibius Cripus 19.4
Vipstanus 2.180
Volcanus 3.93
Vologesus 6.122
Volumnius, Publius 2.147
Vulcan 2.240; 6.187

walls, *see* environment
water 2.161–175, 212–238; 31.4–30
weapons 2.101; 6.176, 194; 7.200–202; 16.159–162
weather, *see* climate
wells, *see* environment
wind 2.114–134; 3.94; 4.89; 5.55; 6.57–58, 100, 106; 19.3–6
wine 2.230; 3.60, 127; 6.161; 7.199; 14.59–76; 31.16, 20; 37.202
women 2.137; 4.62; 5.45, 73; 6.35, 54, 76, 186, 200; 12.54, 84; 19.8–9; 31.8–10; 37.30, 37, 44, 201
wool, *see* cloth
world, *see* universe

Xenagoras Bk. 4–5 source; 5.129
Xenocrates 37.37, 40
Xenophon Bk. 3–6 source; 4.95; 6.200
Xerxes 4.37, 43, 75–76

Zenothemis 37.34
Zeus 4.2, 14, 21
zodiac 2.9, 30–88, 177, 188
zones, earth's 2.172, 177, 216, 245; 6.211–220; 18.216